ENERGY AND PEOPLE

Social implications of different energy futures

Edited by

Mark Diesendorf

Assistant editors

Roger Bartell, Carlie Casey, Alice Day, Lincoln H. Day
Roger M. Gifford and Hugh Saddler

SOCIETY FOR SOCIAL RESPONSIBILITY IN SCIENCE (A.C.T.)
Canberra, 1979

First published 1979 by
Society for Social Responsibility in Science (A.C.T.)
PO Box 48, O'Connor A.C.T. 2601, Australia.

North American distributors:
International Scholarly Book Services, Inc.
2130 Pacific Avenue, Forest Grove, Oregon 97116, USA.

Energy and people.

ISBN 0 909509 14 x hardback
ISBN 0 909509 12 5 paperback

1. Power resources — Social aspects — Addresses, essays, lectures. I. Diesendorf, Mark Oliver, ed. II. Society for Social Responsibility in Science (A.C.T.)

DC 18: 333.7
DC 19: 333.79

Printed in Australia by Southwood Press Pty Ltd,
80-94 Chapel Street, Marrickville NSW 2204.

CONTENTS

Foreword

H. C. Coombs

Centre for Resource and Environmental Studies
Australian National University, Canberra

The issues faced in this book are central to the decisions which society must make in the decades ahead of us. The papers I have read in this volume suggest that clarity and integrity of mind and intellectual innovation are being brought to bear.

Some years ago when scarcity of resources seemed a much more academic issue than it has now become, I wrote a paper called 'Matching economic and ecological realities'. In it I was concerned to consider how far the economic system could assist mankind faced by increasing scarcities to move to patterns of consumption and production which would be ecologically acceptable. I pointed out that there were many widely different life-styles all capable of providing health, security, stimulus and dignity for men and women and that it was possible that human values, never immutable, could change so that the chosen life-style did not 'kick against the pricks' of ecological necessity.

I envisaged this change of values being encouraged and stimulated by changing relative prices until the preferences required were so 'internalised' as the psychologists say, that man would perhaps not be aware that his nature had been changed by the operations of the economic system.

I went on light-heartedly to speculate about what effects this change might have on 'homo economicus' — whether the bustling innovating intrepreneur of industrial society might give way to a worshipful, fearful self-limiting creature.

Looking back on that paper in the light of harder information I believe I was right to place emphasis on the importance of changing values but excessively optimistic about how far the economic system, through the price mechanism, would contribute to their emergence. I almost entirely ignored the effects of price changes on the distribution of wealth (and therefore of power) and on the way the social product is divided — between proprietors of resources and those who wish to use them, between entrepreneurs and those who own the capital they need, between employers and those they employ or fail to employ.

I almost wholly ignored the human tendency to fight relentlessly to retain or to recover a right or privilege once enjoyed and the way that tendency can blind those concerned to the worth of other competing rights — or even to the claims of simple humanity. This tendency, taken together with the fact that the power to defend or restore privilege is very unevenly distributed among the world's population, seems likely to prove a veritable guarantee of social strife and bitterness as established privileges are encroached upon or threatened.

This is no place to continue these second thoughts. I mention them merely to illustrate how important it is that in our deliberations we take account — not merely of the objective components of the problems of energy in a context of increasing scarcity — but of the structure and ideologies of the institutions to which we must look for decisions. The functions which these institutions were established to perform, the sources of their authority and the influences to which they are and have been exposed, will have moulded the men and women who compose them, to their institutional purposes. It may be necessary to ask whether they will be capable of making the decisions required.

Introduction

Mark Diesendorf

Scientists and technologists like to picture themselves as rational pragmatic, realistic beings who are unswayed by emotion or dreams. Yet paradoxically, they have been responsible for a large number of technological fantasies: the project to land the first men on the moon; enormous colonies in space (Gerard O'Neil); an aircraft powered by a nuclear reactor; a space ship for interplanetary travel powered by thousands of atomic bombs (Ted Taylor); a star ship powered by millions of hydrogen bombs (Freeman Dyson); psychosurgery to "cure" hyperactive children, homosexuality and criminal tendencies; a cure for cancer; genetic engineering; biological weapons; controlled climate; the "Green Revolution"; feeding the poor of the world by hydroponics (Herman Kahn). Some of these dreams have been achieved, at a price; others are flawed in their conceptions of the real world (e.g. the needs of the poor countries) and may never be successful.

Of all the technological fantasies, the quest for an unlimited source of energy has probably received more funding and stimulated more science fiction than any other. As Goeller and Weinberg have pointed out, its success would allow all the raw materials of an advanced industrial society to be produced or substituted for, with the exception of one element essential for life — phosphorus. The belief that the age of unlimited energy was imminent probably began with the successful explosion of the first atomic bomb. It was encouraged by the apparent ease of the historical transitions between principle energy sources — from wood to coal to oil.

It was eagerly believed that the next step, to tap the binding energy of the nucleus, would yield the ultimate goal. Unfortunately, this step has introduced problems which are qualitatively and quantitatively different from those experienced previously: nuclear power produces electricity instead of a fuel which is readily stored and transported; its development requires massive inputs of capital and fossil fuels and also many decades of time; and the magnitude and lifetime of its hazards are unique amongst energy sources. Nevertheless, there are still enthusiasts who believe that nuclear fusion will compensate for the disappointments of nuclear fission.

Along the long narrow road to the nirvana of unlimited energy, the technologists recognise a pitfall or two, called by some an "energy crisis". This problem they describe as follows:—

"To produce economic growth, a high standard of living and hence a high quality of life, energy consumption in the rich countries must double, or even triple, by the year 2,000. What existing energy industries can be expanded and what new energy conversion technologies can be introduced to fill the gap?"

Close scrutiny of this type of assessment and the assumptions underlying it has revealed some weaknesses. Energy demands, wants and needs are not necessarily identical; an increase in GNP does not necessarily require an increase in energy use, nor does it necessarily entail an enhanced quality of life.

A broader framework for understanding the energy "crisis" was proposed by Barry Commoner in *The Poverty of Power*:—

"In the last ten years . . . the most powerful and technically advanced society in human history . . . has been confronted by a series of ominous, seemingly intractable crises. First there was the threat to environmental survival; then there was the apparent shortage of energy; and now there is the unexpected decline of the economy. These are usually regarded as separate afflictions, each to be solved in its own terms: environmental degradation by pollution controls; the energy crisis by finding new sources of energy and new ways of conserving it; the economic crisis by manipulating prices, taxes and interest rates."

Commoner shows that energy, environmental and economic problems are closely interrelated and cannot be separated from political questions. A similar broad, holistic approach to energy and society is followed by the Swedish Secretariat for Future Studies and by C. A. Hooker (see *Recommended Background Reading*).

Energy and People attempts to carry forward the discussion of the social implications of different energy futures. Earlier versions of the papers of this book were presented at a National Conference on *Energy and People*, organised by the Society for Social Responsibility in Science (A.C.T.) and held at the Australian National University, Canberra, on 7-9 September 1978. Following the Conference, the papers were revised by the authors and edited.

The first three parts deal with energy strategies in general, transportation in particular and the problems of urban and suburban areas. Their viewpoint is in general that of the scientist, technologist or architect concerned about future directions. John Dick and Chris Mardon point out that energy growth is often assumed to be exponential when a closer examination of the data reveals that it is actually only linear or even saturating. Mark Diesendorf argues that the timescale over which coal-to-oil conversion or fast breeder reactors could make a significant contribution to energy supply is much greater than the timescale over which severe energy shortages would develop if current lifestyles are maintained. Ian Sykes suggests that even before a major energy drought could occur there may be a major economic collapse of the Western oil-importing nations.

Parts IV and V commence by focusing on policies and the planning process. Alan Roberts' paper on the reasons for harmful technologies leads naturally into the analyses by Louis Arnoux and John Price of the relationship of our consumer society to energy use and environment. In particular, Price suggests that the transition to a stable low-energy society would be assisted by permitting people to choose to work shorter hours and be paid on a pro rata basis.

Ethics and equity (Part VI) has common ground with economics, jobs and entropy (Part VII). For example, the choice of interest rates or discount rate has implications for future generations, as Bob Rutherford shows.

The environmental and health implications of energy use already have an enormous literature. So Part VIII considers two areas which have received insufficient attention — the carbon dioxide problem and nuclear fusion — and also offers a cross-cultural perspective from Diana and Ian Maddocks.

Apart from energy conservation, the renewable energy sources which are probably nearest to commercial utilization are solar heat, the production of alcohol from crops, windpower, and solar photovoltaic cells, in that order. With the exception of solar heat, these are discussed in Part IX. (Passive solar design is considered by Tone Wheeler in Part III.) Windpower and, with some minor technical advances, solar cells, are of particular interest in building up a self-sufficient energy future for a community because they can become true "solar energy breeders".

The papers in this book are diverse in their content, viewpoint and presentation. However, they share the conviction that the major problems of our energy futures are not technological, but institutional, political and ethical.

Acknowledgement: The editor wishes to thank the participants at the 1978 *Energy and People* Conference for making this book possible. Ms Jenny Hannan kindly assisted with subediting and typing.

ENERGY STRATEGIES — PRESENT, FUTURE AND HISTORICAL

What is the Basis of Current Energy Forecasts?

J. Dick and C. J. Mardon

Conservation of Urban Energy Group
Conservation Council of Victoria
324 William Street, Melbourne, 3000.

Introduction

Unlike the real world of the market place, almost the entire emphasis of official energy forecasts is on the expected demand for a particular fuel, whether it be oil, electricity or whatever, with little or no regard to the likely future availability of each fuel, or for the overall rational use of resources. Occasional statements are made that "Australia is an energy-rich country", "We are a net exporter of energy" or "We are confident that new developments in technology will yield unlimited supplies of cheap energy for the future", but with the exception of oil and gas, there has been no attempt to estimate the likely productive capacity of our domestic energy reserves over the next 20-30 years, the fiscal and environmental implications of the development of these reserves, or the likely availability of imported oil. We are told that more reserves of oil and gas must be found and that alternative sources of energy must then be found to cope with future growth in energy consumption. Considering the not-inconsiderable amounts of capital that will be required to provide all this extra energy and the environmental consequences which will inevitably arise from the increasing scale and concentration of new development associated with its supply, it is surprising that there has not been more discussion about the implementation of energy policies, or for that matter, the true basis of the growth expectations evident in the demand forecasts.

Despite our general lack of knowledge about how much of our energy resources are in fact recoverable, many people still talk glibly of the life of a resource at present rates of consumption rather than how long we can continue to increase the productive capacity of a resource to meet the demand. This tendency is reinforced by the fact that economists tend not to talk to geologists and that economists (and engineers) tend to draw lines on semi-logarithmic graph paper and calculate percentage growth rates (albeit changing with time) with complete disregard for the availability or cost of potential supplies. Their commitment to growth is so strong that they cannot imagine growth stopping — the law of population growth is forgotten and decline is unthinkable. Even when growth is clearly flattening out, they tack on exponential curves for the forecast period to bring it back again. Temporary deviations from the growth curve are blamed on the "Oil Crisis" or a "recession" without any attempt to analyse the underlying long-term trends in any detail.

The importance of all this is twofold. Firstly, the increasingly long lead times required for the planning and construction of ever-larger power stations, and for the production of substitute fuels such as oil or gas from coal, make it imperative for the Governments or corporations involved to make major decisions, plan their investments and allocate resources much further ahead than they are normally accustomed. For example, the minimum "economic" size of an oil-from-coal plant is now estimated to be some 10-20 times larger than it was twenty years ago and it may take another twenty years to develop and build one. Also, the development of future coal-conversion plants may be pre-empted by decisions committing the few coal deposits large enough to sustain such plants over their working life for use by large power stations. Conversely, are all small-medium coal deposits to be rendered "uneconomic" by the rapidly increasing scale of power stations and coal-conversion plants? Hence, the growth of all forms of energy needs to be known reasonably accurately twenty or thirty years in advance if mistakes are to be avoided and future options not foreclosed. An error of only 2% per annum over such a period could make a big difference in the supply capacity required, but the assumption of even 3-4% growth when demand was only growing linearly or beginning to flatten off could mean truly gross over-investment in unnecessary extra capacity. Errors of this sort have led to a 50% excess generating capacity in Britain's Central Electricity Generating Board already.

The second point is that in an era of increasing resource scarcity and expensive substitutes, further growth of the energy supply system, i.e. growth over and above that necessary to maintain the supply of direct or secondary energy at its present level, can only mean the commitment of an expanding proportion of the available investment capital and natural resources to the energy sector at the expense of other sectors. Also, even without further growth overall, a major shift in the fuel mix, such as the production of liquid transport fuels from coal, and/or a major nuclear power program, could have the same effect. Most economists now living have been brought up in an era of declining fuel costs and it is simply assumed that new technology will make new resources available at ever-lower costs as they are needed. This is no longer happening because, with oil, for example, new discoveries are not keeping up with demand (here or world-wide) and the reserves/production ratio is falling. New technology notwithstanding, no cheap substitute of equivalent convenience and energy-density is in sight, even though we are rapidly approaching the point where world oil production must inevitably fall (an R/P ratio of 15).

The basic reason is that cheap energy from existing deposits is now required in increasingly larger amounts to extract and process increasingly inaccessible resources and also that having built up an excessive dependence on cheap liquid fuels, we now expect the institutions which promoted that dependence to defy the Laws of Thermodynamics! Instead, they are doing all they can to diversify by investing in coal and uranium, but they cannot continue to increase the supply of oil and gas much beyond present levels, and then not for very long. The price must inevitably rise and we must now ask whether the enormous cost of even maintaining the present fuel mix and overall consumption rates is justified, let alone the cost of further growth. We must ask *why* the huge expected increases are necessary, who needs them, how long they can be maintained and what the social costs will be? Moreover, we must also ask whether it would not be cheaper to reduce the demand for primary energy by increasing conversion or end-use efficiencies, promoting less energy-intensive industrial processes or modes of transport, planning future urban development to reduce household and transport energy needs, and so on.

Diesendorf, M. (ed.) (1979). *Energy and People*. Canberra, Society for Social Responsibility in Science (A.C.T.)

Even if we had to encourage such alternatives to increasing the energy supply by some combination of taxes and subsidies, the final social cost would be much less than that of our present course. The only losers would be the Energy Industries themselves. We see current political struggles in the light of obvious pressures from Mining and Oil interests, as well as the Public Electricity and Gas utilities, to maintain their established position of power and use our dependence on what they provide as a means for further increasing their power. If the community were to decide that the cost no longer justified the means of maintaining their present lifestyle, and they had the political will to bring about the necessary changes to those institutions obstructing the changes which we must ultimately make for thermodynamic reasons, then a smooth transfer to a sustainable low-energy lifestyle could be made.

The basic problem is one of values rather than resources per se and we will have to educate economists and politicians alike that no amount of new technology can defy the Laws of Thermodynamics and that a barrel of oil from Saudi Arabia is not the same as a barrel of oil made from coal. Until everybody realises that there are different qualities of energy and that a Joule of steam does not equal a Joule of electricity, then it will be very difficult to arrive at the social and political consensus needed to achieve the institutional changes referred to above.

Energy Availability or Resources In The Ground?

Before going on to examine the energy demand forecasts, we would like to clear up a few misconceptions about energy resources and point out some possible restraints on the likely availability of these resources.

First, there is the frequent confusion between "reserves" and "resources". For example, the Victorian Premier Mr Hamer has often been quoted in the press as stating "We have only used a half of one percent of our brown coal resources and we still have enough to last for a thousand years". While such statements cannot be proved wrong (except by waiting for a thousand years to see what happens), they are misleading to say the least because they assume that all known and inferred brown coal reserves will ultimately be recovered and that brown coal consumption will never exceed the present rate of consumption.

An even less credible example appeared recently in a report of the National Energy Advisory Committee (1). In a table of Australian energy resources, it states that the expected life of our brown coal reserves is 1,300 years at 1975-76 rates of consumption, but only 1,000 years at 1984-85 rates of consumption, a loss of 300 years' supply in only nine years! Even more remarkable, the same table states that the expected life of our crude oil and condensate reserves is ten years at 1975-76 rates of consumption, but it is still given as eight years at 1984-85 rates of consumption when there should theoretically be only one year's consumption left!

These statements illustrate the absurdity of trying to estimate the life of a resource simply by dividing the total amount by the present rate of consumption, but they also raise two other problems. First, the resource base — how much will ultimately be recovered — and secondly, how much will the rate of extraction change in future — will it increase, decrease, or remain the same? The expected future rate of consumption is not a reliable guide to the rate of extraction of domestic reserves because any shortfall must presumably be met by imports and any surplus would probably be exported. Extraction rates depend on a variety of economic, technical and institutional factors which are often beyond the control of the energy forecaster.

The problem with our estimates of the resource base for each of Australia's non-renewable energy resources is that, with the possible exception of oil, the basic data required for the classification of our resources is either inadequate or not available. The Bureau of Mineral Resources is attempting to classify the available data in accordance with the McKelvey system which is used in the United States, but this involves fitting resource data that are not normally classified in this way. For example, total quantities of in situ crude oil and natural gas are never reported; current reserve figures supplied by the oil companies already include their own estimate of the recovery factor and make no allowance for possible additional quantities which might be produced using more expensive secondary or tertiary recovery techniques if economic conditions were to change.

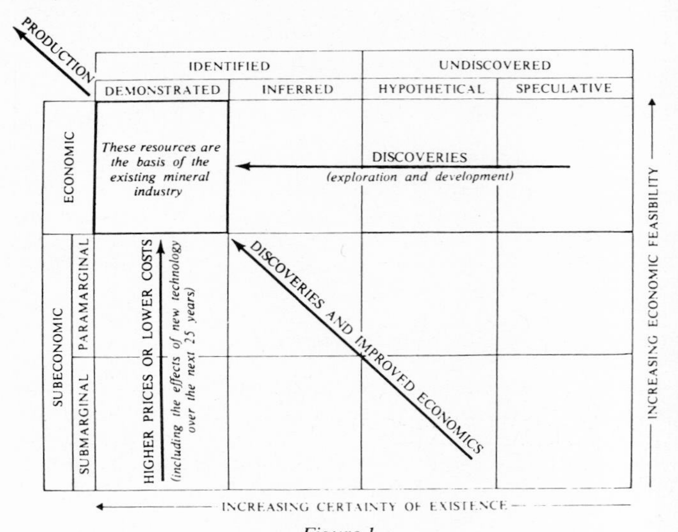

Figure 1

Resource classification diagram showing the general progression of non-renewable resources as a result of exploration, research, development and production

Figure 1 shows the "McKelvey Box" — a system for classifying mineral resources in terms of its certainty of existence and the economic feasibility of its extraction. The total resource can be very large indeed, but the "in-situ" reserves on which our estimates of recoverable reserves are based, are represented by the single square in the top left-hand corner — the "Demonstrated Economic" category. This category includes all resources which have already been discovered and are judged to be economically extractible now *or* potentially economically extractible over an arbitrary 25 year period. It represents an upper limit to the amount of that resource which the mineral industry can realistically expect to extract without better definition and development of inferred reserves or exploration leading to new discoveries. The economically recoverable reserves are often a good deal less than this, sometimes an order of magnitude less, so it leads to the sort of confusion that arises from public statements such as those referred to at the beginning. There is no guarantee at all that the entire in-situ reserves will ultimately be extracted because the total is an aggregate of many individual deposits, some large, some small, some remote or inaccessible, some too deep to make extraction worthwhile, some located underneath towns or cities, some of dubious quality and so on. The average recovery ratio for individual open-cut coal mines is usually 80-90%, but it drops to about 30-60% with underground coal mines and 25-50% with oil wells. Someone will say . . . "But the ratio in Bass Strait is over 60%" . . . which is true, but Barrow Island is only 20%, so that is only another way of saying that there are a few exceptions to the rule and that they can go either way. For a more comprehensive explanation of the methods and problems inherent in resource classification, we would refer you to NEAC's second report (2), D.C.Ion (3) and the WAES Report (4).

The major deficiency of the McKelvey classification is that it gives you no idea of the *likely availability* of resources through time. Additional data indicating sub-categories classified according to the size, accessibility and quality of the resource are essential for the estimation of potential availability at different times in the future.

The very definition of "economic" reserves is open to question because they often include parts of individual deposits which are very unlikely to be exploited in the foreseeable future because they are too thin or too deep as well as other deposits which by themselves are too small or too remote to warrant separate development. This question of the minimum economic size of individual deposits is a complex one, but critical to an understanding of the likely availability of a particular resource.

A number of examples readily come to mind. For instance, a few years ago, a small oil field was found at Mereenie near Alice Springs (as well as a small gas field at nearby Palm Valley) and it was proposed that a small refinery be built there to supply petroleum products to the Northern Territory market. However, because that market is small, the refinery would have been below what is considered to be an economic size and because the location was too remote to allow shipment of the crude oil to Adelaide, the oil remains in the ground and is unlikely to be used. Instead, the oil companies are importing refined products from overseas to supply the Northern Territory market. On a global scale, the WAES Report (4) points out that one reason for the inability of new discoveries to keep up with demand is that the most new oil fields are offshore or in remote inhospitable areas where the extraction cost is high ($5-10/bbl). This tends to limit development to very large fields, usually fields with recoverable reserves greater than 500 million barrels, because only these large fields can justify the very high cost of production platforms and pipelines in these areas. Despite the intensive effort put into oil exploration over the last 100 years, only 240 fields of this size have been found and not enough new ones are being found.

Table 1 illustrates the distribution of the various sized fields discovered so far. Note that world crude oil consumption is now so large that even a large new oil province such as the North Sea, with recoverable reserves now estimated at about 40 billion barrels, is only sufficient to supply world oil demand for two years! This means that a new oil province the same size as the North Sea and consisting predominantly of large deposits must be discovered, developed and brought into production every two years. Data presented by Facer (5) show that total world proven reserves of oil stopped rising in the late 1960's and have actually fallen since 1970. The reserves: production ratio, the oil industries' estimate of working reserves, has fallen consistently since World War II as demand rose and it is rapidly approaching the figure which the oil industry regard as a minimum if total production is not to fall (15:1). The ratio was 27.7 in 1976 and still falling.

Table 1
Approximate Size of World Oil Fields (4), (5)

	Number of fields discovered	*Estimated % of WOCA reserves*
A. All fields	30,000	100
B. Fields greater than 0.5 billion bbl. recoverable reserves	240	73
C. Fields greater than 10.0 billion bbl. recoverable reserves	15	34
D. 4 largest fields: Ghawar (Saudi Arabia), Greater Burgan (Kuwait), Bolivar Coastal (Venezuela), Safaniya-Khafji (Saudi Arabia/Neutral Zone)	4	21

	Year 1950	1960	1970	1976
R/P Ratio	90	38	36	28

Another example of the effect of economic size on the extraction of individual deposits can be seen in the following classification (Table 2) of the Latrobe Valley coalfields in Victoria.

The total "Geological" or "in-situ" reserves are quite large, but some fields are much bigger than others, some have thicker seams than others and the estimates of reserves which are recoverable with present open-cut mining techniques are considerably less than the "Geological" reserves. The new power station now being built at Loy Yang is expected to require some 36 million tons of coal per annum at full output, or 1,000 million tons over its useful life of 30 years. Another 600 million tons will be required for the existing power stations

Table 2
Coal resources overburden < 30m

COALFIELDS	RESOURCES $X10^6$ tonne Geological	Mining	COAL / OVERBURDEN	THICKNESS METRES
Loy Yang	12650	3450	4:1	250
Yallourn	16650 (Yallourn and Morwell)	2350	3·5:1	60
Morwell		2250	4:1	100
Gormandale	3250	800	1:1	200
Rosedale	1850		2·5:1	
Gelliondale	1000+	300	3:1	50
Holey Plains	1000			

at Yallourn and Morwell over the remainder of their useful lives. Hence about 20% of the recoverable reserves in the three largest deposits *has already been committed.* According to the Victorian Government Green Paper on Energy (6), two or more plants may be built in the Latrobe Valley in the 1990's to produce oil from brown coal. Depending on the actual technology adopted, and assuming that the minimum economic size of these plants is 100,000 bbl/day (the same capacity as a large modern refinery), then they will require about 40-90 million tons of brown coal per annum *each*, or a total of 2,000-5,000 million tons over the useful life of the two plants. Hence these two plants alone would commit most of the remaining recoverable reserves in these three largest fields, perhaps leaving enough for one more large power station. Of the other fields, only Gormandale has sufficient recoverable reserves for even *one* large power station, let alone a coal-conversion plant. If a solvent-refined-coal plant for metallurgical char production were to be built on the Gelliondale field, as has been suggested in recent press reports, then most of the recoverable brown coal reserves would be committed. Other small fields exist, but they are no longer large enough to sustain a modern power station or coal liquefaction plant *individually,* and until means have been found to transport large quantities of wet brown coal long distances, most of these deposits will probably stay in the ground indefinitely.

A similar explanation can also be given to show why not all of the "in-situ" black coal reserves will be recovered. It is true that black coal is more economically transported than brown coal, but the size of individual deposits will still be important for much the same reasons. In addition, with underground mined coal, there are the problems of reduced recovery due to increased mechanisation of the mines — the machines are more economic because they dig out the coal much more rapidly than men with picks and shovels, but they are less flexible and remove a smaller proportion of the coal because they require thick seams with no faults or other discontinuities — and that due to "sterilisation" of deposits (or parts of deposits) by flooding, roof collapse or other factors causing an extended interruption to production and possibly leading to permanent closure of the mine. There are also institutional, financial, environmental and numerous other factors which could prevent the development of many deposits still classified as "economic", so we should try to allow for them in coming to a realistic estimate of the resource base and the likely future availability of a resource. Aggregate figures for a few broad categories is not enough.

Many people would then say: "Yes, but when we run out of readily recoverable reserves, we can start using the marginal reserves around existing deposits or other deposits now regarded as sub-economic". Figure 2 gives an indication of what would happen in the extraction of Victorian brown coal. Not only does the calorific value of the coal decline as more and more poorer quality marginal deposits are brought into production — the energy required for mining and pollution control will start to rise rapidly as the economically recoverable deposits are worked out. The result of this is that the *net* energy produced per ton of coal mined falls off sharply as the geological reserves are approached. It is difficult to quantify these changes because we have only mined about 500 million tons so far and present mining energy requirements are only a few percent of the calorific value of the coal mined. It is further complicated by the fact that present mining practice is to dump overburden over underlying seams of marginal coal, thereby increasing the future mining energy requirement. As Lovins (7) has stated so eloquently, changes in energy *quality* are important because the ability of marginal sources to contribute a net yield goes down as the other sources of subsidy become poorer. That is, energy technologies which now yield only a small energy profit will be even less profitable, or even net consumers, as the high-grade (low-cost, accessible energy) available to subsidise them declines. The increasing capital intensity in the energy sector means that even where the quality of a particular resource is not declining significantly, the energy cost of obtaining useful (secondary) energy is increasing. Hence the net energy yields from gross energy resources are declining. This turns the usual assumption of most economists, that marginal energy sources will become economic when the rich sources are gone, completely on its head. Until we recognise the consequences of declining energy quality, we will continue to over-estimate our future availability of resources.

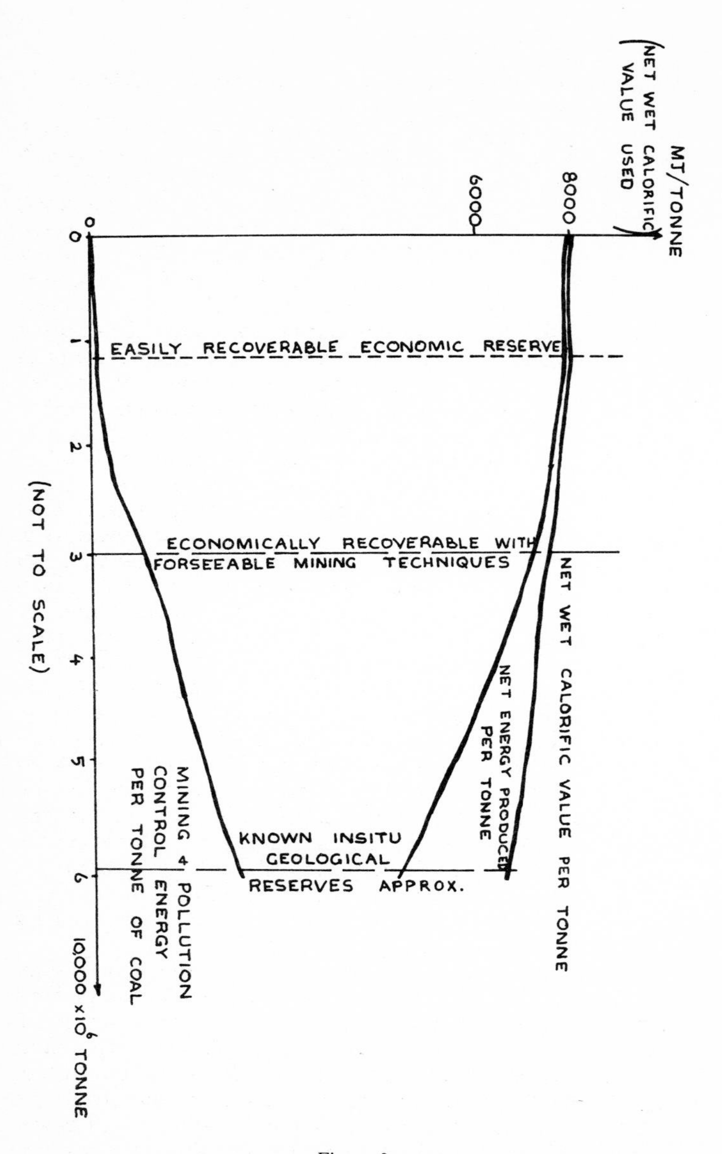

Figure 2
Theoretical plot of available energy with increasing resource size for Victorian brown coal

Returning now to our point about the manner in which the rate of extraction of resources will change in the future, it appears that there is even more confusion about this than there is about the resource base. Judging by the verbal battle that has been raging through the pages of *Search* (8), most correspondents believe that historical energy consumption has been growing exponentially and that it will continue to do so, even if at a reduced rate. As we will show later, even this assumption is not quite true, but the confusion arises when these correspondents make the mental leap from energy consumption to the production of energy resources. Some, as we have already noted, simply divide the total resource by the present rate of consumption. Quite naturally, they conclude that our resources are so vast that we have no need to worry for the forseeable future. In the other camp, Maglen (9) and Rigby (10) have drawn attention to the fact that *sustained* exponential depletion of even very large resources at quite modest percentage growth rates (e.g. 5%), dramatically reduces the life of the resource. For example, if Victorian electricity demand were to continue growing at 6.5% per annum as predicted by the SECV, then the recoverable reserves of brown coal in the Latrobe Valley would be exhausted in about 50 years, even without the construction of coal liquefaction plants! In fact, the rate of extraction of energy resources is neither likely to remain constant nor to grow exponentially. Obviously, if energy consumption were to grow exponentially and the rate of extraction remained constant, then imports would rise exponentially! You cannot have it both ways.

The Importance of Availability

Having explained the difference between the actual amount of energy available and the total amount of resources in the ground, we must now explain why the availability of a particular resource over a given period of time should be considered together with the likely demand for that resource over the same period. There are five basic reasons, all of which have important implications for the future.

1. Possible supply shortfalls

If the demand for a particular resource is likely to exceed its availability at some time in the future, due allowance must be made for the lead time required to discover and develop new deposits or develop *and* produce substitutes. Two examples come immediately to mind. First is of course oil. The best information we have been able to obtain on the likely range of Australian oil demand and domestic oil production forecasts (Fig. 3) has been reproduced from the 5th Report of the Royal Commission on Petroleum (11). All of these projections show continued growth in consumption, despite a fall in domestic production from 1980 onwards. They imply a rapid increase in oil imports, quadrupling by 1990. Similar scenarios can be found in the forecasts of many other countries, but against this unbounded optimism, what are our prospects for increasing our oil imports? As discussed in the WAES Report (4) and in our book *Seeds For Change* (12), our oil imports will essentially depend on how much oil the OPEC countries decide to produce and on what proportion of their oil exports we can obtain. For a variety of reasons which we will not discuss here, it is unlikely that total OPEC production will increase very much above present levels, still less quadruple, by 1990. The WAES Report (4) expects production to level out somewhere between 33-44 million barrels per day by the mid-1980's, compared with current production rates of about 30 million barrels per day. At present, we receive about 0.7% of OPEC's total exports, but in view of the fact that we already have an unfavourable balance of trade with the Middle East, that the USA and Europe will be needing more OPEC oil as the North Sea and Alaskan fields reach their peak in the mid-1980's and that the Soviet Bloc may switch from being a net exporter of oil to a net importer in the early 1980's, it does not seem at all likely that we could significantly increase our share of OPEC exports between now and 1990. If that happens to be the case, then our oil imports are unlikely to increase very much at all (see Fig. 4) and it is quite obvious that unless our share of OPEC exports can be drastically increased, there is no hope whatever of meeting the

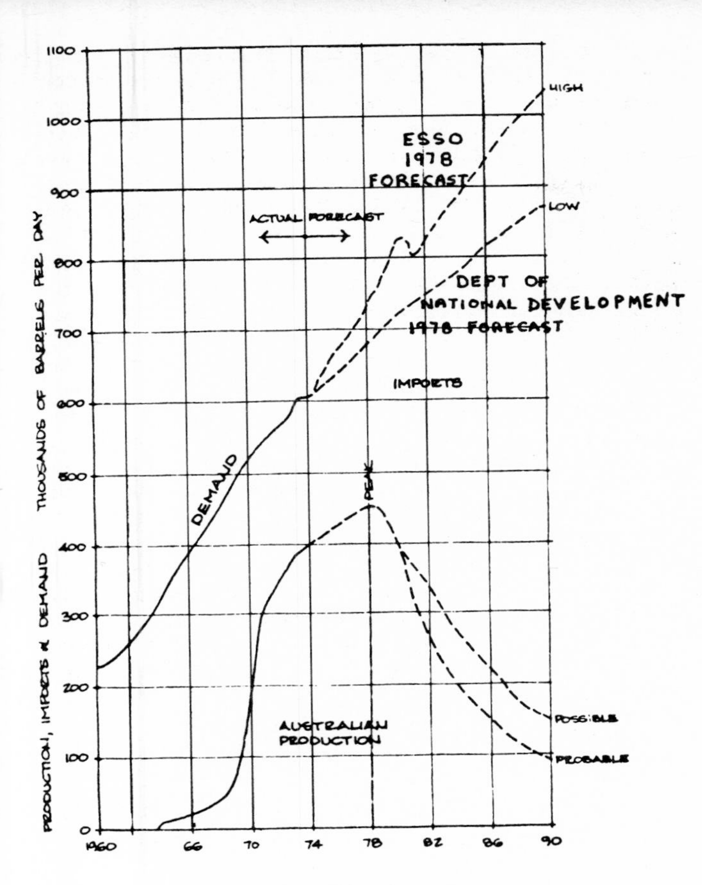

Figure 3
Australian crude oil production, imports and demand

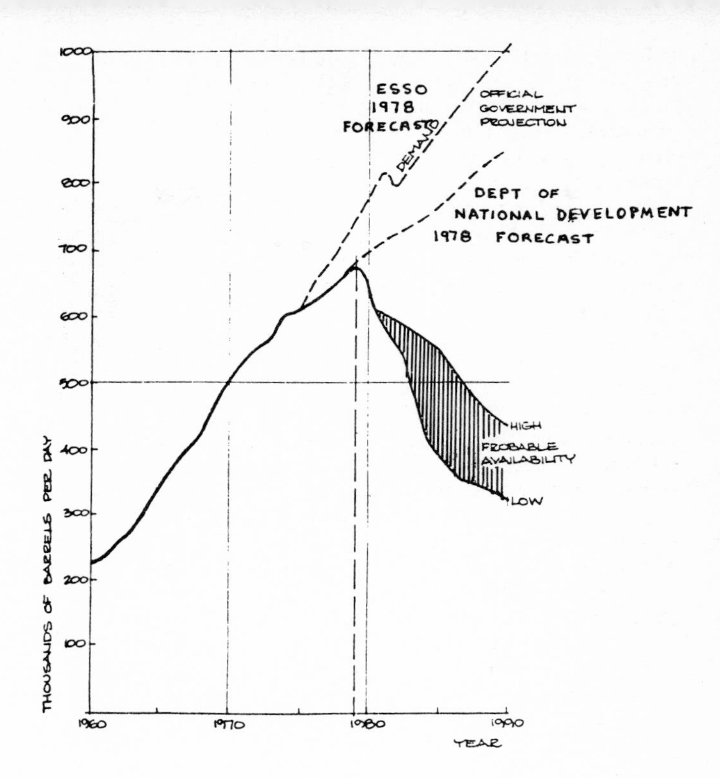

Figure 5
Projected Australian crude oil demand and likely availability

official forecasts, even in the short term, so the forecasts must be revised downwards. If our share of the available OPEC exports (assumed to continue at its present level) is added to the forecast domestic production, we obtain the likely availability of crude oil in Australia, as in Fig. 5. It shows that despite the fact that Australian and World oil reserves are still large, a shortfall in *our* oil supplies is possible quite soon. While it is true that there is a temporary glut in the world oil market and that oil imports might well be increased for another 2-3 years, there is no guarantee that imports can continue to be increased as the market tightens again just because our official forecasts say that we will need more. Hence the problem of oil imports is not just a problem of cost or of foreign exchange. No amount of coal or uranium can buy oil which is not available. The usual assumptions of a free market simply do not apply to the price of oil because it is subject to political manipulation and the influence of power politics. The price of oil bears no relation whatever to the actual cost of production in the Middle East — rather, it is geared to the costs of possible substitutes for oil. It is now apparent that synthetic oil manufactured from coal is unlikely to be a serious competitor to Middle East oil at least until the 1990s and there are few other potential competitors in sight. In Australia, there are few additional oil fields which could be brought into production in time, so the only short-term substitute of any consequence is LPG. However, most of the LPG we now produce is exported, so we cannot use it as a substitute for oil unless we cease exporting it and even then it would only be a partial substitute. We have another four years to wait before the export contracts come up for review again, so we should use that time to consider possible distribution and pricing arrangements. We must also consider ways in which the demand for oil can be reduced so that actual shortfalls will not arise. Possible strategies for Australia are discussed in some detail in *Seeds For Change* (12). Above all, we must not meekly accept official demand forecasts and oil company assurances at their face value — they are wooing us into a false sense of security so that they can exact a high price for accommodating our needs when the time comes.

The natural gas supply situation in Australia is hardly better than that for oil. In its System Study Report (13), the Pipeline Authority predicts that gas shortfalls may occur in South Australia, New South Wales and Queensland by 1985 and in Victoria by 1990. The Victorian situation has since been further aggravated by continued rapid growth in industrial gas consumption and decisions to go ahead with gas-fired power stations at Newport and Jeeralang. Figure 6 shows the effect of adding the SECV gas demand to the Gas and Fuel Corporation demand forecast. Note that instead of being a problem for the

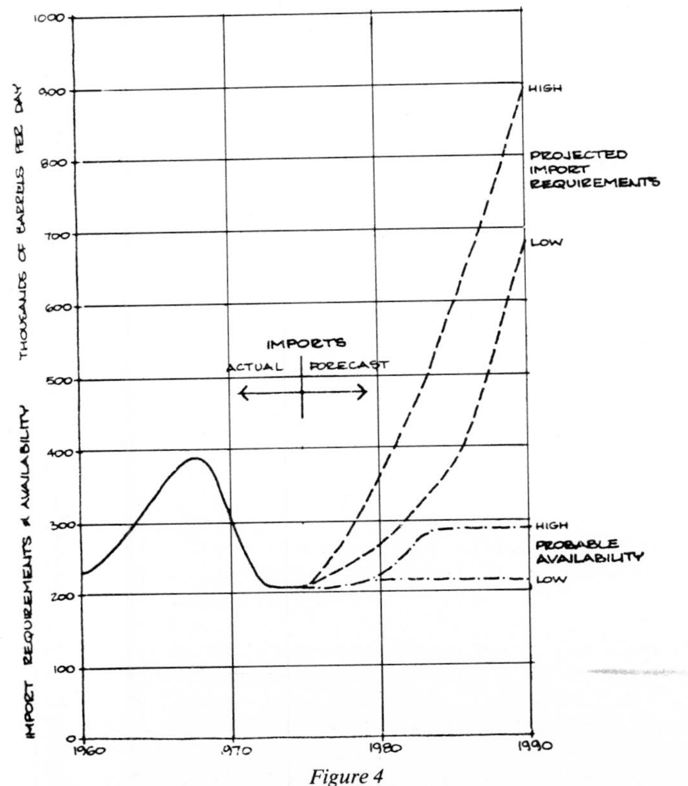

Figure 4
Australian crude oil import requirements and probable availability

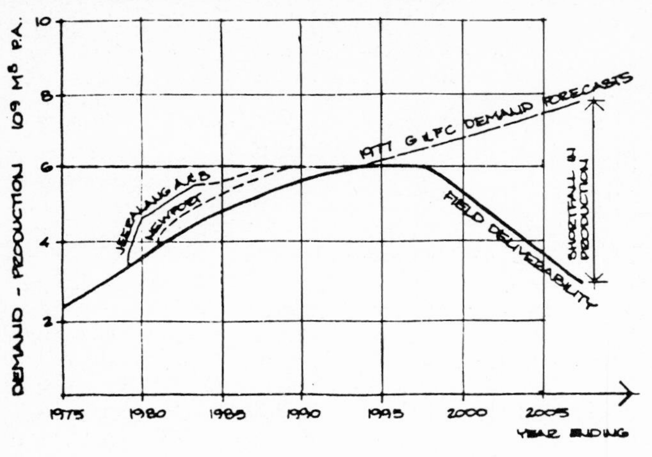

Figure 6
Supply and demand forecast — low growth

The demand in this Figure is based on 1977 Gas and Fuel Corporation demand forecasts. The field deliverability curve is based on the Australian Pipeline Authority's estimates.

mid-1990's, we can now expect gas shortfalls in the late 1980's. While it might be possible to increase the overall field deliverability by drilling more wells, building more production platforms and by storing gas in large underground reservoirs, it can only be done at a cost and something will have to be done within the next ten years if shortfalls are to be avoided, not in the next century as is usually suggested. If further local supplies of gas cannot be found (it appears that initial estimates of recoverable reserves were inflated), then a decision must be made whether to build an LNG terminal in Victoria to receive liquefied natural gas from the North West Shelf or to build a coal gasification plant in the Latrobe Valley. Either alternative will be very expensive and will take many years to plan and construct, so if we accept the demand forecasts as inevitable, we must also be prepared for the costs and environmental consequences of the alternatives. Again, we must not be lulled into a false sense of security by assurances from the producers that all will be well for the forseeable future. If we do not want to be stampeded into expensive or environmentally disadvantageous decisions in the future, we must demand access to the relevant information and demand the right to participate in the planning process long before firm commitments are made to a particular course of action. As these examples show, we are talking of periods up to ten years in which we have to act, not 30 or 40.

2. *Fuel mix problems*

It is important to remember that while oil and gas are vital components of the fuel mix, they only account for just over half of our primary energy supplies. There are a number of other primary fuels and the proportions of each vary considerably with time. Figure 7 shows that immediately after World War II, just over half of our primary energy came from black coal, and that wood, not oil, was the second largest fuel. Black coal reached its peak in the late '40's and then slowly declined, though absolute quantities have actually increased at about 3% per annum. During this early post-war period, however, wood was in even more rapid decline and petroleum products tended to substitute for wood rather than coal. By the '60's, this process of substitution reached the point where wood had become an insignificant proportion of our primary energy and coal was clearly in decline. In its turn, petroleum products reached their peak in 1970. Since then, natural gas has steadily increased its share at the expense of petroleum products and this trend is expected to continue. However, there is nothing inevitable about this succession process and it is reversible. For example, if most of the North-West Shelf natural gas is exported and few new reserves are found, then the natural gas sector will decline and it will be substituted by something else. If synthetic oil or gas are produced from coal, we can expect the coal sector to grow at the expense of natural oil and gas. Similarly, if there is a return to renewable fuels such as wood, straw or wind power, then this sector will increase again. As you can see, nuclear power does not appear on this diagram and there is no necessary reason why it should. It should not be forgotten, however, that these substitutions are generally fairly slow — in our case, it took 20 years for oil to take over from coal as the dominant primary fuel, so we cannot expect miracles to happen over the next few years. With our transport system totally dependent on petroleum fuels and many industries at least partly dependent on them, it will take at least 10-15 years to make any significant shift to an alternative fuel. Therefore, our substantial coal reserves will indeed be useful in the event of oil or gas shortages, but no immediate changes can be made until new supply infrastructures are built up and appropriate investments made in new transport vehicles, industrial boilers and furnaces, etc., so our coal will not help us at all in the short term.

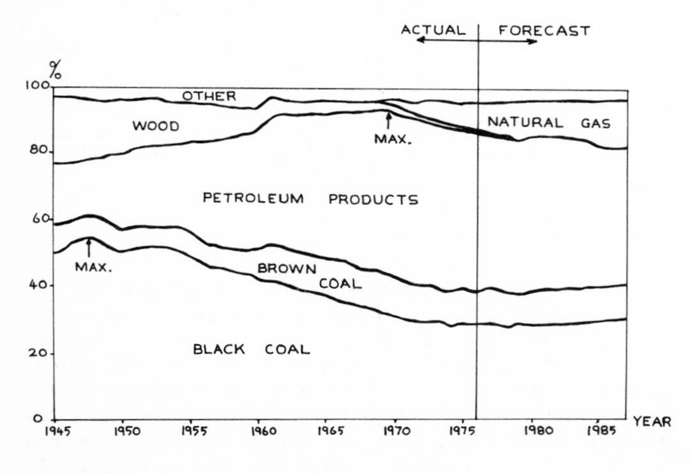

Figure 7
Australian primary fuel mix

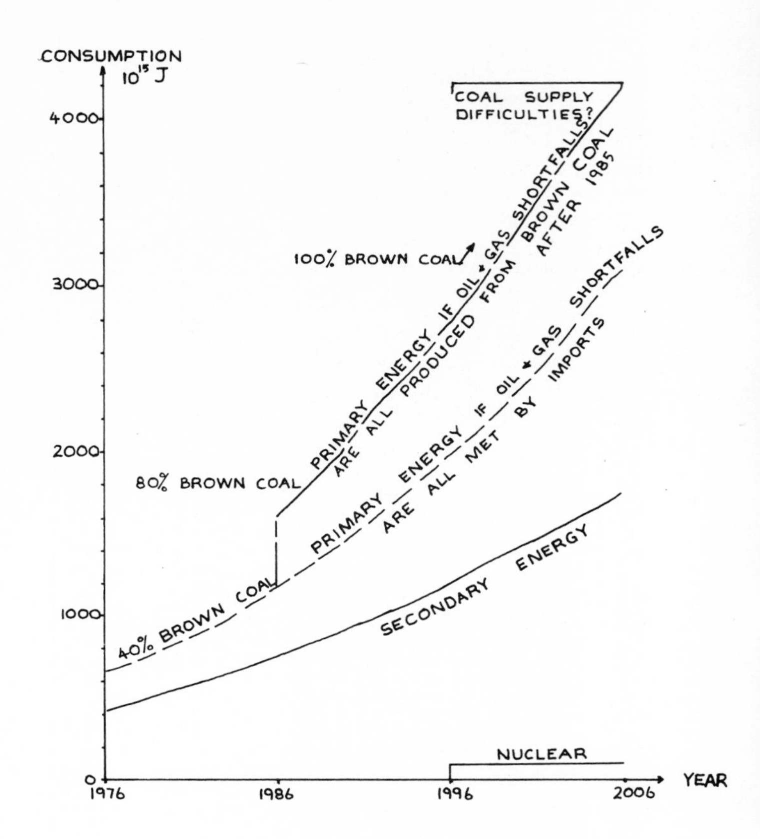

Figure 8
Comparison of primary & secondary energy (Green Paper high-level forecast)

3. Need for substitution
Figure 8 shows that if all the oil and gas shortfalls anticipated by the Victorian Green Paper on Energy were actually met by imports, the proportion of Victoria's primary energy accounted for by brown coal would remain close to its present value of 40%. However, if we were to assume that oil and gas imports to Victoria were unavailable from the mid-1980's and that sufficient production capacity for synthetic oil and gas had been established by that time (this does not seem very likely, but Lurgi spokesmen have claimed that a coal liquefaction plant could be built in Victoria by 1985 if a decision to go ahead were made now), then there would have to be a very rapid expansion of mining activity in the Latrobe Valley coal fields and the proportion of brown coal in the primary fuel mix (Victoria only) would immediately double to 80% and then slowly increase towards 100% as local oil and natural gas production declined. As will be explained in more detail later, such a scenario would soon run into coal supply difficulties, but the important point to be made here is the very rapid increase in brown coal production required by such a scenario if imported oil and natural gas were not available in the large quantities expected to be required in the mid-1980's. Any "fail-safe" energy policy would require that steps be taken very soon to either reduce the demand for oil and natural gas in line with local productive capacity, or build plants for the production of synthetic oil and gas and increase brown coal production to suit, or more probably, some mixture of the two.

4. Increased conversion losses
Figure 8 also shows that both scenarios, but especially the coal-conversion scenario, involve faster increases in the production of primary energy in order to supply the expected growth in secondary energy because the conversion losses associated with the production of electricity, synthetic oil and gas from coal will be higher than the conversion losses associated with the present fuel mix. The conversion efficiencies of synthetic oil and gas plants will probably be in the range 30-60% instead of 85-90% for the refining and distribution of domestic oil and natural gas. It is important to remember too that all of the conversion losses in the coal-conversion scenario will result in the dissipation of vast quantities of low-grade heat, water and air pollutants within the narrow confines of the Latrobe Valley. The demand for cooling water alone would put a severe strain on local water supplies and the greatly increased mining activity would affect local ground-water, cause land subsidence and numerous land-use conflicts. In large, shallow open-cut mines, an increase in the production rate of up to ten or fifteen times the present rate of mining by the year 2000 would require the development of extensive areas and would have very severe effects on the local environment and water supplies. It must be asked whether such a rapid expansion of brown coal production is feasible, and in view of the price of such development, whether it is worth it.

5. Effects of exports
It is frequently asserted that Australia is a *net exporter* of energy. In fact, it has been repeated so often that many people now believe it to be true. However, it is only true if our substantial exports of *coking coal* are included in the overall balance of energy imports and exports. Strictly speaking, coking coal is a *mineral* export rather than an energy export because it is a *raw material* in steel manufacture and the level of coking coal exports is dependent on the demand for steel, not the demand for energy — no-one would waste premium grade coking coal by burning it in a power station! Scarcer energy means higher energy prices and probably also a continuation of the present recession with its depressed demand for steel products, so it is inappropriate to lump coking coal in with other energy resources. The following table shows that in 1975, 60% of Australian black coal was exported, almost all of it coking coal, and that the exclusion of this coking coal from our energy import-export balance makes us a *large net importer of energy*.

Table 3
Australian Energy Production & Consumption, 1975

Primary Fuel	*Production* 10^{15}J	*Consumption* 10^{15}J	*Imports/ Exports* 10^{15}J
Black Coal	2,070	830	1,240
Brown Coal	260	260	—
Natural Gas	190	190	—
Petroleum	909	1,340	(—431)

Obviously, we should differentiate between coking coal and steaming coal consumption within Australia — only a third of domestic black coal consumption is coking coal, whereas almost all of the export coal is coking coal. However, if steaming coal exports were to rise rapidly in future, domestic production of steaming coal would have to rise even if there were no change in local consumption. Too often, local resources are compared with local consumption (Australian or State), completely disregarding the effects of exports. Clearly, the availability of resources for export must be considered together with the availability of resources for local consumption, as indeed must the availability of imports for local consumption.

Methodology of Demand Forecasting

Having raised a few of the problems involved in defining energy resources, we will now concentrate on the methodology of demand forecasting before going on to look at resource depletion policy and the matching of supply to demand.

In Figure 9 (14), we have a typical example of what we choose to call "Extrapolationitis". It assumes uninterrupted exponential growth, i.e. a constant percentage growth rate, and uses a semi-logarithmic plot in order to obtain (conveniently) a straight line. This sort of "linear" thinking is dangerous because, by compressing the upper part of the graph, the logarithmic scale disguises changes in growth patterns and encourages unrealistic expectations of future consumption rates. Even redrawing such projections on a linear scale does not help because historical growth trends vary with changes in economic conditions and are often difficult to classify. Also, many apparently exponential curves have a slowly declining percentage growth rate, making extrapolation hazardous.

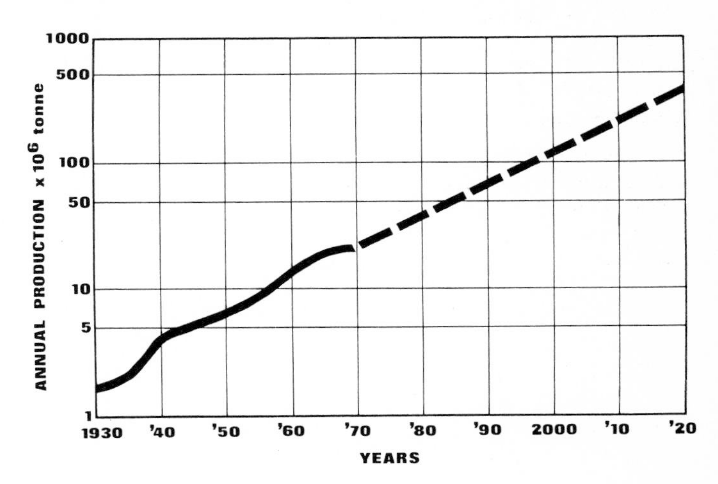

Figure 9
Production of brown coal (14)

If taken over a long enough period, the demand for any product, whether it be petrol or refrigerators, usually goes through a cycle such as that shown in Figure 10 in which a period of rapid growth is followed by an approach towards the saturation point for that product. This cycle, usually known as a logistic curve, is the basis of the theory of population growth (15) and has been found to apply to the growth of bacterial and insect populations as well as to many other growth situations in the natural world and in commerce. As in ecology, the emergence of a competitor may result in substitution of a new fuel or product, leading to a decline in the old one — otherwise, it should stabilise at some constant value.

The problem with most short-term plots of demand is that it is often difficult to tell whether the curve is "quasi-exponential", "quasi-linear" or tending to saturate. In such cases, a much more sensitive test of changes to the growth trend than a semi-logarithmic plot is to plot a differential, or more commonly, difference curve. Figure 11 shows some examples. Clearly, from a consideration of elementary calculus, only an exponential curve will have an exponential differential curve. A constant differential indicates linear growth, while a declining, but still positive, differential suggests that saturation may be occurring. These curves can be drawn without the use of a computer or anything more sophisticated than an adding machine. An approximate curve can be drawn simply by smoothing the original curve visually or by taking moving five-year averages, calculating the difference in ordinate between successive points, and plotting these differences against time. While not mathematically precise, this method does give a good indication of whether the trend in the differences is up, down or constant.

The practice of plotting energy forecasts on semi-logarithmic graph paper has led many engineers, economists and businessmen to overlook departures from the expected trend (or worse, to consider such departures as temporary aberrations) and tack exponential, or at least over-optimistic growth curves onto the end of curves which are clearly tending to saturate. Two examples will be presented here to show the effect of such practices on the expected and likely future trends in the demand for energy. The first, shown in Figure 12, is from a study (16) by the Energy Research Centre at the Open University of the Electricity Industry in the UK. It shows that in fact the trend towards saturation was already occurring at least ten years before the Oil Crisis in 1973, but that this period of time was apparently insufficient for the Central Electricity Generating Board and the Department of Energy to recognise any change from the smooth and regular growth of the preceding years. Coming closer to home, the second example is from the latest (April 1978) Department of National Development forecast (17) for the consumption of crude oil in Australia. As Fig. 13 shows, the forecast points sit rather uncomfortably above what is basically a logistic curve (similar to a Gompertz curve). By non-linear curve fitting techniques, it can be shown that this curve is tending towards a maximum of about 740,000 barrels per day, i.e. not much greater than present consumption, yet the forecast figures pass that mark by 1982. Of course, this maximum level will ultimately depend on whether there are significant changes in such factors as population growth, average personal disposable income, relative energy prices, conservation measures and the extent of substitution for oil by other fuels, but we do not believe that they will be sufficient to raise the maximum level between now and 1985. In fact, they are more likely to reduce it further, even if sufficient oil is available.

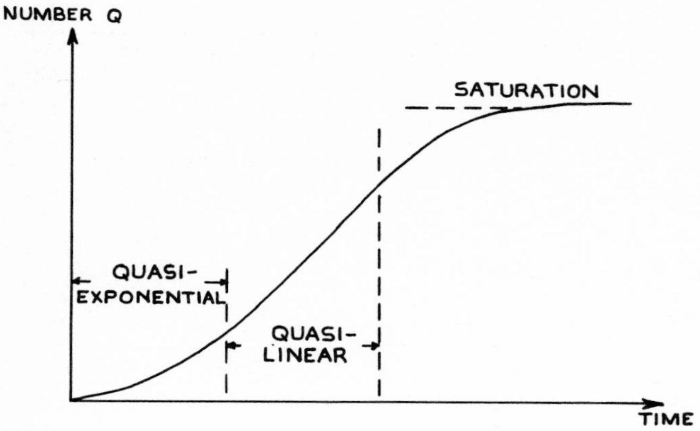

NUMBER Q
SATURATION
QUASI-EXPONENTIAL
QUASI-LINEAR
TIME

Figure 10
Typical demand curve

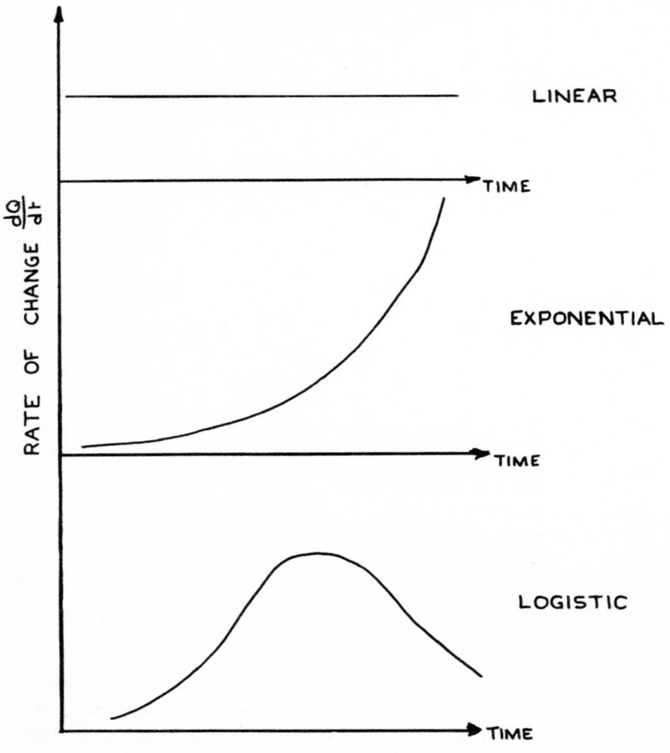

RATE OF CHANGE $\frac{dQ}{dt}$
LINEAR
EXPONENTIAL
LOGISTIC
TIME

Figure 11
Differential curves

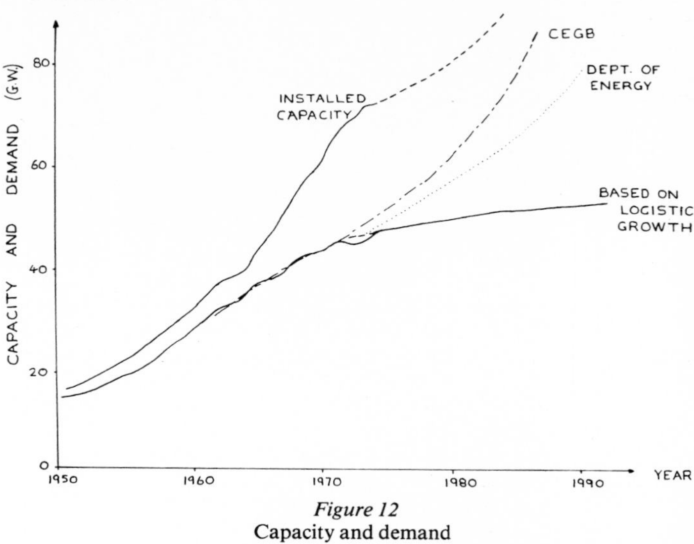

Figure 12
Capacity and demand

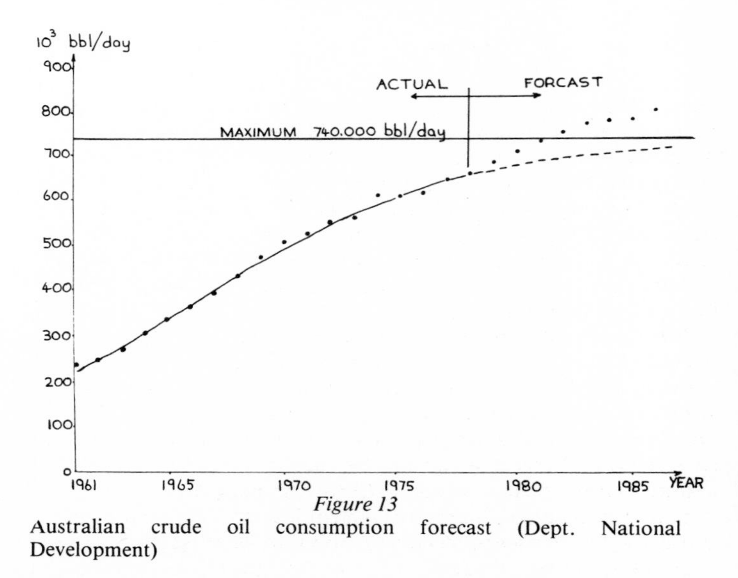

Figure 13
Australian crude oil consumption forecast (Dept. National Development)

In order to understand why the consumption of many forms of energy is now only growing linearly or saturating, we need to look at some of the structural factors affecting the growth in energy consumption. First, let us look at the consumption of primary energy in Australia since World War II (Figure 14). At first sight, growth looks impressive — we now consume six times as much energy as we did at the end of the War. However, the application of the differential analysis technique described above to our primary energy consumption (Figure 15) shows that exponential growth has only even been approached once during this period, and then for only about five years in the early 1960's when there was a rapid expansion of our secondary and mining industries. Before that brief period of rapid growth, the rate of growth was actually declining and we have only experienced a sort of "quasi-linear" growth since. The departure from rapid growth occurred in the mid-1960's, well before the Oil Crisis, the Whitlam regime, or any other popular scapegoat. The real reasons are quite simple and do not require any conspiracy theories. As Figure 16 shows, the growth in *per capita* consumption of primary energy has been remarkably linear since the rapid growth period in the early 1960's. This has probably been caused by the almost linear *decline* in the real cost of fuels (i.e. in constant dollars) over that period and the steady increase in the proportion of *tertiary* industries — 70% of the workforce is now employed in the tertiary sector which generally consumes less energy per worker than the secondary industries. A detailed discussion of these aspects may be found in a very useful paper by Ross and Williams (18). A study by Resources For The Future (19) showed (Fig. 17) that household

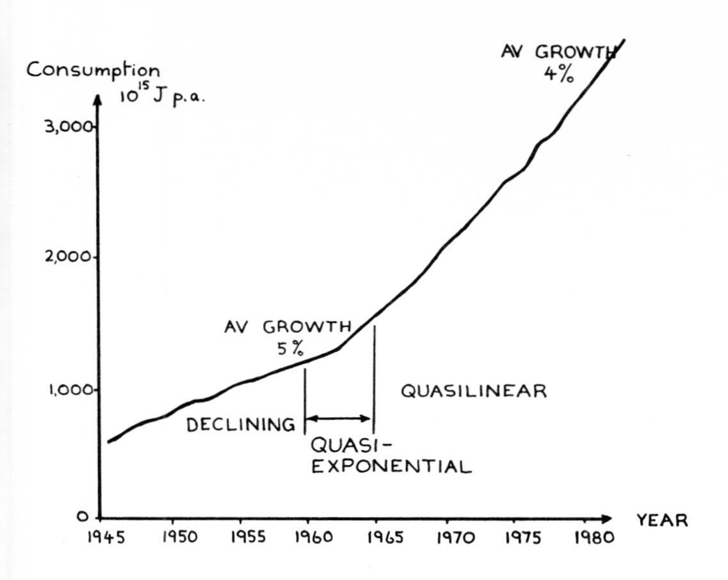

Figure 14
Growth of Australian primary energy consumption

energy consumption in most countries is inversely proportional to energy prices. Given the fact that the real price of energy in Australia has until recently been decreasing linearly with time, it is hardly surprising that the per capita consumption of energy has been rising linearly. It just means that we are spending a constant proportion of our income on energy. However, the recent increases in petrol prices and electricity tariffs may arrest this decline in the real price of energy and lead to a levelling off in per capita consumption.

Also, as Figure 16 clearly shows, there has been a distinct change in the trend of population growth in Australia. Throughout the post-war period, there was a steady and sustained population growth of around 2% per annum. However, in recent years, there has been a decline in both the rate of natural increase and the migrant intake, resulting in a definite tendency since about 1970 for the population to level out. If this trend continues, it too must lead to a flattening out of the total primary energy consumption, so we must be doubly cautious about extrapolating even recent trends into the future.

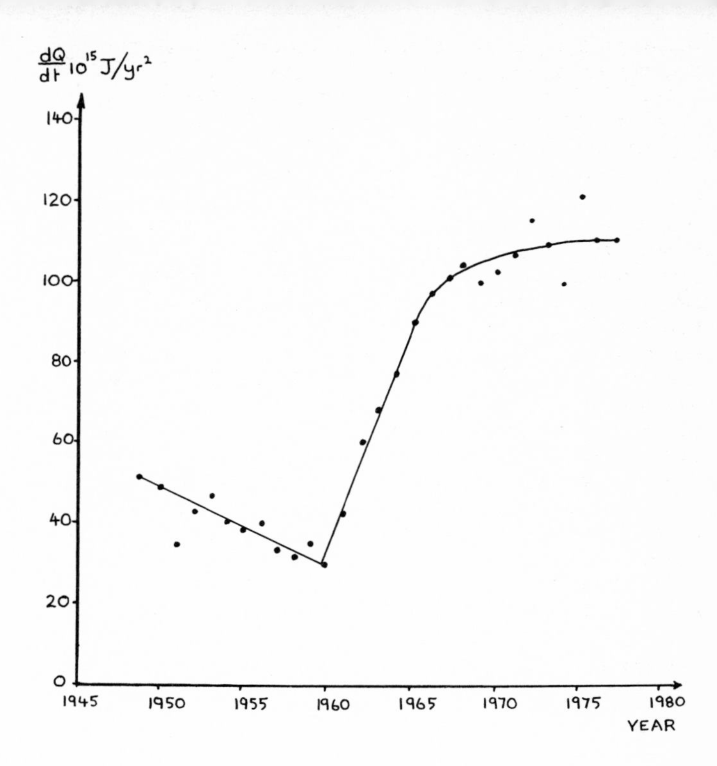

Figure 15
Australian primary energy consumption differential curve

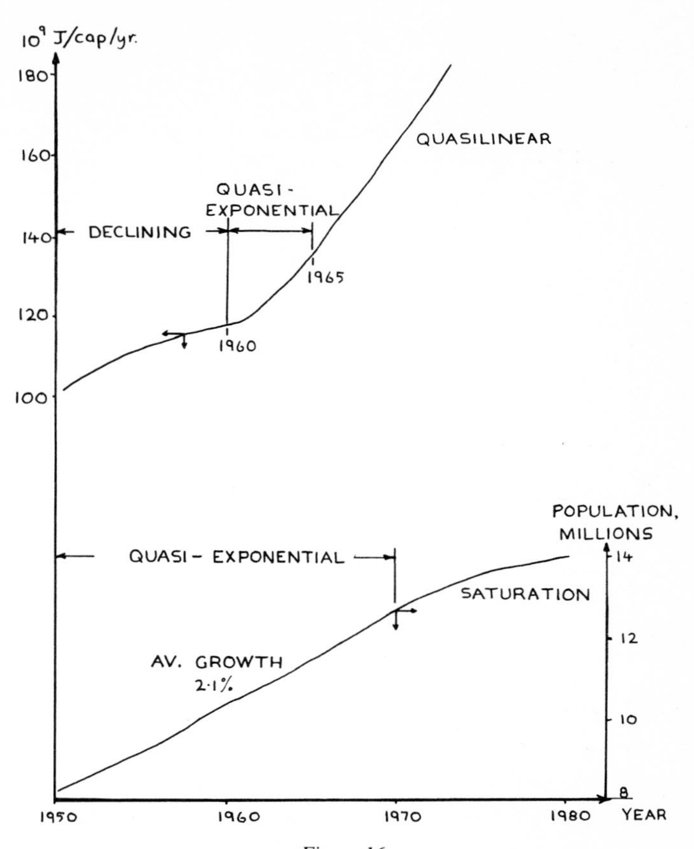

Figure 16
Growth of Australia's population and per capita energy consumption

A final indicator of such structural changes is the so-called "Energy-Output Ratio", the total primary consumption divided by the GDP in constant dollars (Fig. 18). Our ratio has been fairly constant throughout the post-war period, but is still rising in the face of increasing energy prices, whereas the U.S. ratio is falling. For more detailed information on the effects of structural changes in the economy on the demand for energy and labour, it is necessary to divide the economy into a number of sectors and look at the usage of energy and labour per dollar of value added (in constant dollars). The group Environmentalists For Full Employment has begun to look at the Manufacturing Sector in Victoria (20) and has found that the energy/output ratio has increased consistently as a result of Victoria's "Cheap Energy" policy, whereas the corresponding ratio in the U.S. has been equally consistently declined (18). This increase in our energy ratio is over and above that required to account for the substitution of energy for labour which has of course also occurred.

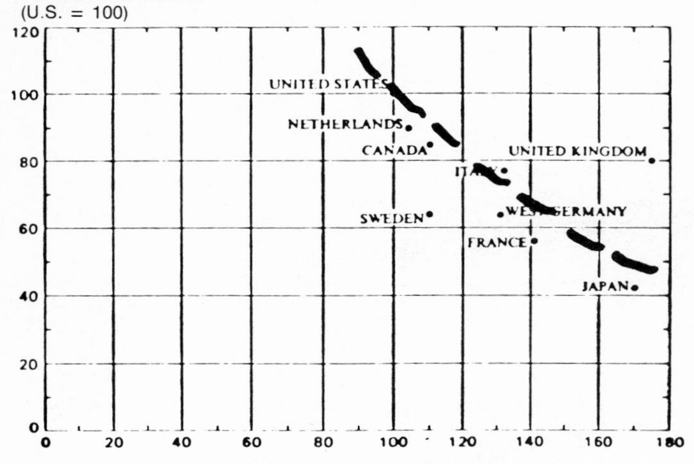

Figure 17
Household energy consumption related to energy prices, 1972

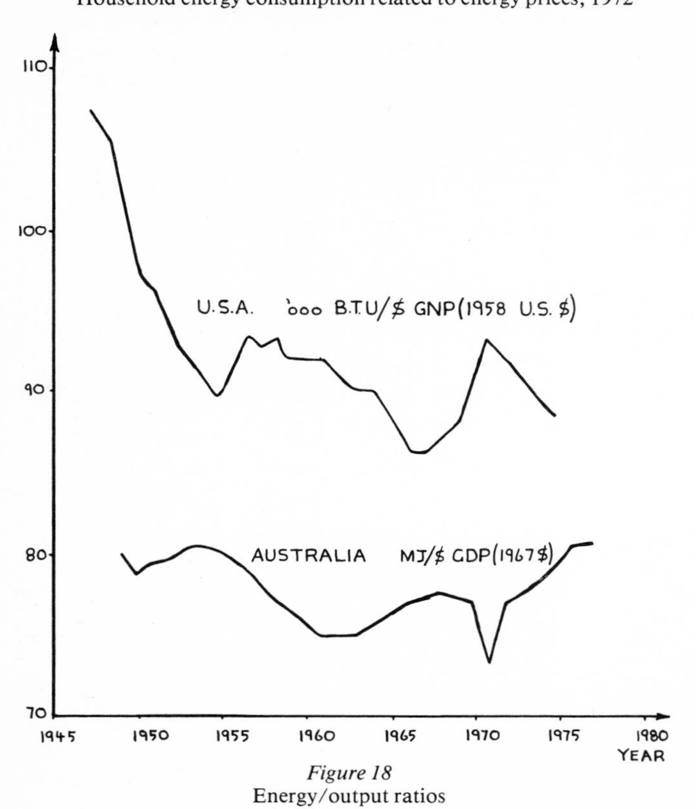

Figure 18
Energy/output ratios

Resource Depletion Policy

This subject has received negligible attention in the debate about energy policy in Australia, but it is being increasingly discussed overseas as awareness of the crucial importance of the actual availability of resources becomes more widespread. It requires an understanding of the time required to bring new resources into production, the pattern of productive capacity over the life of a particular deposit and the aggregate effect of such patterns on the overall availability of a resource. Apart from purely physical restraints, depletion policy also depends on a variety of economic, political, environmental and other factors, some of which are discussed in journals such as *Energy Policy, The Chemical Engineer* and *Resources,* but a detailed discussion of these factors is beyond the scope of this paper.

An example of the pattern of productive capacity for a single deposit is the standard oil production curve (Fig. 19) presented by the Groningen University group in the Netherlands (21) which divides the production cycle of an oil field into five distinct periods:

1. Discovery of the field.
2. Start of production, usually 2-6 years from date of discovery.
3. Build-up of production, usually linear, but fairly rapid.
4. Plateau period, usually 5-7 years, sufficient to recover cost of final increment of production capacity.
5. Production declines exponentially.

Note that only 40% of recoverable reserves are consumed by the end of the plateau period and the maximum annual rate of production is only 8-12% of the total reserves.

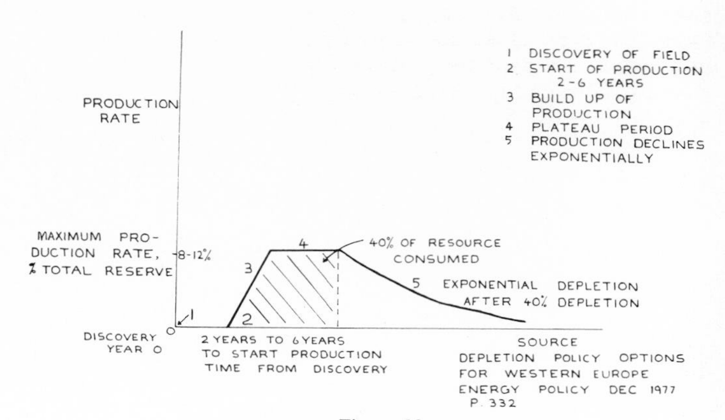

Figure 19
Standard oil production curve

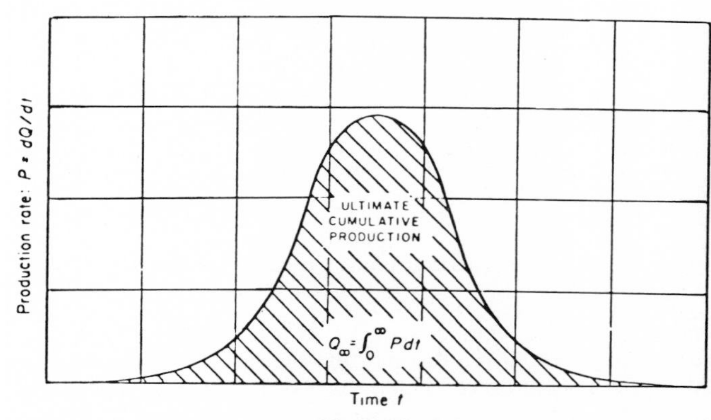

Figure 20
Full cycle of production of an exhaustible resource. (By permission of the American Petroleum Institute, from *Drilling and Production Practice,* 1956, Figure 11, p.12.)

For the analysis of national or world oil production capacity, Hubbert (22) has developed a model similar to a statistical probability curve. It describes the complete cycle of production of a finite resource having ultimate recoverable reserves (ultimate cumulative production) equal to Q_∞ (Figure 20). Although it is only approximate, the importance of this model cannot be over-estimated. It shows that the rate of extraction of a finite resource is limited and that only half of the ultimately recoverable reserves will be consumed when the peak production rate is reached. It also shows that the production cycle can be divided into three distinct phases:

1. Early Phase — characterised by a rapidly increasing rate of production until all of the easily-recovered deposits have been mined.
2. Mature Phase — where production rate flattens out.
3. Decline Phase — characterised by a rapid decline in production.

From a study of World and U.S. oil discovery and production patterns, Hubbert demonstrated the following important relationships:

Cumulative Discoveries (Q_d) = Cumulative Production (Q_p) + Proved Reserves (Q_r)

and

Rate of Discovery (dQ_d/dt) = Rate of Production (dQp/dt) + Increase of Proved Reserves (dQ_r/dt)

Also, Hubbert found the delay $\triangle t$ between discovery and production to be only twelve years, so that if the rate of discovery of new reserves reaches a peak, then the rate of production will probably reach a peak twelve years later. Note also that the rate of increase of new reserves falls to zero midway between the two peaks for the rates of discovery and production, just at the point where the two curves intersect. An approximate estimate of the ultimately recoverable reserves can be obtained by integrating the area under the rate of discovery curve up to the point at which it peaks and then doubling it to account for the fact that only half of the recoverable reserves have been found by that time, i.e. Q_∞ = Area × 2. These relationships are illustrated in Figures 21 and 22. The approximate nature of such estimates does not make much difference to the timing of the peak because it is an inherent

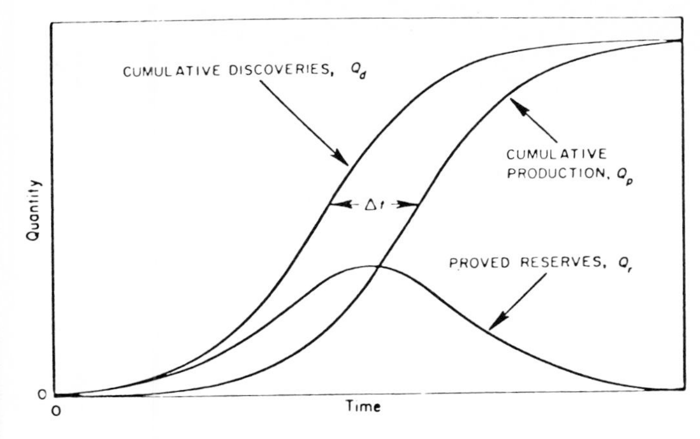

Figure 21
Generalized form of curves of cumulative discoveries, cumulative production, and proved reserves for a petroleum component during a full cycle of production. $\triangle t$ indicates the time lapse between discovery and production. (From Hubbert, Figure 22, p. 55.)

property of this bell-shaped curve that most of the area lies in the middle part. A 50% increase in the estimate for the world's ultimate recoverable reserves would only delay the peak production rate by about 10 years.

Hubbert has suggested that a more accurate estimate of Q_∞ can be obtained by comparing the rate of discovery of new reserves with the cumulative footage of exploratory drilling. There has been some controversy about the use of this method to estimate Q_∞; for a detailed discussion of the pro's and con's, see the informative paper by Harris (23). The fact remains that right through the 1960's, Hubbert was almost a lone voice in suggesting that U.S. oil production would peak in 1969 and that a world oil crisis could occur soon after. History has shown that he was right. U.S. oil production peaked in 1970 and the oil crisis occurred in 1973!

Strange as it may seem, none of our energy authorities in Australia appear to appreciate the importance of the

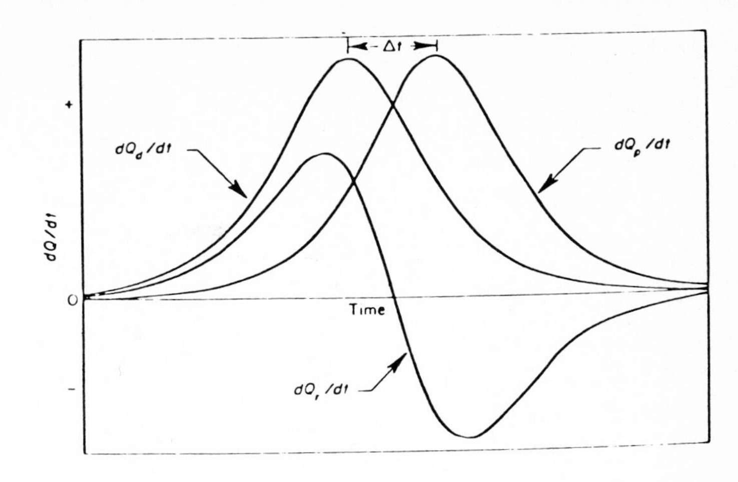

Figure 22
Relations between rate of production *(dQ_p/dt)*, rate of proved discovery *(dQ_d/dt)* and rate of increase of proved reserves *(dQ_r/dt)* during a full cycle of petroleum production (From Hubbert, 1962, Figure 24, p.56.)

production cycle and resource depletion policy and none, including the Bureau of Mineral Resources, collects and analyses data for such things as the rate of discovery of new reserves (the oil companies do provide sufficient information for such figures to be calculated, but no attempt has been made to do so). These methods of analysis are not limited to oil alone — they can be adapted to any finite resource — and in the absence of detailed mine-by-mine data, can provide at least some estimate of the possible availability of a resource through a given period of time and warn of possible shortfalls or indicate the timing of new increments of supply. Although the Hubbert model is strictly only applicable to countries with a large number of oil fields or mines, such as the U.S., or to the overall world situation, the form of the curve is somewhat

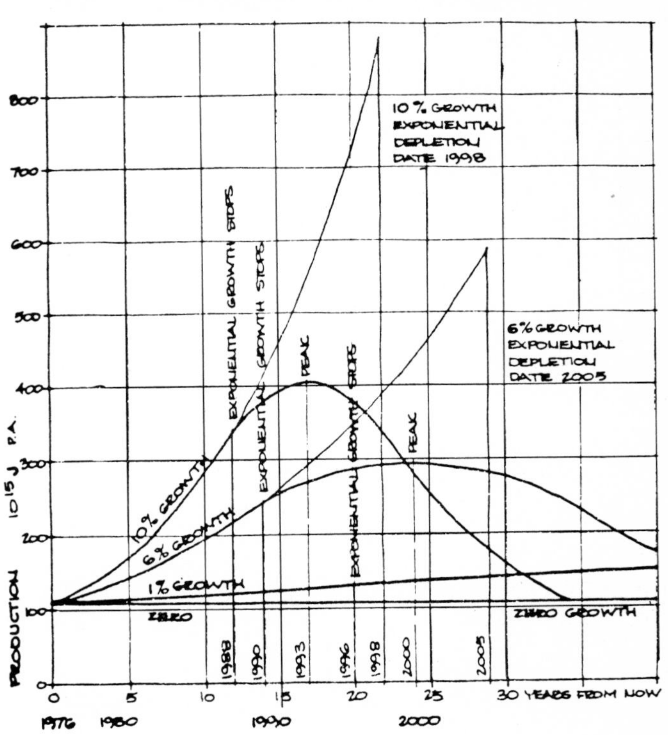

Figure 23
Theoretical production cycle of Victorian natural gas production

This Figure shows the consumption and production curves corresponding to four different demand growth rates. In each case, an energy gap appears when the demand curve deviates from the corresponding production curve

similar to the standard oil production curve shown earlier for the development of individual oil fields and it can give a good indication of productive capacity as long as undue attention is not paid to the precision of the result.

Matching Supply and Demand

An example of the application of Hubbert's method to the production of natural gas in Victoria is shown in Figure 23. Whereas the life of the total recoverable reserves (assumed to be equal to the present level of proved reserves) would exceed 80 years at 1976 production rates, it is 29 years with 6% annual growth in demand and only 22 years with 10% annual growth in demand.

However, as we have indicated earlier, both of these approaches are simplistic. Application of the Hubbert model shows that with a 10% annual growth rate, exponential growth ceases within 12 years, production peaks within 17 years and then declines. Hence, exponential growth at 10% per annum can only be maintained for 12 years with this level of reserves and a new source of gas would be needed thereafter to supplement the existing reserves. As we saw earlier in Figure 6, the actual production curve for Victorian natural gas is complicated by the sharp, discontinuous nature of the SECV demand for gas and field deliverability problems, but the conclusions are much the same — we are likely to need a new source of gas by 1990! The fact that *some* of our gas reserves will still be there in the year 2000 does nothing to ensure that supply continues to meet the demand.

Our second example of the application of the Hubbert model is to the production of brown coal in Victoria. It is very clear from the Victorian Government Green Paper on Energy and the Mines Department graph shown earlier (Figure 9) that great store is being put on the brown coal resources in the Latrobe Valley and that if necessary, electricity, oil and gas may one day be produced from our "vast" reserves of brown coal.

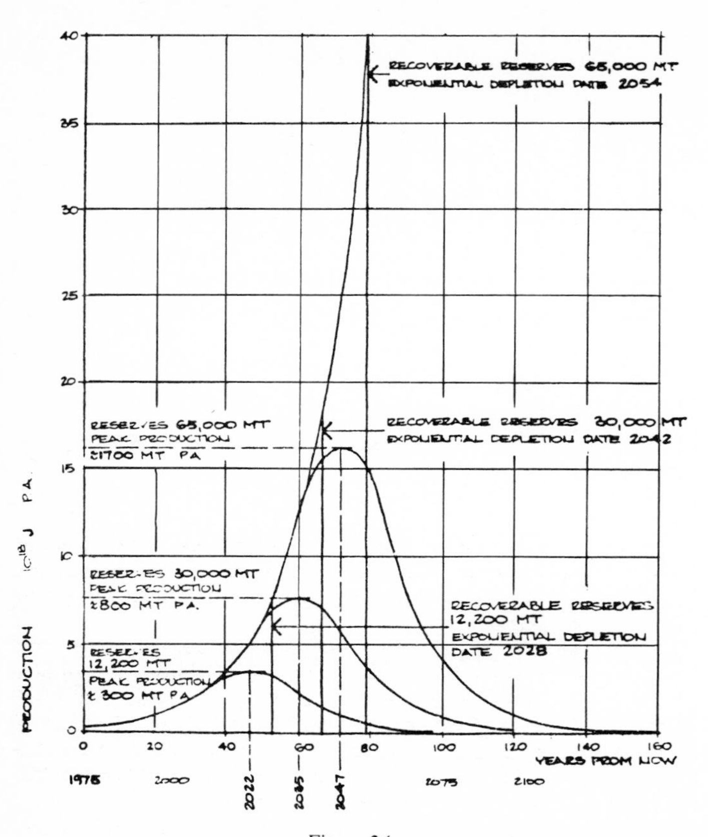

Figure 24
Effect of changes in reserve size on productions of Victorian brown coal

This Figure is based on the use of brown coal for electricity generation only at a growth rate of 6.6% per annum, the official SEC projection. The use of brown coal for other purposes such as oil and gas synthesis would increase depletion rates

Figure 24 shows the effect of assumed recoverable reserve size on the maximum rate of brown coal production and the timing of the peak with a 6.6% annual growth in the demand for electricity only. The assumption of a 12,000 million ton reserve, that which is reckoned to be recoverable under present economic conditions and present technology, enables us to reach a peak production rate of about 300 million tons per annum in about 47 years. Increasing the reserve level to 30,000 million tons, that which is said to be recoverable with "forseeable" mining techniques, enables us to reach a peak production rate of about 800 million tons per annum in about 60 years and further increasing it to 65,000 million tons, the total geological reserves, enables us to reach a maximum production rate of about 1,700 million tons per annum in about 72 years. So much for our "brown coal to last 1,000 years"!

Moreover, if in addition to assuming a reserve level of 12,000 million tons, we also assume that oil and gas shortfalls will be produced from brown coal after 1985 (see our earlier Figure 8), then coal production rises extremely rapidly and would reach its peak around 2007, just one year after the end of the forecast period of the Green Paper on Energy! Moreover, the supply of synthetic oil, gas and electricity may fail to keep up with demand even before the year 2000 because of coal supply problems (see dotted line)! Figure 25 also shows that most of this increase is brought about by the need for coal liquefaction, although the increase in coal requirements for electricity generation is also quite substantial. It shows quite conclusively that we have no reason at all for complacency and that increasing the level of recoverable reserves to 30,000 million tons only delays the peak production rate by 13 years. Further examination of these graphs will show that once a substantial proportion of the total reserves is committed, say about a third, then a reduction or even elimination of further growth in demand does little to delay the inevitable decline in production.

Conclusion

We would like to leave you with two points:

1. We need to take a critical look at the shape of our energy forecast curves, question the need for growth at any cost and look for effective ways of reducing growth in demand, not ways of increasing the supply.

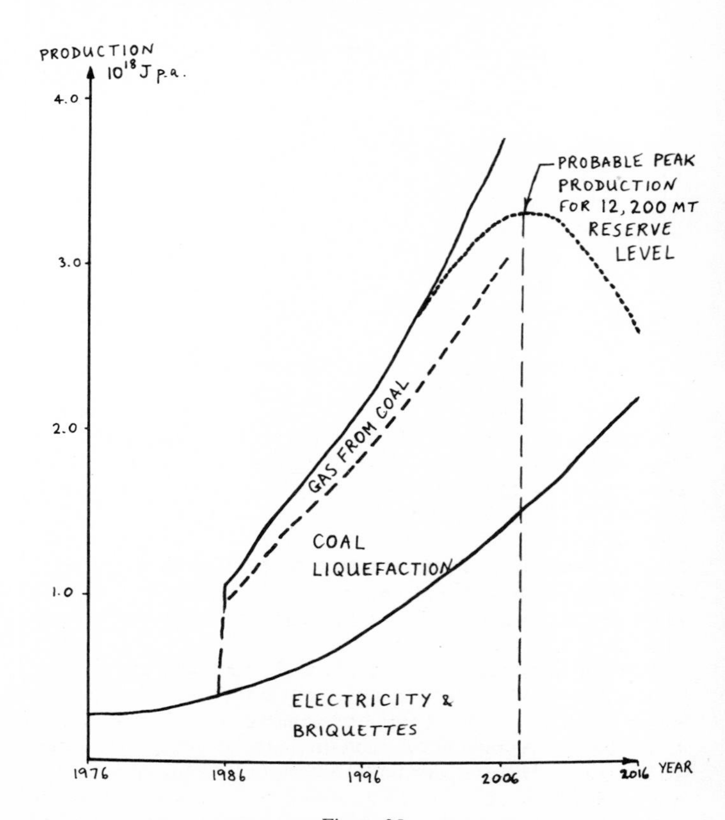

Figure 25
Likely effect of coal conversion on brown coal requirements

2. We should invest fossil fuel energy in the production of renewable fuels now and make the necessary social adjustments so that we can achieve an equitable society which can be sustained long after the oil derricks turn to dust.

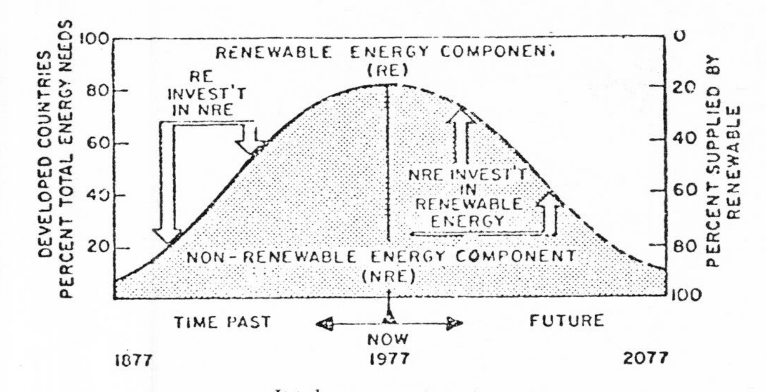

It takes energy to get energy

Figure 26
Energy investment curve (24)

References

(1) National Energy Advisory Committee (1978). *Proposals For a Research And Development Program For Energy*. NEAC Report No. 3, A.G.P.S., Canberra.

(2) National Energy Advisory Committee (1977). *Australia's Energy Resources: An Assessment*. NEAC Report No. 2, A.G.P.S., Canberra.

(3) Ion, D. C. (1975). *Availability Of World Energy Resources*. Graham & Trotman Ltd, U.K.

(4) WAES (Workshop on Alternative Energy Strategies) (1977). *Energy: Global Prospects 1985-2000*. McGraw Hill.

(5) Facer, J. (July 1978). Forecasting and Planning — The Clouded Crystal Ball. *The Chemical Engineer*, 568.

(6) Victorian Government, Ministry of Minerals and Energy (1976) *Green Paper On Energy*. Victorian Government Printer.

(7) Lovins, A. B. (1975). *World Energy Strategies — Facts, Issues and Options*. Friends Of The Earth.

(8) Strahan, R. (1978). The Time To Start Is Yesterday. *Search*, 9(4), 105.

(9) Maglen, L. R. (1977). Non-Renewable Resources and the Limits To Growth: Another Look. *Search*, **8**(5), 158.

(10) Rigby, B. J. (1977). How Much Coal Have We Got? *Search*, **8**(10), 348.

(11) Royal Commission On Petroleum (1976). Fifth Report: *Towards A National Refining Policy*, AGPS.

(12) White, D. et al (1978). *Seeds for Change*. CCV — Patchwork Press, Melbourne.

(13) The Pipeline Authority (1976). *System Study*. (3 vols. + Appendices).

(14) Bowen, K. G. (1975). Victorian Brown Coal Resources, in Victorian Mines Department Geological Survey Report No. 1975/2 — *Hydrocarbons From Coal*, 23.

(15) Lotka, A. J. (1956). *Elements Of Mathematical Biology*. Dover, New York.

(16) Energy Research Group (1978). *A Critique Of The Electricity Industry*. The Open University, Milton Keynes, U.K.

(17) Department Of National Development (1978). *Demand For Primary Fuels: Australia 1976-77 to 1986-87*. AGPS.

(18) Ross, M. H. and Williams, R. H. (1977). *Energy and Economic Growth*. A study prepared for the sub-committee on Energy of the Joint Economic Committee, Congress of the United States, U.S. Government Printing Office, Washington D.C.

(19) Darmstadter, J. et al (1977). *How Industrial Societies Use Energy: A Comparative Analysis*. Resources For The Future, John Hopkins University Press, Baltimore.

(20) Andrews, J. (1978). *Energy Use, Employment & Economic Growth In Victorian Manufacturing Industry, 1969-70 to 1974-75*. EFFE Working Paper No. 2.

(21) De Man, R. et al (December 1977). Depletion Policy Options For Western Europe. *Energy Policy*, 319.

(22) Hubbert, M. K. (1969). *Resources and Man*, Chapter 8. W. H. Freeman, San Francisco.

(23) Harris, De Verle P. (1977). Conventional Crude Oil Resources Of The United States: Recent Estimates, Methods For Estimation and Policy Considerations. *Materials and Society*, **1**, 263-286.

(24) The Biomass Energy Institute Inc. (1977). *Bioconversion: Fuels From Biomass*. The Franklin Institute Press.

The Transport Fuel Dilemma — Patterns of Consumption we cannot Sustain

L. A. Endersbee

Dean, Faculty of Engineering
Monash University, Clayton, Vic. 3168

In the period from the end of the Second World War to the OPEC oil embargo in 1973, the consumption of oil around the world increased six-fold. This made possible a pattern of economic growth that we came to regard as normal, but which was shown to be highly sensitive, even fragile, to the increasing cost of oil.

It was not only the ready availability of oil and its low price that supported such economic growth, it was also the ready transportability of oil. Japan for example is almost entirely dependent on the super-tanker.

It is emphasised that the immediate concern is the adjustment of our society to a constant supply of oil rather than a steadily increasing supply; the major mechanism of constraint is likely to be the increasing cost of oil. The upper level on price could be the cost of the next available alternative, e.g. oil-from-coal, but alternatives can only restrain future increases in price of natural oil if the alternatives are available in sufficient quantity at the time required.

In view of the huge investments and long lead times required to produce synthetic liquid fuels, it must be expected that the price of oil will continue to rise over the time of development of these alternatives.

The economic impact of the oil embargo, and the subsequent higher prices for oil, was quite substantial almost everywhere around the world except Australia, and it was not until 1976 that world energy use again reached the 1973 figure. The rate of growth in energy use has remained low; prior to 1973 it was 5.5 per cent per annum; it is now about 3.9 per cent, and in the coming decade may not exceed the present rate.

Although increasing use of coal features prominently in the energy policies of many countries, the fact is that the production of coal has been declining in most of the major producing nations. The expanded use of coal presents major challenges, particularly in the costs involved in opening up new and often deep mines and new fields, and also because of the environmental impact at both mine and place of use. These

Diesendorf, M. (ed.) (1979). *Energy and People*. Canberra, Society for Social Responsibility in Science (A.C.T.)

prospective limitations in the rate of exploitation of coal are a major factor in the rate at which coal can be developed as a source of synthetic fuel, and add to the already formidable problem of the capital costs of such plants. In fact the attractiveness of oil-from-coal seems to diminish the more we comprehend what is involved.

Energy resources are not at all evenly distributed amongst nations, and many are almost totally dependent on imported energy, as for example, Japan. Fuels are by far the largest item in international trade, and the withholding of supply for political or other purposes (e.g. oil or uranium) could have serious repercussive effects.

Because of the limitations of known Australian resources of oil, Australia seems destined to become more dependent on imported oil. Australia, of course, will also be exporting other forms of energy, i.e. coal and uranium.

The Carter Energy Policy

The United States has passed the peak in domestic oil production, and will also be increasingly dependent on imports. The high prospective U.S. demand for oil imports is a matter of profound concern for all the other oil importers, particularly with prospects of limited oil supplies in world trade.

A revised projection of U.S. energy demand and supply (1) was released recently by Exxon (Fig. 1). It is of interest to note the downward revisions in total projected demand, and the increasing dependence on oil imports.

President Carter, in introducing his National Energy Plan (2), stated that the U.S. faced a situation which was the "moral equivalent of war". This dilemma is illustrated in part in the accompanying diagram (Fig. 2) from one of the policy documents, which shows the dramatically changing pattern of U.S. oil production and consumption.

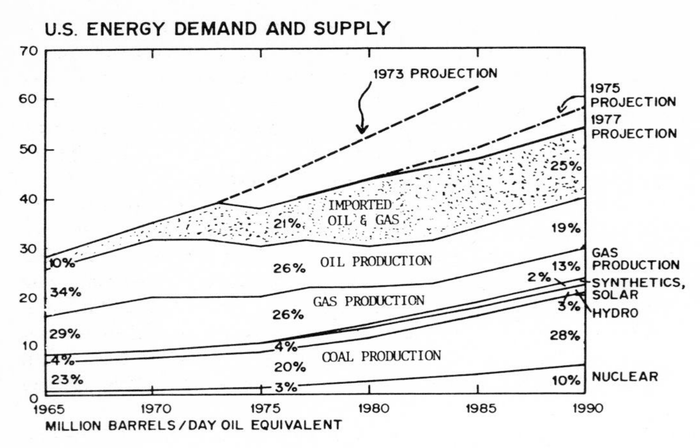

Figure 1

A recent projection by Exxon (1) of U.S. energy demand and supply. Note that the projections of total energy demand have decreased radically since the oil embargo of 1973, and that the projected demand for imported oil and gas is substantially less. The projected growth in energy supply arises mainly from increased coal production and nuclear power

In 1970 the U.S. was able to supply most of its needs from domestic production supplemented by 20 per cent imports. By 1980, oil imports to the U.S. are likely to be about 60 per cent of the total oil consumption, according to a recent estimate by Exxon.

It is of interest that the present U.S. oil consumption is equal to the entire world production and consumption of only about twenty years ago. Even so, the Carter Energy Plan is based on the increasing availability of imports of oil; however, it seems quite likely that within five to ten years further increases in supply may not be possible.

The increasing dependence of the U.S. on imported oil must erode, to a degree, the strategic credibility of the U.S. While this is not a matter of concern for the oil industry, which is mostly composed of multinational companies anyway, it must be a matter of increasing concern to the U.S. Government. For this reason it is to be expected that the U.S. Government will play an increasing role in securing future oil supplies within the U.S., including the early entry into synthetic liquid fuels.

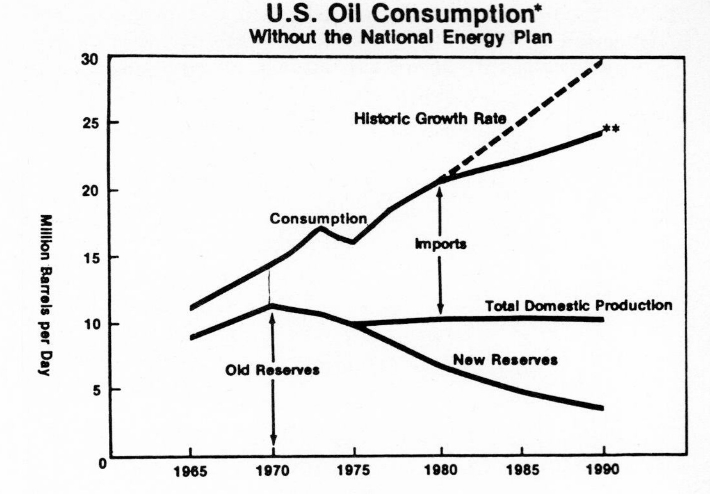

Figure 2

The U.S. National Energy Plan (2) draws attention to the increasing demand for imported oil to offset declining domestic production. Even so, the plan involves accelerated development of new reserves and it remains to be seen whether these are available and can be developed at the rate indicated

Projections of Future Supply and Demand for Oil

In a report of an International Workshop on Alternative Energy Strategies directed by Professor Carroll Wilson at Massachusetts Institute of Technology (3), it was indicated that the oil-producing countries are not likely to continue to match production to meet projected world demands, but are likely to husband their remaining resources. The oil-producing countries are steadily raising prices and limiting production so that their supplies last for a longer period. This is reasonable from their point of view and reasonable for the world, but it places a great responsibility on all countries to manage the transition from such strong dependence on oil to greater reliance on other fossil fuels. The gap between projected demands and available supplies must be met from these other sources, or simply not at all. It is emphasised that long lead times and huge investments are required to produce new fuels on a scale large enough to fill this prospective shortage of oil.

In addition, it must be emphasised that it is quite unlikely that substantial discoveries of oil will be made to soften the impact. Our pattern of oil consumption is already too high, and the best we can hope from new discoveries is a few years' grace. This is illustrated by the short impact of North Sea Oil and gas on British energy supplies.

The international oil companies are looking very closely at international energy strategies. The accompanying diagram (Fig. 3) is part of a study by British Petroleum (4), and shows U.K. primary energy uses and sources of energy. It should be noted that in this case they have estimated that the energy growth will fall to zero by the year 2000, a case which they regard as credible; i.e. possible of achievement. Cases with energy growth above zero at the year 2000 showed difficulties in energy supplies which they considered could not be met within the available time. In this diagram it should be noted that ongoing and probable North Sea oil meets U.K. needs for only a relatively short time, and that by the 1990's the U.K. will again be resuming oil imports. The plan also shows substantial new investments in nuclear power and further development of coal and gas from coal.

A projection of world energy demand (excluding centrally planned economies) is shown in the accompanying projection by Exxon (1) (Fig. 4). The projections carried out in the years

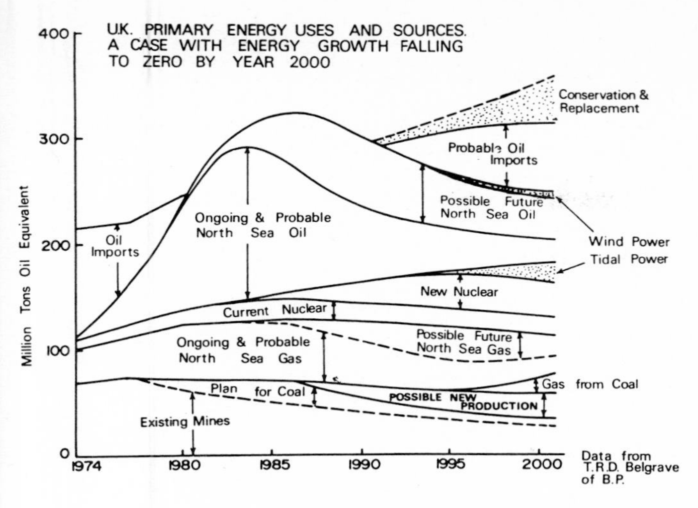

Figure 3

A British Petroleum (4) projection of U.K. energy supply and demand. Note the limited contribution of North Sea oil and gas, and the continued decline in coal production.
The projection of zero energy growth by the year 2000 was regarded as capable of achievement; anything greater showed difficulties in supply

1973, 1975 and 1977 are all shown on this diagram. In 1973, the projected world energy demand in 1980, estimated by Exxon, was 127 million barrels per day oil equivalent; in 1977 that projection was revised downward to 103 million barrels per day oil equivalent, a decrease of 23 per cent. This simply underlines the political, social, economic and technical uncertainties involved in present energy planning.

A major uncertainty is prospective demand outside of the developed nations. The developed nations in North America, Europe and Japan collectively consume some 77 per cent of the present total world energy. The present low use of energy in the developing countries tends to obscure the enormous potential for increased demand, and improvements in communications in developing countries is leading to other economic development and reinforces expectations.

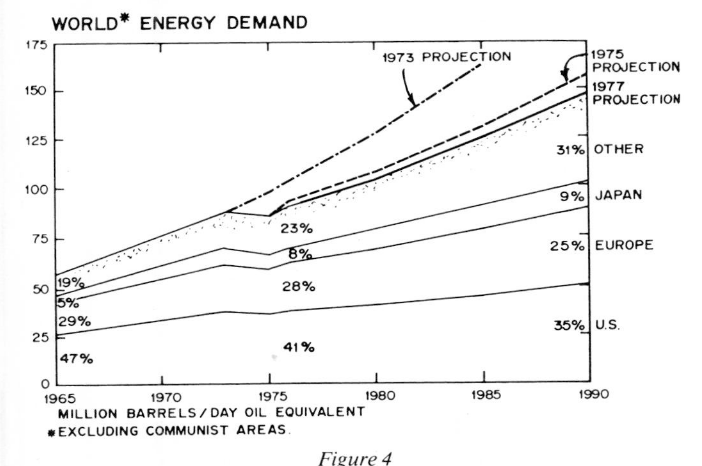

Figure 4

An Exxon projection of world energy demand (1). Note the downward revision of total demand since 1973, and the projected increase in total demand in developing countries

Australian Supply and Demand for Oil

Australia will be dependent to a much greater extent on oil from OPEC countries at the time when those countries are limiting their production and continuing to raise prices, and we will need substantially increased funds to buy imported oil. It must be emphasised that the probable timing of these difficulties will vary according to the extent oil production is varied to suit world demands, and the extent to which the world adjusts its demands for oil to the available supply. These matters are of course conjectural, and we can only simply prepare for what we reasonably expect is likely to happen. The most prudent measure is international and national action or conservation of oil.

It should be emphasised that the transition from a world economy dominated by oil to one of greater energy diversity will take time and money. Progress overall seems likely to be limited by the supply of capital for major investments to offset dependence on oil. Non-oil energy sources, particularly coal, must later increase in relative importance, but in the short term an increasing proportion of the available crude oil is likely to be devoted to transportation fuels. It is to be expected that in time oil, and later natural gas, will be restricted almost exclusively for transportation and petro-chemicals.

Ultimately, liquid fuels produced from coal, tar sands and oil shales, alcohols and other organic fuels will provide a greater proportion of the world's liquid fuel needs. Because of the time lags involved, however, they are not likely to be significant contributors to meeting energy requirements until the next century, even given favourable economic relationships. Thus conservation of oil and more economic use of liquid fuels is the only immediate response available while time is gained to develop, fund and construct the liquid fuel plants to extend and supplement the present supplies.

Energy Policy in Australia

Responsibility for energy matters in Australia resides largely with State Governments.

Under the Australian Constitution, only specified powers are conferred on the Commonwealth Parliament, the remainder being retained by the parliaments of the six States. Energy was not one of the powers conferred on the Commonwealth Parliament. Primary responsibility in this field has remained with the States, which control energy production and distribution through various government departments and public authorities. The Commonwealth Government has a direct role in controlling the export of energy resources arising from its responsibility for overseas trade. The State Governments in Australia are already active in promoting conservation of energy, and are co-operating in plans for an Australian Conservation of Energy Programme.

In recognition of the needs for national co-ordination in energy matters, the Commonwealth Government is co-operating with the States through a Council of responsible ministers, the Australian Minerals and Energy Council. The Commonwealth Government has also formed the National Energy Advisory Committee, which advises the Minister for National Development, and thereby the Commonwealth Government, on matters of energy policy, including energy conservation.

The first report of NEAC was on energy conservation (5), and was prepared at the urgent request of the then Minister, Mr Anthony. This report was primarily concerned with measures for the conservation of oil, and was issued in September 1977. Subsequent reports have been prepared (6-10).

The Commonwealth Government has also recently constituted the National Energy Research, Development and Demonstration Council to support increased research in energy, for which additional funds were provided in the budget for 1978-79.

Energy Conservation through the Market Place

In the recent budget, the Federal Government determined that Australian oil would be taxed to bring the price into refineries up to world parity. This had been one of the strong recommendations of The Institution of Engineers "Task Force on Energy" (11). Prior to the rise to world parity, motor spirit prices in Australia were virtually the lowest in the world, and they are still comparatively low. These low prices have encouraged the relatively inefficient use of petroleum fuels in transport and industry, and distorted the relations between oil and other fuels, supporting, for example, the use of oil where coal or gas or solar energy may be the preferred longer term source of energy.

Conservation is an essentially relative concept: the objective in conservation of oil is to modify patterns of consumption of oil in relation to other fuels, in the anticipation of further increases in the price of oil relative to these other fuels.

These objectives could of course be met by the normal operation of the market system; and the move to world parity pricing in Australia is simply an adjustment to the world market price.

The need for introduction of conservation measures for oil, in addition to normal market adjustments in oil pricing, reflects the unique importance of oil for transportation and a certain lack of faith that the world markets will be able to adjust sufficiently rapidly to the depletion of natural oil and its replacement by other sources of energy.

It is noteworthy that some of the major oil companies are outstanding advocates of conservation of oil; the next major alternative to natural oil is at least two to three times the price, and there is little incentive to invest in synthetic fuel plants until the price approaches that level.

Conservation of Transport Fuels

In the studies of the Task Force on Energy, one Working Party devoted its attention to "Energy Conservation and Transport" (12).

In the final summary report of the Task Force, the various discussions on energy conservation in transport were brought together and specific recommendations listed. The Task Force noted that,

There are many ways in which energy can be saved in transport. In fact energy costs have hitherto not been a significant factor in transportation planning or in design of transport vehicles including motor cars. If we were to take account of energy factors in transport, it is estimated that it would be possible, in time, to save 30% of transport energy consumption without serious inconvenience to transport users. Insofar as this saving would mainly be shown as a reduction in demand for imported oil, it can be regarded as a net economic benefit to the community.

In the longer term, we must have regard to the overall efficiency of transport and the energy costs of our present urban society.

In the short term, we must initiate several programmes of energy conservation in transport and this will allow time to:

- *create more efficient forms of transport;*
- *plan for more energy efficient city forms;*
- *introduce alternative fuels.*

Immediate measures for energy conservation in transport should be directed primarily at improving the energy efficiency of private motor cars. Our present use of motor cars is characterised by:

- *low occupancy rates;*
- *high fuel consumption;*
- *stop/start traffic flow;*
- *short trips.*

Conservation actions need to be reinforced by a publicity campaign focusing attention on and explaining the various conservation measures being adopted.

Publicity needs to be given to the energy implications of vehicle characteristics. Australian motor car manufacturers are already active in planned developments for future models having improved fuel economy. In the past, motor car buyers have been more influenced by appearance, comfort and speed. There is need for Government action to reinforce these planned developments by the motor car manufacturers for improved fuel economy.

In the short term, smaller and more fuel-efficient vehicles need to be stressed. In the long term, alternatives to oil-fuelled vehicles may need to be encouraged by Government initiatives.

The Task Force also made a number of specific recommendations on energy conservation in transport, noting that it would be a major task of Governments to prepare desirable targets for the conservation measures, to develop priorities and to assess effects as measures are implemented.

Fuel Economy Goals for Motor Vehicles

The Task Force on Energy noted that the fuel economy of new motor vehicles in Australia could be improved by about 30 per cent with little inconvenience to drivers and passengers. It recommended that motor vehicle manufacturers should be encouraged to design and build motor vehicles of much improved fuel economy, and suggested that for this purpose it may be necessary for the Commonwealth to prescribe future fuel economy goals.

NEAC also recommended that the Commonwealth Government undertake a study directed towards prescribing such targets (5).

The present average consumption of motor spirit by Australian cars is about twelve litres per one hundred kilometres of travel. The U.S. and Canadian Governments have recently introduced sales/weighted fleet/average fuel economy targets for 1985 for manufactured and imported vehicles of 8.6 litres per one hundred kilometres, and year by year targets up to that date.

In response to the NEAC proposals for Government regulation in this field, the Federal Chamber of Automotive Industries has recently completed a study (13), to which all Australian motor car manufacturers contributed, and proposed "A Uniform Code of Practice for Passenger Car Fuel Consumption Reduction". The Ministers for Transport and National Development have welcomed this move by the automobile industry, noting that self-regulation by the industry is preferable to regulation by Governments. In essence, the FCAI has proposed that the Australian motor car industry meet essentially the same standards of fuel economy as prescribed for North America for 1983 and beyond (Figure 5). The FCAI proposals are now being reviewed by NEAC and the Federal and State Departments of Transport in the light of other parallel proposals for conservation of oil, including the proposed increase in octane rating of standard gasoline from 89 to 92 R.O.N., and the review of lead levels and auto-emission controls. Of course, the FCAI proposals for self-regulation can be reinforced by government measures such as higher sales tax on larger cars and higher registration fees, in addition to the market response to higher fuel costs.

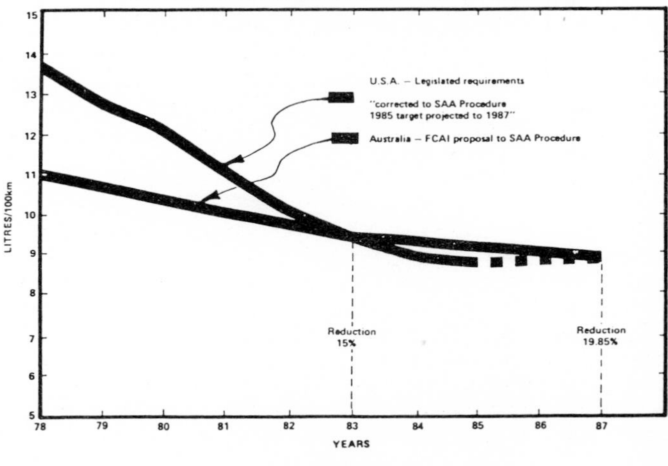

Figure 5

Federal Chamber of Automotive Industries proposals for new passenger car average fuel consumption and comparable U.S. legislated requirements

The design of passenger cars can be modified in several ways to improve fuel economy. These include:

- Reduction in aero-dynamic drag.
- Reduction in vehicle weight.
- Improvements in transmission.
- Improved efficiency of engines and use of higher compression ratios.
- Reduction in use of auxiliary drives (e.g. air conditioning).
- Control of engine adjustment.

In summary, there is no doubt that there are substantial prospects for improvement in the fuel economy of Australian motor vehicles, and that the motor vehicle will continue to play the major role in transport in Australia. It should also be noted that until recently, fuel economy has not been a major factor in the design of Australian automobiles. Cars have become progressively larger and more powerful since the introduction of the first Holden car. Styling, comfort, speed, and recently safety, have been the major market factors in the last 20 years.

Review of Motor Vehicle Exhaust Emission Standards

The introduction of the first stage of motor vehicle exhaust emission standards led to an increase in fuel consumption, estimated to be 7 per cent or more on the heavier, larger cars. In addition, there is increased fuel consumption in the refining process, as more energy is required to produce super-grade fuel at the same octane number but with a lower lead content.

The introduction of fuel economy targets for motor vehicles provides an opportunity to review this question and to study the whole system of motor car, and refinery, and environment and to work towards a more rational solution overall. It is recognised that there may be a need to limit lead levels in the major urban centres, but not necessarily throughout most of Australia.

In the long term the possible introduction of electric vehicles and the possible use of LPG and alcohol-gasoline blends will assist in reducing air pollution levels in cities.

Use of Alternative Fuels

(a) LPG (Liquid Petroleum Gas)

Australia is well-placed to make substantial use of LPG for transportation as we are likely to have a continuing surplus by virtue of the further development of natural gas resources, particularly the substantial gas fields in the Northwest Shelf.

While LPG can be used in place of motor spirit for cars, it is especially well suited for urban buses and urban vehicle fleets. In Melbourne a large number of taxis are using LPG, and the Commonwealth Government is converting part of the Commonwealth car pool.

(b) Alcohol

There are also technical prospects for direct use of alcohols as transportation fuels and for augmenting supplies of motor spirit by blending ethanol or methanol with motor spirit up to a proportion of about 15 per cent. During the Second World War a 10 per cent ethanol blend was used in Queensland. At the present time the Government of Brazil is pursuing an ambitious and heavily subsidised programme to produce ethanol from crops as a means of reducing the demand for imported oil.

The large discoveries of natural gas in the Northwest Shelf have drawn attention to the possibility of converting natural gas to methanol to use as an alternative liquid fuel for transport. At the present time the cost of methanol produced from natural gas is significantly greater than the world market price for oil, and higher still on an equivalent energy basis. However, continued escalation in the price of natural oil may make this methanol alternative an attractive proposition for additional supplies of liquid fuels in Australia.

(c) Diesel Engines

Diesel engines offer advantages over petrol engines for car operation through substantially lower fuel consumption, despite a penalty of some additional engine weight. The use of diesel engined motor cars is growing very rapidly in Europe, presumably reflecting the higher cost of fuel in most European countries. Most automotive diesel engines operate with compression ratios approaching 20-1, and these give higher thermal efficiency and low fuel consumption. The higher thermal efficiency favours the diesel engine as far as exhaust emission is concerned, and there is no lead additive to diesel oil.

It is expected that diesel powered motor cars will increase in popularity, especially for fleet use.

(d) Electric Vehicles

The National Energy Advisory Committee has recently submitted a small state-of-the-art report on electric vehicle development and battery technology; and has suggested means of encouraging the development of electric vehicles in Australia.

Electric vehicles offer a further means of reducing dependence on imported oil, and also of reducing air pollution in central city areas. However, there are definite limits to the development of electric vehicles using present-day lead acid batteries, and it must be anticipated that the major development in electric vehicles is likely to come with the development of more advanced types of batteries. Battery systems of main interest are:

Lead-Acid, Nickel-Zinc, Nickel-Iron, Zinc-Air, Sodium Sulphate and Lithium Sulphate.

These advanced batteries do not appear to be suitable for supplying high motor currents required during acceleration. Hybrid systems are now under study in Japan where by advanced lead acid batteries are used to supply instantaneous high power with other types of batteries being used to store most of the energy, and to re-charge the lead acid battery during travel.

It is anticipated that the major motor car manufacturers will be including electric cars within their normal production range within the next decade, and the timing of this development will depend on progress in new battery development in relation to the changing cost of motor spirit.

More Effective Use of Motor Cars

A large proportion of transport fuel in Australia is consumed by privately owned cars being used for travel to and from work with only one occupant. It is estimated that this amounts to 25 per cent of the fuel consumed by privately owned cars. It is considered that increased fuel costs will encourage people to travel together to common areas of employment.

An interesting development in the United States is the use of small buses for work-related travel for groups of people travelling to a common destination. One example is the Commute-a-Van Programme of the 3M Company in Minnesota. The Company operates a 96-vehicle fleet, providing a home-to-work transport service for over one thousand employees. For urban travel, such buses are over 5 times more efficient than cars with only one occupant.

Energy Conservation in Transport Planning

Motor cars and small buses with considerably improved fuel economy seem likely to continue to meet the greater part of our urban transport needs for some time. This focuses attention on the present problems of traffic congestion in our cities, and the fact that this congestion will become increasingly expensive to the community. This prospective increasing cost of traffic congestion should be taken into account by public authorities in planning improvements to road and freeway systems and in planning extensions to public transport systems. There are many cases in our cities of part-completed freeways which cause particular problems in traffic congestion.

In some of our cities there are several examples of part-completed and unconnected freeways. These cause serious traffic congestion and are a discredit to our urban road authorities and to State Governments who have encouraged such haphazard construction programmes.

Governments must accept the fact that cars will continue to be built, and will be smaller, and that traffic congestion will be an increasingly serious cost to the community.

It is important that governments accept the need for long-range planning of urban transport facilities, including freeways, and that construction should proceed on a rational basis.

Improved traffic management, including co-ordination of traffic lights can also assist with lower fuel consumption.

Electric Powered Public Transport

Urban rail transport based on electric power can be expected to continue to meet major peak-hour travel in city areas, although increasing energy costs can also be expected to lead to curtailment of services in off-peak periods. The further development of electric trolley buses seems to be an economic way of extending urban rail transport, and the Task Force on Energy recommended the further study of the possibility of using trolley buses on major feeder services.

The trolley bus was once highly regarded as a public transport vehicle, and became unpopular because of the complicated overhead network and the greater flexibility and lower cost of diesel-powered buses. With rising fuel costs and problems of air pollution, the trolley bus is now being reconsidered and further developed for medium density routes, particularly in Europe.

Most of the new urban development in Australian cities since the Second World War has been in areas away from fixed rail transport; the rail systems were mostly completed in their present form by the time of the First World War. For the past few decades the urban rail transport in Australia has been running at substantial losses, and governments have not been encouraged to commit funds for further development. However, the prospects for improvement of urban transport through the use of electric trolley buses on feeder services may justify demonstration routes in one or more of our major cities.

Energy Efficient Land Use

The pattern of development of our cities reflects the cost of energy just as much as it reflects the customs and living habits of our society. With increasing cost of fuels we may see an end to the urban sprawl and some reduction in demand for commuter travel as people move closer to their work.

Energy costs and related costs of transportation seem likely to be a matter of increasing interest to industry, and will influence location of new industry, as it does not. For these reasons there is advantage to be gained in the early publication of long-term transportation strategy plans for each of our major cities.

Aviation

The Jumbo-Jet and the wide-bodied jets are already reasonably efficient in overall fuel consumption, with figures of 4 litres per 100 seat km on long haul and 5 litres/100 seat km on short haul.

Conservation in use of aviation fuel is most likely to come from improved management of aircraft operations rather than continued improvements in aircraft efficiency. In addition, Middle East crudes yield a higher proportion of jet fuel than Bass Strait crudes, and supply of jet fuel seems secure in the short short term at least. But there are other competing uses for this fraction, and so jet fuel prices are expected to follow the overall movement in crude prices.

It is therefore reasonable to anticipate further development of civil aviation for passenger transport, and further improvement in operating efficiency.

Shipping

The increasing price of bunker fuel will direct attention to substitutes such as coal/oil slurry and pulverised coal. Even now the relatively low price of coal offers cost advantages, but not yet sufficient to offset the convenience of oil and to justify the design of new ships for coal firing.

Conclusion

It seems inevitable that a tight oil situation will develop over the next decade or so, depending on the level of economic growth around the world, and that the price of oil will rise to limits determined by its value as a premium transport fuel and the cost of substitutes.

The time scales for development of alternative transport fuels based on oil shale and coal are very tight indeed. The high costs of synthetic fuel plans inhibit early investment decisions, and the environmental aspects may impose further restraints. The coal and oil shale resources of Australia favour the early development of synthetic fuel plants and there seems to be a need for new approaches in federal-state relations to ensure appropriate government support and participation in these new energy initiatives.

Provided of course there is no dislocation of the international oil market by such problems as trade embargoes, and provided early investment decisions are made on substitute transport fuels, there could well be a smooth transition from major dependence on oil to increasing dependence on substitutes.

In these circumstances, the pattern of transportation in Australia will be able to continue along present lines, but with greater emphasis on fuel economy in transport, especially motor cars, and on more efficient use of transport services overall. The prospective increasing cost of fuel should be taken into account in long term planning of transportation facilities.

References

(1) *World Energy Outlook,* April 1978, Exxon Background Series, Public Affairs Department of Exxon Corporation.

(2) *The National Energy Plan,* April 1977, U.S. Government Printing Office.

(3) *Energy — Global Prospects 1985-2000,* Workshop on Alternative Energy Strategies.

(4) *An Analysis of the Energy Balances for the UK and Western Europe to the Year 2000,* Lecture to Royal Institute for International Affairs, March 1977, by Robert Belgrave of BP.

(5) *An Australian Conservation of Energy Program,* N.E.A.C. Report No. 1, September 1977.

(6) *Australia's Energy Resources,* N.E.A.C. Report No. 2, December 1977.

(7) *A Research and Development Program for Energy,* N.E.A.C. Report No. 3, December 1977.

(8) *Motor Spirit: Octane Ratings and Lead Additives,* N.E.A.C. Report No. 4, February 1978.

(9) *Electric Vehicles,* N.E.A.C. Report No. 5, June 1978.

(10) *Exploration for Oil and Gas in Australia,* N.E.A.C. Draft Report, June 1978.

(11) *Recommendations for an Energy Policy for Australia,* The Task Force on Energy, October 1977.

(12) *Energy Conservation and Transport,* Kneebone & Wilkins, The Task Force on Energy, October 1977.

(13) *A Uniform Code of Practice for Passenger Car Fuel Consumption Reduction,* Federal Chamber of Automotive Industries, June 1978.

The Petroleum Industry and the Economic Instability of the Western World

Given as a talk without notes

Ian G. Sykes

Chairman, XL Petroleum Pty Limited
317 Queensberry Street
North Melbourne, Australia, 3051

The Petroleum Industry was established by the coincidence of three events in 1900. The first of these was the invention of rotary drilling in the United States. This method allowed drilling at about 100 times the rate, and at a fraction of the cost, of the existing cable tool methods. Rotary drills can penetrate up to 400 metres of rock per day and reach 10,000 metres below

Diesendorf, M. (ed.) (1979). *Energy and People.* Canberra, Society for Social Responsibility in Science (A.C.T.)

the surface. Second, a group decided to drill on a tiny hill in a swamp near Bakersfield in Texas. In the following year that hole, Spindletop, gushed oil at incredible rates, and although the United States was already the largest producer of crude oil, Spindletop produced more oil than the 35,000 or so wells, located mainly in the Pennsylvania area, that were already active. Third, in 1900, some odd man from Australia residing in London, William Knox D'Arcy, decided that there could be oil in the Middle East, and started securing concessions for the Western world in that area. After his well in Persia gushed oil in 1908, other companies followed suit.

The oil-automobile age was founded by these events, but the ramifications were not felt at first: even in 1929, coal was still responsible for about 80% of energy use, oil for 15%, natural gas for 4% and hydro for 1%. By 1961 however, petroleum products, particularly oil, came to a position where they supplied approximately half of the world energy. In that year, oil supplied one third of the energy used. By 1971, natural gas, especially in North America, contributed a general order of one-third of the energy used compared with the order of 20% for the world. So the petroleum industry replaced the industrial structure that we had before and transport, city design, army and agricultural methods all changed with it.

Of course, you are sitting in a city which is absolutely designed on the assumption — I think it is a wrong assumption — that cheap energy goes on forever; and this city is not only designed on cheap energy but, in particular, on cheap energy from oil. Half or more petroleum energy is used in transport in Australia. For travel to Canberra I used to take an overnight train service from Melbourne (in terms of moving people around, trains are the most economical form of transport based on petroleum energy). However, the train gets into Canberra without passing one house in the A.C.T. If I move around Canberra, I am forced to use petrol-driven taxis or diesel buses, and now, because the overnight train service from Melbourne does not operate anymore, I am forced to arrive in an aeroplane. The city is designed with a central business section which is not easy to walk to from wherever you live, and its housing is widely spaced. Not only are its suburbs strung out, but there is a long distance between the residential areas and the industrial area at Fyshwick, making walking to work or shopping centres almost an impossibility. Children must often be taken to school in cars or in buses. Canberra's design is untenable.

The industrial structure that revolves around the petroleum industry is also generally untenable. That includes the way we build buildings, the way we heat and ventilate them, and so on. We use high quality energy such as heating oil in many Australian cities, including Canberra, to heat our under-insulated and energy squandering homes. We should not, in my view, look at the details of how to save 6% of energy here and 5% of energy there, over several decades (this is how I understood the previous speaker was approaching the problems), because I do not think we have, in a realistic sense, got the option of making the gradual change at all. I do not say that because I think there is a limitation on the quantity of ordinary fossil fuels available; I say it along the economic lines that I think there is a funding problem in the Western world to actually buy all their oil needs and to change in time. Unfortunately for America, it developed its oil industry first: by the early 1900's it was a major exporter of oil, and it fuelled a couple of world wars. Now its energy position is that its crude oil is largely depleted and, except for Alaska, its production is in rapid decline. The U.S.A. can still produce about 60% of its oil consumption, but this self-sufficiency level must soon be lost. Yet, even now, in balancing their international accounts we can see a broad general capital deficit accruing at the rate of $20,000 million a year for the first three quarters of 1978.

The problem can be epitomised in observing that Spindletop is virtually depleted with only 3 million barrels of oil yet to recover; the Middle East has 500,000 million barrels of oil available for export and the OPEC countries are rapidly acquiring the Western world's technology. Including Alaska, the U.S.A. and Canada have only 40,000 million barrels of known oil remaining and it is clear, from exploration results in the last decade, that most of their oil has already been used.

Australia's case is similar to America's except that our oil industry, as far as exploration is concerned, developed much later. However, we also see a general inability to balance our trading account and we lost about $2,000 million of capital on our balance of payments last year. That is the basic factor that is going to cause change. I do not think that, when a country the size of Australia economically is that far in debt and that far out of balance, you have got any more time to make structural readjustments at the rate of 1% or so a year. The time for that has passed. Another alarming fact is, I think, that the West has been lulled into a position of absolutely false and unjustified security. We have had a very successful and powerful age based on industries and a power structure that rests on oil and gas. We assume, however, that although we are now seeing some reversal since the last decade, perhaps we can extrapolate from this glorious past to a glorious future and that these reverses are just temporary. I am putting the view that they are not temporary reverses: they are the beginning of a structural readjustment. Therefore it matters in the most primary sense to me as to how quickly the entity — it is a nation in our case, and a nation in America's case, or a group of nations if you consider the Western world — can adjust to the new circumstances.

I want to dwell for just a moment on Australia's political background, which might help explain why we are taking the slowest steps towards readjustment of any country in the Western world. We can be singled out, I think, as the nation with the worst record of adjusting to obvious changes. I believe that the reason for this is that we have been so closely bound to the power structure of the large oil companies and the items of industrial structure that rest on them, such as the automobile industry and so on, that we just cannot move. The political structure will not look beyond the oil and automobile industries to see what a dangerous position it has already put the country in.

Of course, looking at the Menzies era of politics we see the rise of power of the Liberal Party which, I think, evolved probably on three basic things: first, a cessation of the idea of nationalising the banks which was proposed by the Labor Party (the idea — leave finance problems to those outside the Government); second, the sale to BP of the Commonwealth Government Oil Refinery in which the Government owned just over 50% of the shares (the idea — leave the international corporations alone); and third, the cessation of petrol rationing (the idea — resources are plentiful and we can build up a nation relying on unrestricted access to oil etc.). Two of these things specifically rest on the petroleum industry while the third concerns the banking system which itself is so enmeshed in the oil and automobile age. Throughout the Menzies era we simply went along with the Prime Minister, who had formerly represented Shell on a Royal Commission into the Petroleum Industry and in legal work involving the oil companies in the 1930s. We went along with more or less what the international oil industry, and car industry and its bankers wanted.

In that period we saw rapid growth and a rapid industrialisation of Australia; everything looked rather good. And we saw, as I said, the designs of our cities and additions to existing cities being based on the assumption that ubiquitous motor cars would be forever. In that period public electric or coal transport was frequently changed to private petroleum or was closed down altogether. So we have seen a progressive decrease in the use of steam trains, electric trams and similar systems.

The whole jerry built structure is persisting with the Fraser regime, which is dominated by a backward looking politician whose forebears could never understand change, believing in privilege and patronage. This is illustrated by the 1978 Budget where we see a very late adjustment of petroleum prices to world parity; at least four years after there was a clear need to readjust the prices and achieve some sort of recognition that the era of cheap oil had ended forever in 1973. The Budget also decreased the price of some motor cars through some sales tax adjustment to try to stimulate car production. Those sorts of moves do not meet the requirements of a nation that really is making adjustments to its real future. Progressive moves would have been removal of sales tax on solar heaters and insulation, a subsidisation of changeovers from oil home heating to other forms of heating, removal of sales taxes on diesel cars and increased taxes on petrol; removal of sales taxes on diesel engines; a steep tax on fuel oils and heating oils; a tax on refineries burning fuel oils instead of using natural gas as a fuel; a subsidy to buy refinery plant for the conversion of the extra fuel oils into transport fuels; and laws to stop the export of natural gas and the use of natural gas in power stations.

The Western World's ideas, used by the status quo, are promoted now through the large banks in America, like the Chase Manhattan Bank, which are heavily financing the oil industry. Their public philosophy is that the Arabs do not really want to cause some sort of financial problem in the Western World because this is not in their interests. That is a very dangerous assumption, and I think it is absolutely wrong. If you take the Old Testament idea that the borrower is the servant of the lender — and I think that is still true — you do not assume anything so lethal as that. The fact of the matter is now that the Western World cannot balance its capital or income accounts at all, and with the lack of balance, it is losing power, both military and economic, at an extraordinarily fast rate. I calculate the rate in real terms as twenty times the rate that applied to capital accumulation by the Western world during the Industrial Revolution. The relative rate of loss of capital now by the Western world is the fastest economic change that the world as a whole has ever seen. If you look at the quantity of reserves available to balance the U.S.A.'s position, you can look at America's deficit in terms of its remaining gold reserves. It can only balance, for about 2½ years, if it sells all of its gold at the current market value. What is the use of the American Treasury trying to bluff people by saying "We'll stop the run on the American dollar, and we'll help balance our accounts by selling half a million or a million ounces of gold"? That sort of statement is like threatening a bushfire with a water pistol. The U.S.A. Government's statements are unreal. It is terribly dangerous because it symptomises that the power structure in our side of the world is incapable of adjusting. This is because the bulk of the power structure has accepted, incorrectly, the assumption that we can continue on our way because some magical source of new energy might become available and capital deficits overall do not really matter.

Some of the national restructuring to which Australia specifically needs to turn its attention include:

1. We must cut down on our reliance on crude oil.
The Government Fuels Branch persists in extrapolating an ever upward course for the figures on crude oil use in Australia. This is nonsensical. Sooner, if we decide to cut down on our crude use, or later, if a decline in crude use is forced upon us by a financial crisis, crude oil consumption must decline absolutely. We can no longer tolerate using crude oil for heating boilers and homes. The whole of our reserves must be set aside for transport, agricultural and industrial use. It is tragic that already five years have been wasted since the need for these changes became clear.

2. We must stop our refining industry squandering oil.
Refineries in Australia squander oil in two ways. First, they burn oil to heat their refineries instead of using natural gas. Second, they fail to use natural gas to convert fuel oils into transport fuels. In the U.S.A. the best refineries yield 110 barrels of products for each 100 barrels of crude oil, whereas in Australia refineries produce only 92 barrels of products for each 100 barrels of crude. In spite of the fact that natural gas is available at every refining centre in Australia, only one refinery uses gas. Refineries in Australia are obsolete tin kettles which are gobbling up huge volumes of crude oil unnecessarily.

3. There is a need to try to find what our oil and gas reserves really are.
Currently only about eight rigs are looking for oil and gas in Australia. Six are drilling gas extensions on known fields and two are wildcatting, which is drilling exploration holes. Canada has over 300 rigs and the U.S.A. has over 2,000 rigs at work. Both these countries have a similar area of sedimentary rocks — rocks in which oil may occur — as has Australia. We would need 100 rigs wildcatting — that excludes development or follow up drilling after a discovery — for 20 years or more to get a fair idea of our petroleum prospects. That program would involve some 20,000 wildcats, compared with 1,000,000 which have been drilled altogether in the U.S.A. and Canada. So far, only about 2,000 exploratory wells have been drilled in Australia.

Australia's exploration efforts are pathetically small in relation to our good chances of finding more oil. We are now reliant for over 90% of our local production on two incredibly productive fields, Halibut and Kingfish, which yield over 360,000 barrels of oil a day. This is over 60% of Australia's needs. These fields will be virtually depleted in 10 years. This month Mr Dixon, who owns Drilling Contractors Pty Ltd, shipped the last of his three rigs from Australia and closed his office. This leaves only two land based contractors of any size in Australia.

The large oil companies had their lives made very easy by a Prices Commissioner of South Australia who awarded high selling prices on the basis of the International Companies' cooked accounts. The oil companies in Australia imported oil at higher prices than the actual prices from their international associates. This caused poor "apparent" returns in Australia, caused the Prices Commissioner to raise prices, and maximised the international profits of the majors. This led to an over-developed marketing system with too many petrol outlets and a run down exploration industry. For the international oil companies, there could have been nothing worse than local production sold at real prices. The arguments about the sale of Moonie oil, found by an independent, illustrated this in the 1960s.

4. The Prices Justification Tribunal must be brought to a realistic prices body.
Since 1973, the Prices Justification Tribunal has kept on adding declared costs by large oil companies to wholesale allowed petroleum prices. As a general rule the submissions of the oil companies have been untrue and unfair. The result of this, and of the chronic and weak behaviour of the Tribunal, has been enormously rapid increases in wholesale prices above what the actual wholesale prices really are. For example, in 1973, XL was purchasing petrol at arm's length (that is, independently) for the order of one cent a litre below the P.J.T. justified wholesale prices, and this disparity has now risen to about three cents per litre. Blind Freddie can see now that every major oil company is selling retail, in every major city in Australia, far below the justified P.J.T. wholesale prices. The industry is currently nagging away at the P.J.T. to award it price rises of about three cents per litre, without a public hearing. Petroleum prices have by far the largest effect on the price structure of this nation.

Besides bringing dishonour to the court system, the Tribunal's failure to make its wholesale prices what the wholesale prices really are, prejudices Australia. Presently the large oil companies can charge very high prices in areas where there is no competition, and oil must be used. They use some of these profits to subsidise the sale of fuel oils to compete with gas and coal. In Sydney, fuel oil is now dumped at about half the allowed P.J.T. prices. The present system means that the refineries need not invest to modify their output patterns to change fuel oil into transport fuels. While this may be in the interests of the refiners to maximise profits or minimise investment, it is very much against the interests of Australia. An honest system of wholesale pricing would obviate the national disadvantages stemming from the P.J.T.'s unreal price system.

5. Diesel engines should be encouraged.
Diesel engines in vehicles are about 30% more efficient than petrol engines. Cars must be switched to diesel engines as quickly as possible in Australia.

6. There are also needs to improve public transport and rearrange incentives as already outlined in my comments on the 1978 Budget.

7. Australia must devalue its currency.

In 1974 I believed that Australia's currency was overvalued by about a factor of two compared with the OPEC and Southeast Asian currency levels. The Australian Government should reduce our currency value to at least dollar for dollar parity with the U.S.A. currency, and this should fall to parity with the Canadian dollar as quickly as possible.

As the factors impinging on the North American currencies are likely to cause them to weaken further, adjustment to about half the 1974 value may then be achieved by market forces as these raise the relative values of other currencies.

Australia cannot sustain its present rising indebtedness, and its present indebtedness has largely risen from the money used to build an economic structure which may soon be worthless; while present rising borrowings are being used to sustain a structure which ought to be changed. Given a continued period of borrowing without readjustment, our credit worthness will be approaching exhaustion at about the time there is a general realisation of the vast investments necessary to change our system. Much investment in overseas goods will be needed to implement the necessary changes.

8. There is a need to realise that the price of oil is going to rise in real terms.

After the 1973 OPEC price rises on crude oil the United States Government made inane predictions that the U.S.A. could somehow become self sufficient in oil again and that the OPEC countries would lower prices. Quite definitely, oil prices will rise to the cost of synthesising oil from coal or making oil from oil shales or tar sands, and that price, in real terms, is at least double the present OPEC price level.

9. Australia must not countenance exporting any natural gas.

The only known significant unutilised petroleum resource in Australia is the North West Shelf natural gas. The Government has recently granted export permits and tax concessions to allow and to encourage this gas to be exported at virtually no profit to Australia. In the next decade, the gas now being flared in OPEC countries will be committed and then gas export prices will rise steeply. Gas can be changed into liquid transport fuels very much more easily and cheaply than can solid materials. All the North West Shelf reserves will be needed in Australia for use within the next three and a half decades. To export them now, on a subsidised or marginal basis, would lead to tragic consequences for Australia as well as for the shareholders owning the gas. This conclusion cannot be altered by any ordinary discount rate.

Problems of High- and Low-Energy Futures

Mark Diesendorf

CSIRO Division of Mathematics and Statistics, Canberra

1. *Introduction*

This paper is a preliminary attempt to assess some crucial problems of high-energy and low-energy futures for Australia. We focus on problems which have previously received little public attention: the timescales required for the development of the very large, capital intensive, commercially unproven technologies of a high-energy future and the new policies and socio-economic changes which would be required for a smooth transition to a low-energy future.

After a summary of the pattern of present energy use (Section 2), the problems of timescale, technologies and economics in a high-energy scenario are considered (Section 3). An alternative moderately low-energy scenario is presented (Section 4), together with some of its advantages — rapid implementation and "solar energy breeders" — and its difficulties — the need for planned institutional, legal and economic changes to permit this pathway to unfold.

2. *Pattern of Energy Consumption*

Figure 1 (1) shows (at the top) Australian sources of primary energy in 1975, and a breakdown into end-use categories (at bottom). Oil is the major source of primary energy (48%), closely followed by coal (40%), with a small but growing contribution from natural gas (7%). The miscellaneous sources are hydroelectric power (2%), bagasse (the waste of the sugar cane industry) (2%) and wood (1%).

Heat. The main category of energy end-use is for non-electrical heat (47% of end-use energy), most of which is produced in industry by burning coal and, to a lesser extent, oil and natural gas. Natural gas is now gradually replacing some of the industrial and domestic heat obtained from oil. Solar

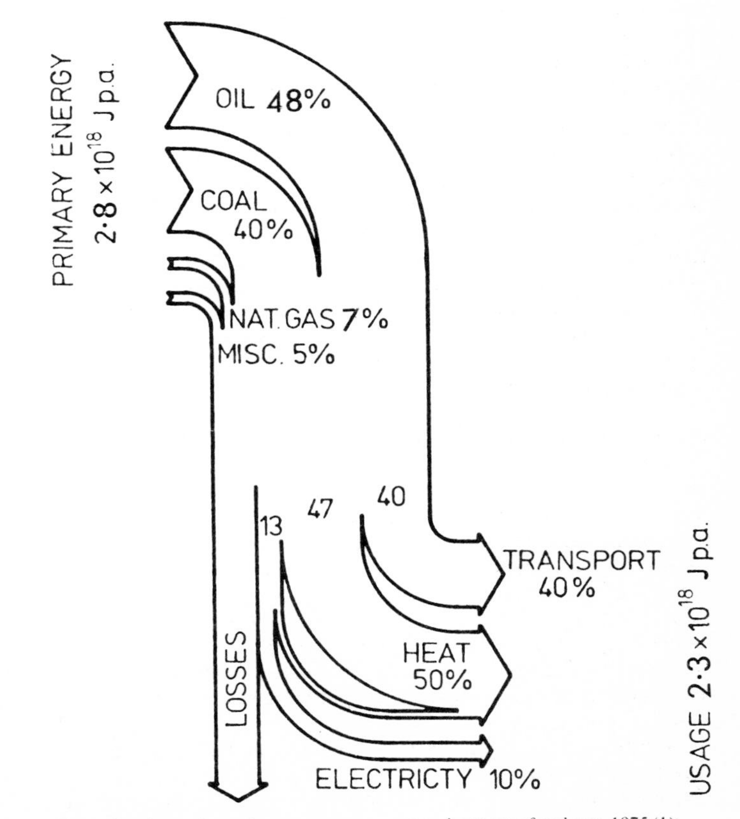

Figure 1. Australian primary energy sources and pattern of end-use, 1975 (1).

Diesendorf, M. (ed.) (1979). *Energy and People*. Canberra, Society for Social Responsibility in Science (A.C.T.)

energy has the potential for substituting for a significant proportion of low-temperature heat. The exact amount is unknown because the temperature distribution of heat use in Australian industry has never been measured. Although a large proportion of industrial heat is required at high temperatures for the production of iron, steel and other metals, there will be several industries which only require low-temperature heat. For example, in the particular case of the food processing industry, 89% of energy requirements are for heat, almost all of which is required at temperatures below 150°C (2). This temperature lies within the range of solar collectors being developed at Sydney University.

Transport, which takes 40% of end-use energy, is currently run almost entirely on oil.

The third largest output of the energy flow diagram comprises, not electricity, but the *losses* which amount to 18% of the primary energy. Almost all of the losses which occur in the conversion of primary to end-use energy result from the generation of *electricity*, the smallest proportion of end-use energy (10%, or 13% if one includes electricity used inappropriately to produce low-temperature heat).

It may be surprising to some that electricity makes such a small contribution. However, to produce one unit of electrical energy in a thermal power station, three units of fossil-fuel energy have to be burnt. The missing two units go into the cooling water and up the stack of the power station as waste heat. Therefore, if low-temperature heat is required for domestic or industrial purposes, it would generally be far more efficient, thermodynamically speaking, to burn the fuel on-site to produce the heat directly, instead of first converting it to electricity. As a corollary, it follows that the conservation of one unit of electrical energy would save three units of fossil fuel energy.

This is a particularly important consideration for natural gas, which could play a vital role as a source of heat (and also as a feedstock for the chemical industry) during the coming period of transition to new energy technologies. Yet natural gas is still being used in electric power stations in South Australia (and shortly, in Victoria), to generate electricity at a thermal efficiency of 35% or less.

Australia's basic energy problem is now well-known: although the nation is currently producing about two-thirds of its oil consumption from indigenous oil-wells, within ten years (unless there is a major new discovery in Australia), these wells could be so depleted that, if the motor transportation lifestyle is to be preserved, it may become necessary to import the equivalent of two-thirds of current consumption.

In exploring different scenarios for solving this problem, it is important to consider the *timescales* for the depletion of the available energy sources, and for the deployment of new energy technologies. In this paper, the *short-term* future will cover the ten years 1979-89 over which substantial depletion of Australia's known oil reserves will occur. The *medium-term*, 1990-99, could witness the substantial depletion of those Australian natural gas fields which are already being tapped (i.e. the North-West Shelf is excluded), while the *long-term* future will start in the year 2000. In the next section it will be shown that the timescales of development required by the technologies of a high energy future pose fundamental problems for high-energy scenarios.

3. *A High-Energy Scenario for Australia*

It is difficult to find a coherent written presentation of high-energy scenarios for Australia. The basic elements of the scenarios, which are part of the conventional wisdom of the Goverment's energy advisors, are visible but the logical connections between them and the problems of implementation are rarely displayed. The reports of the National Energy Advisory Committee (NEAC) (3), the Australian Science and Technology Council (ASTEC) (4), CSIRO Energy Review Committee (5), the Institution of Engineers (6,7) and the Senate Standing Committee on National Resources (8) are all concerned with policies for energy research and development in Australia. Their common orientation appears to be towards high-energy high-technology pathways; this is not surprising since, with the possible exception of ASTEC, their authors are, or rely for advice on, a common core of individuals, mainly engineers, employed in high-technology and high energy-use fields (9). The treatment of energy policy tends to be fragmented into different topics — coal, oil, gas, uranium, electricity, renewables and a small measure of conservation — and it is difficult to detect integrative policies and explicit statements about goals. More important, there is rarely any discussion of choices between *different* goals; yet such choices can be readily found in publications from New Zealand (10), the USA (11,12) Britain (13) and other countries.

The main axioms or presuppositions of the Australian energy policy reports, in order of priority, seem to be:

(i) that the dominant role of the individual private motor vehicle should and will be maintained;

(ii) that energy use will increase by a factor of 2 to 3 over the next 20 years;

(iii) that there should and will be a shift in the pattern of energy consumption to give electricity a greater share of the cake (14);

(iv) that Australia should play a "watching brief" while most of the R & D for renewable sources is performed overseas (3,4), even though much of that R & D is relatively inexpensive.

The result of applying axioms (i)-(iv) to Australia's energy problem leads to the following general policies:—

First, an overwhelming emphasis is placed on the development of industries to convert Australia's vast coal reserves into oil.

Second, there is high priority on the further development of the mineral export industries — such as (a) the production of bauxite, alumina and aluminium and (b) uranium mining, enrichment, possibly reprocessing (15) and even storage of the world's radioactive wastes (16) — justified by the hope of using these exports to pay for part of the oil imports while coal liquefaction is developed. The uranium industries, excluding nuclear power plants, are now seen as the main justification for the disproportionate investment of our energy R & D funds ($26m p.a.) into the Australian Atomic Energy Commission (17,18).

Third, the option to develop nuclear power in the long-term (post year 2000) is maintained, despite the belief that "it is expected to have negligible impact on Australia's fuel supplies for the foreseeable future" (18). The possible role of nuclear power in the high-energy scenarios will become clearer after the discussion of the problems of the coal liquefaction scenario. For the moment, it is simply noted that, for unstated reasons, nuclear power is often assigned a "high priority" in official energy policy reports (19). Since stocks of uranium are inadequate (20) to provide a substantial long-term contribution to electricity supply from the current generation of nuclear power plants (the "burner" reactors), nuclear power will be treated in the context of the proposed second generation of nuclear power plants (the "fast breeder" reactors (21)).

I should like to focus on the coal liquefaction → fast breeder scenario because it appears to be the current conventional wisdom for Australia (6,7,54). However, there may be other high-energy technologies with shorter development times which may offer a greater chance of maintaining the motor car lifestyle. The production of methanol from North-West Shelf gas is one possible example, but at present it seems that the development of this very large resource would be contingent upon the export of most of it as liquefied natural gas. Methanol from coal may contribute. Oil shale is another possibility which has received little discussion in Australia; the severe environmental impact of its development could rule it out. The market for private electric cars might be limited to the second family car (6), until such time as metal-air batteries are developed which permit a longer distance to be covered between charges.

The high-energy proponents claim to be conservative, in the sense that they wish to preserve the present lifestyle. A closer examination reveals that although the high-energy scenarios encourage an existing trend towards more private cars, centralised capital-intensive industry, growth of suburbs, energy and resource wastage, and pollution, they would, if successful, lead to massive changes in the physical structure, economics, industries and lifestyle of our existing society (49). Furthermore, a number of fundamental problems emerge which challenge the feasibility of a successful transition to a high energy future. These problems involve serious gambles about:

- technologies
- timescales for development
- economics and the economy
- environment and health.

Since the environment and health implications of an increase in the use of fossil fuels and nuclear power have been receiving growing attention over the last three years, this paper will focus on the first three problems, which have received scant attention in energy policy reports and public discussions.

Gambles on technologies

Tables 1 and 2 summarize the main problems of coal liquefaction and fast breeder reactors respectively, the principal energy sources proposed for most high energy scenarios. In particular, it is not generally realised that these technologies are not simply available "off the shelf", but still require considerable developmental work.

From Germany's war-time experience and South Africa's current heavily-subsidised production via *Fischer Tropsch synthesis*, we know that it is at least possible to produce liquid fuel on a large scale from coal. However, the thermal efficiencies of the two processes currently favoured in Australia, *pyrolysis* and *hydrogenation*, are uncertain, as are the costs of commercial production.

In the case of fast breeder reactors, the technological uncertainties are even greater. In the usual scenario, breeder reactors would be initially fuelled from plutonium which would be obtained from the spent fuel of the current generation of burner reactors (which are fuelled on uranium-235) and from military wastes. Around the core of the fast breeder is placed that isotope of uranium which is usually discarded in fuelling burner reactors, uranium-238. This is gradually converted to plutonium by the intense stream of neutrons emitted from the breeder's core. In this way, more nuclear fuel is "bred" while, at the same time, the breeder generates electricity. Thus, it turns out that the breeder can, in theory, extract about 60 times more energy from the original uranium than the burner alone.

However, in practice the situation is rather different. A key step in the "recycling" of plutonium is *reprocessing*, the extraction of plutonium from the spent fuel of both burners and breeders. Although the reprocessing of *military* wastes has been performed on a large scale in several countries, there is no large-scale reprocessing of the rapidly accumulating commercial wastes. For a start, commercial wastes have almost 100 times the radioactivity per unit volume of military wastes (which are left in the reactor for shorter periods) (22) and commercial reprocessing has to make some concessions to economics. Even so, the US taxpayers could be forced to pay heavily for at least one of the three commercial scale reprocessing plants which have been built and are not operating (23). The French taxpayers are already paying heavily for the commercial reprocessing plant at La Hague, which is currently operating well below its design capacity.

There are no commercial fast breeder reactors in the world, although one, the "Super-Phénix", is under construction in France. Therefore, much of the technology, safety problems, economics and the doubling time (24) for breeding more plutonium, are unknown or uncertain. However, work with experimental and prototype breeder reactors suggests that the doubling time, originally believed on theoretical grounds to be less than 10 years, may turn out to be 25-50 years, i.e. longer than the expected working life of the reactor.

Gambles on timescales

An important implication of the longer doubling times for the plutonium production in a fast breeder industry (Table 2) is

Table 1

Problems of the Coal Liquefaction Pathway

Technology
The two processes most favoured for Australia (6), *pyrolysis* and *hydrogenation*, are only at a pilot plant stage overseas and there is no experience in converting local coals. A third process, *Fischer Tropsch synthesis*, operates on a commercial scale in South Africa but is very inefficient and very expensive.

Yield
One commercial *pyrolysis* plant would consume 30,000 tonnes/day of black coal* (or 90,000 tonnes/day raw brown coal) to produce 2,000 MWe of electricity and 30,000 bbl/day of synthetic crude oil, which is just 5% of current crude oil consumption.

One commercial *hydrogenation* plant would require 40,000 tonnes/day black coal† (or 100,000 tonnes/day raw brown coal) to produce 100,000 bbl/day syncrude.

Total coal requirements
In 1976, Australia produced 68m tonnes black coal (of which 31m tonnes were exported) and 31m tonnes brown coal (no exports).

To produce all of the current crude oil demand (660,000 bbl/day) from black coal (which is unlikely because specific types of black coal are required — see footnotes) would require Australian black coal production to rise to between 1.4 and 3.5 times the 1976 level, depending on the process used. Alternatively, to produce the same oil demand from brown coal would require 8 to 23 times the 1976 level of raw brown coal production.

Lead times
The construction time, from initial decision to commercial production, for one 30,000 bbl/day plant would be about 11 years (6). The time, from the decision to build the first large pilot plant, to production of Australia's current oil demand, would be *at least* 30 years‡. This estimate allows little time for gaining experience with local coals and assumes that the capital will be available.

Lifetime
Despite Australia's large coal resources, known reserves of coal preferred for pyrolysis and hydrogenation and suitable for low-cost mining would only meet our current liquid fuel demand for 17 years (6).

Costs
Very uncertain. Estimates for capital costs of plant, coal mines, water supply, refining facilities and intrastructure for a 30,000 bbl/day pyrolysis plant with electricity operation are in the range $1,500-2,000m, while a 100,000 bbl/day hydrogenation plant could be in the range $3,000-4,000m. The cost of syncrude has been estimated at $25-35 per barrel, at least twice current price of petroleum before excise.

Thermal efficiency
Estimated optimistically to be 65-70% for pyrolysis, 60-65% for hydrogenation and is in practice 33% for Fischer Tropsch. Thus the overall thermal efficiency of motor vehicles would be reduced significantly below the current low value obtained with natural petrol.

Water requirements
Very large (10,000 m^3/hr) amounts of water are required for the pyrolysis plant, which uses process water plus cooling water for electricity generation from the char and gas produced. For hydrogenation, process water at a rate of 1,000m^3/hr is required.

Pollution
Coal liquefaction releases CO_2 and the same carcinogenic hydrocarbons as produced in fossil-fuel power-stations, though possibly less SO_2, NO_x and fine particles. There is also a risk of polluting the cooling water.

* More precisely, these calculations were performed for a specific type of black coal known as "high volatile, bituminous" (6).

† In this case, black coal which is "high vitrinite, plus exinite, low ash; low to medium rank", is preferred (6).

‡ Actually, the Institution of Engineers' Report (6) suggests 30 years to produce a smaller amount, 430,000 bbl/day.

Table 2

Problems of the Fast-Breeder Reactor Pathway

Technology
Although there are several operating prototype fast reactors overseas, each of several hundred MWe capacity, there is no commercial (1,000 MWe) fast breeder. One commercial-size plant is under construction in France.
There is currently no commercial-scale reprocessing of spent oxide fuel: a reprocessing plant is operating at La Hague, France, but well below its design capacity; none of the three commercial-scale oxide reprocessing plants built in the USA is operating, as the result of safety, regulatory and economic problems.

Lead times
Construction time (from initial decision) for a single reactor would be about 10 years. Even if breeders were built very rapidly, they could not make a significant contribution to electricity supply for the next 50 years or so, primarily because
i) their initial fuel charge would have to come from the reprocessed spent fuel of the current generation of burner reactors;
ii) the doubling time for "breeding" new plutonium from uranium-238 is much longer than expected — now believed to be 25-50 years.
In addition, there are constraints on growth of the industry resulting from the problems of
iii) creating a commercial reprocessing industry at a time when it is not clear that this is the best route for the disposal of high-level wastes (51);
(iv) putting in more energy (mainly in the form of scarce fossil fuels) than is generated (as electricity) during a period of rapid growth of the reactor industry (52).

Safety
In addition to having similar safety problems to those of burner reactors, fast breeders lead to
i) greater risks of the proliferation of nuclear weapons because of the larger quantities of plutonium produced;
ii) greater risks of serious reactor accidents: because of the denser core, the possibility of a limited nuclear explosion (which is impossible in a burner reactor) cannot be ruled out; furthermore, if the liquid sodium coolant came into contact with water, a large conventional explosion would occur; in each case the huge radioactive inventory of the reactor core could seriously contaminate enormous areas;
iii) greater risks from terrorism and sabotage.

Civil liberties
As a result of (iii), the constraints on civil liberties of nuclear workers and members of the public living in a "plutonium economy" would be considerable (53).

Economics
Very uncertain. Huge amounts of capital would be required for the reactors and reprocessing.

that the rate of construction of breeders is limited by the rate at which plutonium becomes available from existing burner reactors. As a result, it has been pointed out by Merrick (24) that the 60-fold theoretical improvement in uranium utilization by the breeder would not be realised until well into the 21st century. The breeder could make no contribution to the energy crisis in the short- or medium-term. Indeed, Merrick has argued that even in the UK, which already has significant stocks of plutonium, breeder technology cannot ensure an indigenous supply of electricity for at least 50 years (24). There is one (theoretically) possible way of speeding up the growth of a breeder industry. If all the plutonium in the world's nuclear weapons could be "melted down", it would be more than sufficient to fuel a large breeder industry.

In Australia, it would be technically possible to use highly enriched uranium as an initial fuel for breeders. However, this would be very costly and the reactor performance would not initially be in the "breeding" regime. Hence a long timescale for development of the industry would also be expected in this case.

In the case of coal liquefaction (Table 1), the timescale problems are even more acute. It is thought that to construct one plant, capable of providing for 5% of current crude oil consumption, would take about 11 years (6). Assuming that there were no major problems with the first plant, a crash program to provide just 2/3 of *current* oil consumption from coal could well take 25-30 years (6). Yet, unless there is a major new indigenous discovery, Australia could face a major oil crisis within 10 years!

Even if an industry, producing enough oil to meet *current* demands, could be built up over the next 35-40 years (and the high-energy scenario envisages a considerable growth in oil consumption beyond the current level), by then it would have already exhausted most of the known reserves of coal preferred for pyrolysis and hydrogenation and suitable for low cost mining (6). Thus, it would seem that the coal liquefaction solution will provide too little, too late, for a high energy society.

What role for nuclear power?

The timescale problems of coal liquefaction provide the clue to the mystery of the inclusion of nuclear power under "high priority" by proponents of a high energy future. A major commitment to coal liquefaction, taken together with an increasing rate of coal exports and increasing supply of electricity, could lead to a coal shortage in the first decade of the 21st century. For example, the Ranger Uranium Environmental Inquiry (25) estimated the lifetime of recoverable reserves of Victorian Brown Coal, at the 1975 consumption rate, to be 444 years. Yet a mere 6% per annum of exponential (compound interest) growth in coal use would reduce this lifetime to 55 years; 8% growth would reduce it to 45 years and 10% to 38 years. Major problems with supply, resulting from competition between different uses for the coal, shortages of particular types of coal, etc., could occur well before these time-periods.

Thus, in the high-energy scenarios for Australia, coal could only play a transitory role until nuclear fission (or fusion) could take over. This explains, at least in part, the role of nuclear power in the high energy scenarios, as well as the pressure to shift over to a greater use of electricity.

The coyness in official energy policy reports about the forthcoming demands on coal could perhaps be explained by anxiety about the public reaction. The public are only just beginning to recognise that they will probably have to face a major oil shortage in the near future. They would hardly be pleased to learn that current energy policies may lead their children into a major coal shortage as well.

In attempting to reduce the timescale problem on nuclear power, to the extent that a viable fast breeder industry is in existence before the coal shortage develops, the high-energy proponents may be tempted to ignore the major technological, economic and safety problems of fast breeders and reprocessing plants and try to develop a nuclear power industry in Australia over the medium-term.

On one hand, those States which are short of fossil fuels — W.A., S.A., Tas. — have electricity grids with maximum demands in the range 800-1500MWe. These grids are much too small for the incorporation of a 1,000MWe (or even a 600MWe) nuclear power plant, the size which is generally regarded as being the most economic (or least uneconomic). These States also have considerable windpower potential (26,27). R & D on large-scale windpower is proceeding rapidly in Denmark, Sweden, the USA and several other countries; this research may prove relevant to the needs of W.A., S.A. and Tas.

On the other hand, in those States whose grids might be large enough to incorporate nuclear power plants in the medium-term — N.S.W. and Vic. — nuclear power cannot compete in economic terms with the cheap (in the short-term and probably the medium-term) local coal.

It has been suggested (28) that one means of introducing nuclear power would be to begin with the introduction of the other industries in the nuclear fuel cycle — uranium mining and enrichment, in particular. Nuclear power might then be "justified" as a means of ensuring the financial viability of the large investment in uranium enrichment and also to provide on-site electricity for the project.

The notion that the uranium industry could make a significant contribution to Australia's foreign exchange earnings rests on the uncertain assumption that there will indeed be a big world market for uranium in the medium-term. However, as the result of the economic, environmental and safety problems of nuclear power, and a growing public opposition, the industry is almost at a standstill in all countries except France, the Soviet Union and South Korea. In addition, with the new Canadian uranium mining developments, there will shortly be a temporary glut of uranium on the world market. Even if a market can be established for Australia's uranium, a second uncertain assumption pointed out by Ian Sykes (29) remains. There is no guarantee that the Organisation of Petroleum Exporting Countries will be willing or able to provide Australia with petroleum, even if we could pay for it with the earnings from mineral exports.

Gambles on economics

The capital costs of the high-energy technologies are very large. One 100,000 bbl/day plant to produce liquid fuel from coal by hydrogenation, together with associated coal mines, water supply, refining facilities and infrastructure, could be in the range $3,000-4,000m. One 1,000MWe commercial fast breeder reactor, *without* reprocessing plant and other facilities, could cost $1,500-2,000m, at today's prices. Since the technologies are unproven at a commercial scale, and the timescale problems are substantial, massive investments into these areas would entail a considerable financial risk. Serious economic difficulties on the national scale could not be ruled out.

Even if an extreme outcome could be avoided, a likely result of following such a pathway would be that the major part of available investment capital might become tied to the energy industries. Thus, the economy would be greatly distorted, the development of renewable energy sources pre-empted, and the funding of more labour-intensive areas such as health, education and welfare might suffer. Commoner (30) has noted that already for 1970-3, the capital required to support US energy production amounted to one-quarter of all capital invested in US business as a whole and that this fraction was expected to reach one-third within the period 1975-85. Some contemporary figures would be valuable.

A less obvious area of economic doubt consists of excessive reliance by Governments on traditional economic theories in a high-growth, high-consumption society. Several of these theories have recently been called into question. For example, Blatt (31) has shown that the classical result of Kemeny, Morgenstern and Thompson — that, in an economy with exponential growth and ultimate consumption, the physical growth rate of the economy is equal to the equilibrium rate of profit — is incorrect. In another example involving investment evaluation under conditions of unpredictable events in the future (i.e. a real world situation), Blatt (32) has shown that the well-known "discounted present value" method is inadequate; he has proposed a new method which is more in accord with practical business requirements.

Summary — from oil crisis to coal crisis

A realistic assessment of a high-energy scenario for Australia reveals a number of fundamental problems. Since luck appears to be the only way of surmounting most of them, I have referred to them as gambles. The gamble on timescales is possible the biggest plunge of them all. Unless there is a major oil-strike in Australia, indigenous supplies could be low by 1990. If oil consumption is allowed to grow through the 1980s and hopes are focussed on coal liquefaction, serious national economic difficulties could occur if OPEC restricted its supplies around 1990. Coal liquefaction is unlikely to make a major contribution before 2005-2010, while the "all-electric" plutonium economy could not readily take over from coal before 2030 at earliest. Yet a major coal shortage could well occur in 2010-2020.

To the uncertainties about timescales we must add the gambles by taxpayers and business of enormous amounts of capital on commercially untried technologies; excessive reliance by the Federal Government on inadequately tested economic theories and practices; and finally gambles on the environment and health impacts of the technologies. Apart from some of the environment and health issues, which are important in their own right, these points have received little public discussion. Thus, options for a stable transition to a low-energy future could be foreclosed.

4. *A Moderately Low-Energy Scenario*

Harsh realities

Serious consideration of a low-energy future should follow from the recognition of these realities:—

- that the world's natural oil and natural gas are likely to be substantially depleted by the year 2000; in particular, unless there are new oil discoveries, Australian oil production could be very low by 1990;
- that prospects for oil imports may become more and more uncertain after 1990;
- that there is at present no general substitute for oil and gas, the only energy sources which can be readily stored and transported;
- that it is very unlikely that coal liquefaction, fast breeders and other high-technology "solutions" could be brought into production quickly enough to permit a continued growth in energy consumption;
- that a last-ditch attempt to attain a high-energy future could foreclose our options to make a gradual transition to a low-energy future based on conservation and renewable energy sources.

Timescales of renewable energy sources

Those renewable sources which would in theory be capable of supplying *very large* amounts of energy — solar satellites, ocean thermal gradients and wavepower — also have very long development times and probably very large capital costs. None of these technologies is likely to make a major contribution during this century.

However, contrary to the beliefs expressed in Australia's many energy reports, there are several technologies for tapping renewable energy sources on a *medium scale* which could make a significant contribution to a moderately low-energy society before the year 2000.

Amongst these are new technologies for producing alcohol fuel from crops and from forest and crop residues which, according to McCann (33), could provide 10-15% of current transport oil consumption by 1990. The rate of growth in alcohol production would slow substantially after 15% is reached, because it would then become necessary to develop marginal land. Still, alcohol from biomass is physically capable, in the short-term, of providing a major part of *essential* requirements: transportation of agricultural produce to rural railways, and urban bus transport for commuters.

R & D into solar heating, at temperatures below 150-200°C, is proceeding rapidly and could be widespread for industrial as well as domestic purposes in the medium-term. If, for instance, 30% of heat demand could be supplied by solar heating, this would amount to 15% of Australia's current total end-use energy. For comparison, in countries heavily committed to nuclear power, this source only contributes 2-3% of total end-use energy, after nearly a quarter of a century of commercial operation.

For electricity, the major load over the next 20 years would still be carried by coal. Nevertheless, two renewable sources — wind generators, for high-power requirements in grids and isolated regions, and solar photovoltaic cells, for low-power requirements — would each reach the point of mass commercial manufacture within 10 and 5 years, respectively. The developmental work still required for windpower is minor, and the main constraints upon development are non-technical ones such as the unrealistically low price for the competing conventional fuels (e.g. natural gas) and high interest rates on the capital required (27). Once these impediments are overcome, wind systems can be constructed and deployed very quickly; for large wind generators the construction time and the time needed for paying back the initial energy investment are both only one year (27).

The development of solar photovoltaic cells still requires some important technical breakthroughs, in order to reduce the cost and the energy inputs: either a cheap process for mass-producing the amorphous silicon cell or a cheap system incorporating sunlight concentrators which does not overheat (34). Once one of these developments is achieved, photovoltaic cells may be able to compete economically with mains' electricity. Since the amorphous silicon cells are expected to have an energy payback time of only 2 years, they too could proliferate rapidly.

A system for utilising renewable energy sources, which can rapidly pay back the energy required for its construction, can expand without placing any burden on existing fossil fuels. Therefore, wind generators and amorphous silicon photovoltaic cells can be considered to be "solar energy breeders" (35). These particular solar breeders produce a high-grade form of energy, electricity, which can be converted to mechanical energy or heat with high efficiencies. But unlike the generation of electricity from fossil fuels or uranium and plutonium, there are no problems with waste heat, CO_2 or radioactivity. The final level of energy conversion by solar breeders would be constrained by the diffuseness of the sources, and so a self-limiting mechanism would assist in protecting the environment from excessive technological growth. Starting from a modest initial outlay, a nation with a relatively small economic capacity but a good range of natural resources could eventually grow to complete energy independence. The timescale for such a transformation, given a short energy payback period for the sources, would be determined mainly by the timescale required to change the nation's pattern of energy use and to match the economy to this pattern — several decades at least.

Returning now to Australia's short-term energy problem, we observe that none of the medium-scale renewable technologies could be developed sufficiently quickly to avert an oil crisis in a high-energy society in 1990, although intensive work on alcohol from biomass appears to be the best technological prospect. The only measure which has the potential to dramatically alter Australia's energy prospects in the short-term is *conservation*. Controlled reduction of energy consumption seems to be the only realistic alternative to the high risk of a sudden uncontrolled reduction. While oil supplies dwindle, conservation, coal and natural gas will act as transitional "fuels" while the medium-scale, medium-term renewable sources are developed. The private motor car, with its legacy of accidents, air pollution and alienation would be allowed to fade slowly from the scene while public transport and local community development would be encouraged. That, in essence, is the low-energy scenario considered in this paper.

Low-energy futures, based on conservation and renewable energy sources, have been discussed previously for Denmark (36), Sweden (37), Canada (38,39), the USA (40,41) and Britain (42).

How much conservation?

The answer to this question depends on one's particular social viewpoint. Most official energy reports envisage the paring off of a few percent of current energy use here and there. For example, the member nations of the International Energy Agency (IEA), which now include Australia, have each set themselves the modest goal of reducing oil consumption by 5% in 1979. If this reduction could be maintained for 11 consecutive years, a 43% total reduction in current consumption would be achieved in 1990. Although the IEA has not proposed the latter measure, at least one high-energy oriented body, the Australian Institution of Engineers, seems to believe that such a reduction is possible:

> "Under emergency conditions, even with the severest rationing, more than 50 percent of current liquid fuel supplies will be required to maintain normal industrial activity" (43).

Although "emergency conditions" (i.e. the timescale of the reduction and of the preparation for it) and "normal industrial activity" were not defined, the statement gives an indication of the kind of goal which might be possible.

Until recently, Australia's primary energy consumption was doubling every 12 years. It follows that halving consumption need not necessarily take society back to its primitive beginnings, providing that such a measure does not occur suddenly. Halving our *per capita* energy consumption would take us back to the early 1960s, a time with higher levels of employment and lower levels of inflation than the late 1970s. However, there is a real problem: during the intervening period the outer suburbs of the major cities have grown into areas which are often served poorly by public transport, yet are distant from the workplace (44). The existing stock of houses and industries is an important constraint on the types of conservation measures which can be implemented in the short-term.

An alternative method of judging how much energy conservation may be really necessary, would be to start by facing squarely the likelihood that imported oil could be cut off shortly after 1990. What measures should be taken in the interim period to permit *survival*, rather than "normal industrial activity"? Once these requirements are determined, how can we build on them to create a viable economy based on *essential* industries? Such a study is vital for planning purposes. It would require a small multidisciplinary team of people with skills in e.g. geography, economic history, energy analysis, sociology and political science. Only a few tentative notes will be made here: it should be understood that these reflect my personal views on possible social change.

In a warm-temperate country, food is the predominant requirement for survival. Centralised water supply, sewerage and electricity could be listed next. When telephone/data links could be maintained, the transportation of people may not be of such high priority for maintaining a basic level of economic activity as the transportation of goods. Since most of Australia's population is concentrated in a few large cities, which are now almost totally dependent upon the motor vehicle, an expansion and diversification of non-oil transportation modes is urgently needed.

For survival of the cities, it would be necessary to reorganise and upgrade Australia's railway network. Coal-fired locomotives should be re-introduced for rural freight, while the main intercity railway links should be electrified. These would be the most expensive changes of the scenario. In rural areas, alcohol-powered trucks, fuelled (whenever possible) from locally-grown biomass, would deliver produce to the railways.

Within the cities, the introduction of light electric delivery vehicles, interconnecting with rail freight terminals, is probably already economically viable. For the journey to work, buses fuelled on 100% alcohol would be introduced, along with better facilities for changing between different transportation modes. The bicycle could play an important role in providing rapid access by a large proportion of the population to public transport nodes. There are simple, inexpensive and effective schemes for introducing cycleways in urban areas (45,46). Facilities for transporting bicycles on trains and in trailers behind buses are also feasible.

In Table 3, a summary is given of a program for a transition to a moderately low energy future. Whether the transition is smooth (i.e. avoids social and economic disruption) or not depends on the time remaining before a major oil drought and the speed of implementation of the proposed reforms. It is my personal opinion that this transitional society would be more stable (against sudden shocks to the economy), more equitable and healthier than one based on the uncertainties of high-energy scenarios. It keeps open a range of options for the long-term future. For instance, large-scale, centralized, renewable energy sources (wavepower, ocean thermal) might dominate, or decentralisation to a large number of self-sufficient townships and communities might be chosen, or some mixture of both

Energy end-use category	Technologies to be developed and implemented	Socio-economic changes required for implementation
heat	*short-term:* construct new buildings according to passive solar design; fit solar hot water; retro-fit existing buildings with insulation, solar greenhouses, etc.; make existing heating and cooling systems more energy efficient; construct thermal storage in hot-water tanks or rock beds; perform R & D for industrial solar heat to 200°C. *medium term:* commercialize industrial solar heat to 200°C; and thermal storage via solar ponds, phase transitions, etc.	revise building codes; introduce low interest loans and income tax concessions for solar heating, cooling and conservation; remove sales tax from insulation, solar heating/cooling systems; keep the price of oil at world parity and provide incentives to shift from oil to natural gas and coal for high-temperature heat; encourage medium density housing, with community thermal storages, in cold regions.
transport	*short-term:* Start to electrify and up-grade main railway lines, eg. between Melbourne & Sydney; re-introduce some coal-fired locomotives on rural railway lines; introduce light trams or trolley buses to inner city areas; integrate different urban transport modes, e.g. car-train, car-ferry, cycle-train interchanges; construct urban cycleways by blocking some minor roads to motor-vehicle through-traffic; LPG for taxis and other fleets; commence production of alcohol from biomass for mixing with petrol; introduce electric delivery vehicles; use wide-bodied jets for long-distance internal as well as external flights; perform R & D into dirigibles, automatic sailing ships. *medium-term* modify urban buses, and trucks in agricultural areas, to run on 100% alcohol; commercialize electric delivery vehicles; construct pneumatic tube systems for inner city freight, if energetically viable; use dirigibles and automatic sailing ships for much overseas freight; use rail for most long-distance internal freight.	re-organise the railways; in particular, develop a reliable coordinated interstate rail freight system with provision for door-to-door deliveries via truck; increase tax on intercity road transport according to weight to pay for main-road maintenance; ban private cars from inner city; provide incentives for using LPG or diesel fuel in motor vehicles; require trains, urban express buses and inter-city buses to provide trailers or racks for carrying bicycles; strengthen system of priority lanes for buses; increase flexitime, part-time employment and shopping hours to spread transport peaks; fund suburban and neighbourhood community centres with facilities (e.g. visi-phone, computer links, offices) for working at centres; increase registration fees for large cars and automatic transmission (concessions for large families and the handicapped, respectively); reduce registration fee for diesel and LPG cars; introduce petrol rationing.
electricity	*short term:* develop solar photovoltaic cells (amorphous semiconductors) with battery storage for local (non-grid) supply at low power; construct the first large commercial wind generators for use as fuel savers in isolated towns and electricity grids in southern States. *medium-term:* construct pumped hydro-electric storages, based on fresh-water in Victoria and salt-water in S.A. and W.A.; expand large-scale windpower in those States and Tas. — with short-term storage and demand modification (see next column), windpower will have some "firm" capacity (70-80%); keep coal as the main source of grid electricity while additional renewable sources — ocean and tidal currents, wavepower — are developed; develop flywheel and hydrogen storage.	modify demand by introducing inverted tariffs for electrical *energy* used, higher "rental" rates for high-*power* circuits, and cheap rates for circuits with interruptible supply; increase sales tax on electrical heaters and air conditioners; remove sales tax from photovoltaic cells; remove subsidies for oil-fired power in isolated towns but provide low-interest loans and other incentives for wind and solar power; raise price of natural gas and restrict its use in power-stations to peakload only.

Table 3

Program for a Transition to a moderately Low-Energy Future in Australia

approaches could evolve. In the long-term, it is possible that the three categories of energy end-use treated in Table 3 — heat, transport and electricity — would merge. Greater attention would have to be paid to efficiencies of energy conversion as determined by the Second Law of Thermodynamics (47), with the result that any burning of fossil fuels to produce electricity would have to be accompanied by the utilisation of the waste heat by industries. As in the high-energy scenarios, electricity would play a larger role in end-use patterns than it does at present (public transport is an efficient use of grid electricity), but the *absolute* amount of mains electricity generated in *thermal* power stations could be very much less than the present value.

The second column of Table 3 lists the main technologies to be developed and implemented, while the third column indicates some of the socio-economic changes which would be required for the implementation of the technologies.

Socio-economic changes needed for adoption of the low-energy scenario

These would include:—

- *Economic* changes such as price incentives and disincentives. The standard objection that governments should not interfere with the "free" market is invalid in the field of energy, which is dominated by OPEC and the multinational energy companies. Would it not be better for governments to manipulate prices in the public interest than to have them entirely determined from overseas?
- *Legal* changes to facilitate the introduction of conservation and renewable energy technologies. These would replace existing laws and regulations which tend to favour energy wastage and non-renewable sources.
- *Institutional* changes have been stressed recently by Lönnroth (48) and Hooker (49). These would create new policy-making institutions which would be democratic, sensitive to value judgements, capable of learning, economical of scarce resources and integrated across the traditional boundaries of intellectual disciplines (49).

It should be clear that socio-economic changes offer the key to a low-energy future. These changes can spring from new policies by federal and state governments, local authorities, professional groups, employers and unions. They need not be expensive but could be very difficult to introduce in a society with large, existing bureaucratic institutions and a public conditioned to believe in technical fixes.

To meet the requirement for survival, the most important social change is the re-introduction of community life in Australian suburbs and towns. The motor car and its freeways, the construction of huge regional shopping centres, jobs which require people to shift house frequently, nursing homes for the aged, television, etc. have destroyed much community spirit and support and have made people even more dependent upon the motor car and centralised services. An oil shortage could leave millions of Australians isolated and helpless, with no community basis for organising food supplies, transport, employment, etc.

In *Seeds for Change* (46), it is suggested that a return to community involvement — primarily at the local neighbourhood, but also in the suburban and regional centre — could play a fundamental role in facilitating a stable transition to a low-energy future and diminishing the ill-effects of an energy crisis. As a bonus, the general quality of life would also be improved, and dependence on alcohol, analgesics, psychoactive drugs and other "magic bullets" would be reduced (50). Already there are signs of a spontaneous flowering in community activities — childcare, food cooperatives, self-help groups, community buses, craft groups, community workshops, organic growers and learning exchanges — to name a sample. Local councils and governments, State and Federal, can assist in this process by reducing institutional, legal and financial constraints upon such developments. The need for community centres and the growth in cooperative housing projects also offer the opportunity for the development of demonstration models in solar housing.

Conclusion

Both high-energy and low-energy pathways have unsolved problems; both entail major changes in lifestyle and have profound human implications. Both require large investments of capital in new energy sources. However, in the low-energy case, R and D and commercial implementation can generally proceed with much smaller units of investment and hence with smaller economic (and environmental) risk.

The conventional high-energy scenarios for Australia, based on coal liquefaction followed by fast breeder reactors, entail major gambles on the timescales required for the development of huge, capital-intensive, commercially unproven technologies. Conservation, followed by medium-scale renewable energy technologies, can be implemented more rapidly. This pathway can utilize "solar energy breeders" and permits a diverse range of self-sustaining long-term futures to emerge. Its main obstacle is the difficulty of implementation of new policies to permit locally-planned socio-economic changes, in the face of inevitable social inertia.

To maintain our future options, it is in the interest of society at large for demonstration models of low-energy *communities*, as well as the technologies, to be set up. It will be of interest to see which (if any) of the two energy futures finally emerges in Australia.

References

(1) Morse, R. N. (1977). Solar energy in Australia. *Ambio* **6**, (4) 209-15.
(2) Proctor, D. and Morse, R. N. (1975). *Solar energy for the Australian food processing industry*. Int. Solar En. Congress, UCLA.
(3) Australia, National Energy Advisory Committee (1977). *A research and development program for energy*. See also the other NEAC reports listed in L. A. Endersbee's paper, this volume.
(4) Australian Science and Technology Council (1978). *Energy research and development in Australia*. Canberra, AGPS.
(5) CSIRO Energy Review Committee (1977). Report to the executive. September.
(6) Australia, Institution of Engineers (1977). *Submissions of working parties*. Canberra, the Institution.
(7) Australia, Institution of Engineers (1977). *Recommendations for an energy policy for Australia*. Canberra, the Institution.
(8) Australia, Senate Standing Committee on National Resources (1977). *Report on solar energy*. Canberra, AGPS.
(9) e.g. see Crossley, D. (1979). *Energy policy and people*, (this volume).
(10) Maiden, C. J. (1976). *Energy scenarios for New Zealand*. Newsletter No. 5, NZ Energy Research and Development Committee, Auckland.
(11) USA, National Research Council, Committee on Nuclear and Alternative Energy Systems (CONAES) (1978). US energy demand: some low energy futures. *Science* **200**, 142-52.
(12) Wilson, C. L. (Project Director) (1977). Energy: global prospects 1985-2000. Report of workshop on alternative energy strategies. McGraw-Hill.
(13) Chapman, P. (1975). *Fuel's Paradise*. Penguin.
(14) See, for example, (5) p4 item 8 and (7) section 4.6; the emphasis on *primary* energy in (3) section 4 also introduces a bias towards electricity which is inconsistent with its small contribution to *end-use* energy.
(15) Butler, S. T., Raymond, R. and Watson-Munro, C. (1977). *Uranium on trial*. Horwitz.
(16) Sabine, T. (1976). *National Times*, May 10-15.
(17) See (7) section 4.3.
(18) See (5) paragraph 36.
(19) See (5) paragraph 107 and (7) section 4.3.
(20) OECD Nuclear Energy Agency and IAEA (1977). *Uranium: resources production and demand*. Paris, OECD.
(21) e.g. see (7) sections 2 and 4.3.
(22) Krugmann, H. and von Hippel, F. (1977). *Science* **197**, 883-5.
(23) Carter, L. J. (1977). West Valley: the question is where does the buck stop on nuclear wastes? *Science* **195**, 1306-7; Severo, R. (1977). Too hot to handle. *National Times* (May 16-21).
(24) Merrick, D. (1976). *Nature* **264**, 596-8; and his ref. (3).
(25) Fox, R. W., Kelleher, G. G. and Kerr, C. B. (1976). *Ranger Uranium Environmental Inquiry, First Report*. Canberra, AGPS.
(26) Mullett, L. F. (1957). *J.Inst.Eng.Aust.* **29** (3) 69-73; (1976) Report FIES-7604, Inst. for Energy Studies, Flinders Univ. of South Australia.
(27) Diesendorf, M. (1979). *Search* **10**, 165-73.
(28) Hutton, B. (1979). Nuclear power in Australia. *Chain Reaction* **4** (4) 9-13.
(29) Sykes, I. G. (1979). *The petroleum industry and the economic instability of the western world*, (this volume).
(30) Commoner, B. (1976). *The Poverty of Power*, New York, Bantam.
(31) Blatt, J. M. (1978). Consumption in von Neumann growth theory, Univ. of NSW, School of Maths, preprint.
(32) Blatt, J. M. (1978). Investment evaluation under uncertainty. Univ. of NSW, School of Maths, preprint.
(33) McCann, D. J. (private communication)
(34) Trickett, E. S. (1979). *Solar cells and people*, (this volume).
(35) Slesser, M. and Hounam, I. (1976). Solar energy breeders. *Nature*, **262**, 244-5.

(36) Blegaa, S., Josephsen, L., Meyer, N. I. and Sorensen, B. (1977). Alternative Danish energy planning. *Energy Policy*, p.87-94 (June).
(37) Johansson, T. B. and Steen, P. (1978). *Solar Sweden*. Stockholm, Secretariat for Future Studies.
(38) Middleton, P. and associates (1976). *Canada's renewable energy resources: an assessment of potential.* A study prepared for Office of Energy Research and Development, Dept. of Energy, Mines and Resources.
(39) Valaskakis, K., Sindell, P. and Smith, J. G. (1977). *The GAMMA report on the conserver society*, Vol. I, Uni. Montreal/McGill Univ.
(40) Lovins, A. B. (1977). *Soft energy paths*. Penguin.
(41) Hayes, D. (1977). *Rays of hope*. New York, Norton/Worldwatch.
(42) Leach, G. (1979). *A low energy strategy for the U.K.* London, IIED.
(43) See (7) section 4.2.
(44) King, R. (1979). *Energy prices, urban structure and social effects*, (this volume).
(45) Parker, A. A. (19) Bicycles and the conservation of oil reserves. *Polis: a Planning Forum* 5 (2) 26-37.
(46) White, D., Sutton, P., Pears, A., Mardon, C., Dick, J. and Crow, M. (1978). *Seeds for Change*. Melbourne, Patchwork Press/Conservation Council of Victoria.
(47) Lustig, T. (1979). *Planning criteria to cope with the entropy crisis*, (this volume).
(48) Lönnroth, M. (1976). *Swedish energy policy: technology in the political process*. Stockholm, Secretariat for Future Studies; Lönnroth, M., Steen, P. and Johansson, T. B. (1977). *Energy in transition*. Stockholm, Secretariat for Future Studies.
(49) Hooker, C. A. and van Hulst, R. (1977). *Institutions, counter-institutions and the conceptual framework of energy policy making in Ontario*. Dept. of Philosophy, Univ. of Western Ontario.
(50) Diesendorf, M. (Ed.) (1976). *The magic bullet: social implications and limitations of modern medicine, an environmental approach*. Canberra, Society for Social Responsibility in Science (A.C.T.).
(51) Breach, I. (1978). *Windscale fallout*. Penguin.
(52) Lovins, A. B. and Price, J. H. (1975). *Non-nuclear futures*. Friends of the Earth/Ballinger.
(53) Flood, M., Grove-White, R. and Suter, K. (1977). *Uranium, the law and you*. London and Sydney, Friends of the Earth.
(54) Baxter, J. P. (1979). *Sydney Morning Herald*, letter, 2 July.

Note

Any social interpretations expressed in this paper are those of the author and do not necessarily reflect the views of CSIRO.

From Hunting Tribes to Industrial Nations

Dan Coward

Economic History Department
Research School of Social Sciences
Australian National University

"In the long run" J. M. Keynes once wrote, "we are all dead". But that event for all people, does not occur simultaneously. Generations overlap. A period of one hundred years merely spans the period from the birth of a person to the death of his or her first child. These are not remote relationships. But concern, ethically and practically speaking, should sensibly extend beyond an interest in the fate of our next of kin. Processes in train — like world population growth, industrialisation, the growth of cities — force us to focus on the consequences of human activities for our future. Yesterday our habit of shortsightedness may not have been grossly harmful over the long age of our evolution as a species. Today we live in an age of bulldozers, insecticides and nuclear weapons. Our potentially earth-shattering activities might unexpectedly shorten Keynes' "long run" to now.

Our inquiry into the social implications of energy raises in a new form an old question: what is to be the purpose of our industrial-urban society? That question was put and political solutions proposed about "the condition of England" during the early nineteenth century decades, notably the 1830s and 1840s, of that incipient industrial state. The focus then fell on the social impact of industry. Economic growth, partially captured in the contemporary notion of "progress", it was believed, would, in the long run, raise the material condition of impoverished groups. As a makeshift, government intervention, for example, in sanitary regulation in the hope of reducing sickness and disease and in income support via the Poor Laws, would ameliorate the condition of the poor.

About ten years ago the debate opened on the physical impact (and the consequences for collective human welfare) of our growing number of industrial nations on the earth's fragile biosphere. The spotlight fell on the problem of man-made pollutants. What happens when increasing quantities and kinds of man-made wastes over-tax the ability of the air, water and land to absorb and disperse them? We recognise the manifest physical consequences of the process of pollution — dirty air and filthy water — but what of their long run health and genetic effects on human beings? We still don't know the answers to some of these questions. Moreover, the question of biological impacts is often not asked, let alone the answer sought. Wealthier industrial societies, in the main, generate the bulk of the wastes that provoke unease. More particularly, polluting processes are a by-product of city activity. Cities, as prolific consumers of energy and materials, concentrate the accumulation of wastes. Here the intractable problems are to be discovered.

From the point of view of man, energy can be thought of in two ways. First, that needed in the form of food, water and air for existence. Second, the production of energy, either through the use of human muscle or the use of technology, which can be used to procure food and to master other forms of energy production. In short, we can consider energy as intake and output. To appreciate the very recent appearance of high energy-using societies, which have an enormous capacity to control and transform the physical environment, let us briefly survey the past.

We can identify three main phases of energy use: hunting and gathering; agricultural and, very recently, industrial. Successively, as we move closer to present time, these phases become shorter in duration. Of course, historical differences between different societies means overlap in phases: thus, for example, the persistence of aboriginal hunting and gathering economies in contemporary urbanised Australia. Sketching the outline of these phases means also outlining the general picture of our human development.

The human species and our traceable ancestral species, relative to the age of the earth (something like 4500 million years), have lived a brief existence. Despite the accumulating evidence of archaeologists and palaeoanthropologists, much remains to be discovered about our ancient origins. Human and pre-human fossils so far unearthed are rare, fragmentary and scattered in their geographical distribution. What follows, therefore, may be disputed territory among specialists in the field. Pilbeam and Simons tentatively date the evolutionary divergence of hominids (that is, man and his immediate and recognisable ancestors and relatives) as probably 10 million

Diesendorf, M. (ed.) (1979). *Energy and People*. Canberra, Society for Social Responsibility in Science (A.C.T.)

years before the present, possibly even 14 million or 15 million years ago. Long-term climatic changes, so it is conjectured, slowly transformed forests into open woodland. The primate ancestors of hominids, so the theory continues, slowly adapted by changing from vegetarian foragers to being hunters and gatherers. The significance of the transfer from wholly vegetarian to partial carnivore is that our ancestors moved up the food chain. As there is more energy in the proteins and fats of meat, the greater efficiency and economy implied in the change of diet possibly had other very long run effects in man's evolution.

That hunting way of life shaped man. While not fully bipedal, the partial use of "hands" for primitive stone toolmaking, in a complex interactive process, influenced the skeletal development and brain structure in hominids. The oldest stone tools so far discovered are dated at about 2.5 million years old. The pathway for animality to humanity had begun. It seems that hunting was firmly established as a way of life 3 million years ago. Hominids evolved during the onset of cooler climates and the successive "ice ages" which penetrated to the middle latitudes during the Pleistocene epoch (about 3 million to 2.5 million to about 10,000 BP — that is, Before Present). They became large-brained bipeds, tool-making and linguistic creatures. From fossil evidence, the ancestral species of modern man, *homo erectus,* existed 1 million years ago. Hitherto it had been a biological inheritance that had shaped the possibility for change. Now cultural acquisition — imitation, training and learning — became the main force transforming man's way of life. Social evolution, as distinct from biological evolution, opened up unlimited capacity for adaptation. By 50,000 years before the present, modern man, zoologically classified as *homo sapiens sapiens,* had evolved.

Table 1
Theoretical Outline of the Evolution of the Hominid Family

Geological Epoch	*Years BP**	
Pliocene	c. 15-10 million (?)	•first hominids
	c. 3 million	•hunting established
	c. 2.5 million	•oldest known stone tools
Pleistocene	c. 1 million	•*homo erectus*
	c. 800,000-500,000	•first use of fire
	c. 50,000	•*homo sapiens sapiens* (modern human species)
	c. 10,000 (7000 BC)	•beginning of plant and animal domestication
Recent		

Source: Pilbeam (1972)
*Before Present

As has been argued, for example, at an international symposium in 1955, whose papers are published as *Man's Role in Changing the Face of the Earth,* fire was the first great force employed by hunting and gathering man. The oldest use of fire by *homo erectus* so far discovered was at Choukoutien (near Peking). That use is anything from 800,000 to 500,000 years BP old. Control of fire made man more efficient in finding ways to exploit his environment. Fire enabled the cooking of food, provided warmth and could be used to drive animals from cover during the hunt. In much more recent times it aided the clearing of bushland in primitive agricultural techniques of shifting cultivation, and with suitable fuels enabled the smelting of metals.

Summarising, hunting and gathering in small tribes or bands has been the predominant mode of living for the vast part of the existence of modern man. In the twentieth century it barely survives as a way of life. But that "economy" has a prior history in our ancestral species of some million years. Not surprisingly, because of its antiquity, the contrast of hunters and gatherers with today's urban-industrial peoples, living in large industrial cities, has prompted much speculation about the human condition. As Pilbeam puts it, in pessimistic overtones: "To what extent is modern urban man maladapted biologically to living in large, crowded communities?" A lot of popular writing around similar themes is merely that: speculation. Fanciful paperback books flourish because of the mystery shadowing our, essentially, irrecoverable past. ("Irrecoverable" is not to suggest that we cannot discover more. It is clear from the fruitful work of natural and social scientists reconstructing data on the past that we can.) One major intellectual obstacle arising from stressing the conjectural hunting (rather than "and gathering") origin of man is that of determinism. Crudely put, this reduces to the dogma: man is "nothing but" an animal. (Embedded in "animal" is a complex of notions including that of "beast".) Allied to this is the opinion that man is "conditioned" (in the sense of command and dictate) by his animal past. Little, if any, room is left for ethical responsibility — a common theme in this book — in these authoritarian notions. By contrast, an alternative view to "man-as-beast" is that which stresses adaptability, inventiveness and capacity for change in human beings. Man has the *capacity* to acquire ethics and values. The postulate stressed here, in the context of the social implications of energy use, is that one necessary ethical step is to see ourselves as belonging to one species, confined on this planet with its fragile biosphere. It is obvious, so far, that an historical perspective can have important implications for the way in which we perceive our society today. How can we reduce the chances of speculation creeping in and taking root as fact? The historical method: inquiry, the search for, the scrutiny and criticism of available data, is such a way.

The relevance of this digression is that molecular biologists, such as Sir Macfarlane Burnet (following the lead of ethologists and their studies of animal behaviour), theorise that man's hunting past is with us still in socially significant ways. Aggression, characteristically a male attribute, and its manifestations (authority, leadership, prestige, physical courage as well as arrogance, tyranny, cruelty and so on), he declares in his book *Endurance of Life*, are not correlated with intelligence. Further, he argues, "human behaviour is based on a genetically determined neural infrastructure that evolved essentially to its present form in the long hunter-gatherer phase of human prehistory" (p. 164). But, he warns, this statement is a gross oversimplification. In short, we are not necessarily trapped in a one way "deterministic" street by behaviour. We have the *capacity* for ethical behaviour. We are human beings. Here we need to bear in mind that the analogies and inferences drawn from contemporary studies of animal behaviour are disputed. The differences, even in the human infant, qualitatively outweigh our similarities with other mammals. Whatever the theories advanced, on inferential reasoning, for the evolution and function of aggression in our ancient past, we face the fact today that aggressive behaviour (reacting to, and influencing other perceived behaviour), is linked with the "control", by a few nations as yet, over a forbidding energy source, nuclear power. Burnet underlines his concern that our behavioural shortcomings as a species do not match our technological ingenuity. "Like many others", he declares, "I strongly oppose extension of nuclear power, and especially the entry of breeder reactors, not for their danger as such, but because of the great increase in the amount of plutonium that will become available for malevolent use . . . Traffic in plutonium", he continues, "built up by some unholy alliance of paranoia and state policy, could be the trigger that in the end will destroy civilization." (pp. 171-2).

We now come to the second phase in energy use. That period begins with the retreat of the glaciers, conceptually dated by geologists as the transition from the Pleistocene geological epoch to that of the Recent epoch, which is roughly about 10,000 years before the present (c. 8000 BC). Prehistorians, again emphasising the frailty of the evidence, see this as the dawn of the agricultural "revolution". Around this time villages became more numerous in the Syrian-Palestine region, the natural habitat of emmer and einkorn wheat and of the earliest domesticated animals: goats, sheep, cattle and pigs. But it must be emphasised that agriculture has no single, simple origin. Archaeological work in central America, China, India, the middle east and Europe has illumined again the variety of human invention.

Although inefficient and feeble for a long time to come, agriculture freed people from the necessity of searching and foraging for food. The adoption of food-producing behaviour and the gradual assimilation of farming techniques such as the domestication of plants and animals gave mankind the potential to use a more abundant source of energy. Some specialisation did emerge in the two forms of peasant farming and nomadic pastoralism. But agriculture supported larger populations. For example, Leach estimates that subsistence agriculture requires (depending on soil, climate, protein yield of food crops and so on), from 0.1 to 2 square kilometres of land per person to provide an adequate diet. By contrast, some estimates for hunters and gatherers prior to European invasion are 20-25 square kilometres per person for Indians on United States prairies; 30 for Australian aborigines; and 140 for Eskimos of the Canadian northwest. Agriculture did not necessarily imply subsistence farming. Productivity, as always, varied. For example, in certain fertile river valleys of Egypt and western Asia sun, silt and water made farming relatively productive. Surpluses of food made possible the division of labour (that is, freedom from food production) and the development of trading those surpluses. Thus, about 5000 years ago metal working of copper and bronze (implying the application of heat energy) had its beginnings. Moreover urban living became possible. For these reasons the term agricultural "revolution" is used because it conveys the idea of a fundamental shift in economic organisation and the development of social relations in human societies.

Given that most people today live in societies where agriculture is the dominant sector in the economy, it is necessary to stress here, following Leach, that nearly all pre-industrial farmers and food gatherers achieve very large energy returns from their land or territory as measured by the work they put in. This fact becomes particularly clear when compared to the energy-intensive agriculture (fuel, machinery, and artificial fertilizers) of industrial nations. At the same time, the possibility of crop failures and famine and the presence of nutritional deficiency diseases are relatively common experiences in agricultural economies.

Two relative conceptions of the rates of social and economic change during the emergence and consolidation of agriculture must be kept in mind here. First, relative to the long, slow evolution of the hunting and gathering way of life, the pace of change in those regions which adopted the use of plants and animals for food production, was rapid. By contrast with the speed of social, economic and technological change in the nineteenth and twentieth centuries AD, the rate of change in the incipient and developing agricultural economies appears inordinately slow.

In this crude sketch of long run trends it is not possible to do more than mention the approximate dating (bearing in mind the problem of evidence) of major innovations affecting the type of energy available to human societies. In particular must be stressed the inventions of sailing boats (c. 3500 BC), water mills (c. 100 BC) and wind mills (c. 600 AD). Accordingly, human and animal muscle could be supplemented enormously by machines. But there is a warning. There is no correspondence between invention and widespread application of that invention. It is convenient to list some of these innovations in a table.

Table 2

Indicators of the approximate timing of changes in ancient agricultural economies

Approximate date BC	*Innovation*	*Site or region of evidence*
c. 6750	domestic pig, einkorn wheat	Jamo (Iraq)
c. 6500	use of pure deposits of copper	East Anatolia (Turkey)
c. 6000	domestic cattle	Argissa (Greece)
c. 5000	millet	China
c. 5000	flax	India, China
c. 4000	millet	Tamanlipas (Mexico)
c. 4000	rice	China
c. 3500	maize	Central America
c. 3500	smelting of bronze	Ur (Iraq)
c. 3500	sailing boat	Egypt
c. 3000	use of wheeled vehicles	Sumaria
c. 3000	soya bean	China
c. 2700	taming of steppe horse	Ukraine
c. 1000	smelting of iron	Anatolia/Syria/Palestine
c. 100	water mill	Europe
c. 600 AD	wind mill	Persia

Source: *Encyclopaedia Britannica* (1976 ed.).

About nine generations ago we can trace the origin of a period we know as the "industrial revolution". That term is used here in the not so usual sense of indicating long run cultural changes in the recent history of our species. A substantial increase in energy-intensive activities is characteristic of the period. Humans effectively harnessed energy sources by inventing and improving various energy converters. For example, the steam engine, fuelled by coal, converted heat energy into usable mechanical energy. The beginning of the industrial revolution is conventionally and arbitrarily dated at about 1750 in Britain. But, it must be emphasised, the apparent rapidity of economic developments should not disguise the fact that these changes had deep roots nurtured in previous centuries. The first phase of the first industrial nation was based, essentially, on the exploitation of inanimate energy derived from an irreplaceable resource: coal; and to a larger extent than is usually realised, on the harnessing of water power. Industry very gradually supplanted agriculture as the dominant sector in the economy. Intricate demographic trends accompanied this fundamental change in economic structure: accelerated population growth, rural depopulation and the rapid growth of cities. For example, Greater London in the period 1801 to 1861 rose from 1.1 million to 3.2 million people; Liverpool from 82,000 to 472,000; and Manchester from 75,000 to 399,000. Equally, the social consequences in the new industrial cities were massive: overcrowding, illness, unemployment, poverty and accumulating wastes and pollutants. Currently, historians are pursuing regional studies of social change and economic growth of industrialising England in order to clarify how detailed history fits or is different from the national experience.

During the nineteenth century the industrial revolution flowed through to parts of Europe and North America. Again, its cumulative process of expansion touched all aspects of social and economic life. In our own century it has been initiated in most parts of the earth. Not only has the scale changed, so too have the ingredients in the revolution. Britain's industrialisation based on coal energy was added to by other irreplaceable resources: petroleum, natural gas and lignite. Water power is a fifth major energy source. The growth in world energy production is illustrated in the table following.

Table 3

World Production of Inanimate Energy, 1860-1970

Year	*Coal* m tons	*Lignite* m tons	*Petroleum* m tons	*Natural gas* 10^9 cubic metres	*Water Power* million megawatt hours p.a.
1860	132	6	—	—	6
1890	475	39	11	3.8	13
1920	1193	158	99	24,0	64
1950	1454	361	523	197.0	332
1970	1808	793	2334	1070.0	1144

Source: Cipolla (1978).

The growth of urban electric power generation, an innovation less than one hundred years old, has had a significant impact on the consumption of energy by industry, transport and household users. The invention and improvement of the filament lamp in effect demonstrated a useful purpose for electric power. In 1882 the world's first electric power generating plant, Edison's Pearl Street station, opened in New York. This implied the expansion of wire networks to distribute power to urban, and later rural, consumers. In order to safeguard and rationalise large investments, publicly- and privately-owned electric power corporations were granted monopoly rights over electricity distribution for particular districts. Typically today we have huge centralised power stations, fired by fossil fuels, or driven by water power, or impelled by nuclear fission, requiring large scale investments. One significant implication is that decision-making, which has the potential to affect large groups of people, is also centralised. In Australia, the government of Victoria was the first to centralise power generation and distribution under public authority in its State Electricity Commission Act of 1918. Included in its statutory duties the new Commission, headed by General Sir John Monash, was enjoined 'to encourage and promote the use of electricity and especially the use thereof for industrial and manufacturing purposes'.

The use and the impact of industrial technology has penetrated to all human societies. But the geographical distribution of the end use of that energy is greatly unequal. The political economy of energy use, as Earl Cook calculated in 1971, was that the industrial regions, with 30 per cent of the world's people, consume 80 per cent of the world's energy generated each year. More strikingly, the United States, with 6 per cent of the people, consumes 35 per cent of the energy. These levels of consumption are related to the degree of industrialisation. As the industrial revolution spreads to Asia, Africa and Latin America, we now face the prospect of competition over newly-realised scarcities of fuel. These dominantly agrarian economies, among other things, are seeking to raise the standards of living of the mass of people above subsistence level. Moreover, the same industrialising regions are also experiencing a phase of rapid population growth, particularly owing to effective controls over death rates and a relative absence of controls over birth rates. These facts aggravate their problems of reducing poverty. The scale of this growth can be appreciated in our context of energy "revolutions". Thus around 10,000 years ago, at the outset of the agricultural revolution, the human species numbered between an estimated 2 to 10 million people. By 1750 estimates range from a minimum of 650 million to a maximum of 850 million. By 1975 the number had accelerated to over 4000 million. Growth rates are rising. While the annual average growth rate was about 0.7 per cent in 1850-1900; it rose to about 2 per cent in 1965-1970. In short, it has taken thousands of years for the human species to reach a population of 4000 million. But a mere thirty years will add the second 4000 million. As it stands, the human species faces growing demands upon finite energy sources to meet expanding needs. How long can non-renewable energy resources be made to last? How do we decide who gets what? The energy-based industrial revolution is thus at a critical phase.

The industrial revolution has produced, and is producing, two obvious types of physical "debts", which at some future time will affect the human species. We are still too close in time and lack data on trends to determine the shape of that impact. First, the depletion of irreplaceable fossil fuels. Without effective and safe energy substitutes, this means that in the long run the industrial revolution will be undermined. Second, the unknown impact of waste discharges into the environment. Much of the energy consumed ends up as waste heat. Moreover, the huge quantities of waste gases (for example carbon dioxide, hydrocarbons, sulphur dioxide and nitrogen oxides) and dust and particulates discharged have observable effects on urban environments and urban micro-climates. For example, central London, although now clear of smog, is probably today, on average, about 2°C warmer than the surrounding countryside. It is also possible that these man-made waste discharges, although in annual totals infinitesimally small in relation to the volume of the earth's atmospheric gases, have implications for global climates, and hence on agriculture, as well. Thus, by burning fossil fuels we are consuming atmospheric oxygen and replacing it with carbon dioxide. Additionally, deforestation will have added to the level of carbon dioxide. As this gas, together with water vapour, acts as a so-called "greenhouse" (namely absorbing the outgoing heat from the earth's surface), it is supposed that the earth's atmosphere is in the process of becoming warmer than otherwise would be the case.

One response to the idea of fossil fuel exhaustion is to ration, to conserve what we have. This means conservation by the national possessors of energy resources. Interestingly enough, the notion of finite energy resources is not new. The British economist W. S. Jevons published in 1865 a book entitled *The Coal Question*. It is sub-titled: "an inquiry concerning the Progress of the Nation, and the Probable Exhaustion of our Coal-Mines". This perceptive man foreshadows the political and economic questions we are asking today. Given the existing technology, the costs of mining and the limited coal fields, he asks: "Are we wise in allowing the commerce of this country to rise beyond the point at which it can be long maintained?" In the final quarter of the twentieth century, we face the dilemma similar to that put by Jevons to his generation of British people: do we squander our irreplaceable resources for particular uses, and for lucky generations; or do we adopt policies to eke out our finite supply of fossil fuels?* We add today: for lucky nations. The question of energy supply is far more intractable than that of over 100 years ago. In a global context equity has international not simply national implications between generations of people.

A second response, perhaps partly based on ideological conceptions of "economic growth" and the recent historical experience of the "long boom" of the "affluent" 1950s and 1960s, is to search for inexhaustible sources of energy. In particular, the search is for modes of generation that can be adapted to our centralised electricity networks. Thus we have advocates for solar derived energy and for nuclear fission (and ultimately fusion) derived energy. In this essay, sketching in the main phases of social and economic changes influenced by and influencing changes in energy technology, it is appropriate to make some comments on possible social implications of a greater reliance on nuclear energy. Proponents, some passionate, declare that nuclear fission power has several advantages over the burning of fossil fuels in conventional electric power generating plants. For example they argue, essentially, that fission power is cheaper, far less polluting, and more efficient when compared with say, the mining, transport and burning of coal. Let us leave aside the detailed studies which challenge all these propositions. Assuming a *perfectly* working technology their claims might have some substance. The great hazard in this vision is, of course, the ionising radiation emitted by the fission process. Ionising radiation occurs, not only from the reactor and from spent fuel reprocessing but also, importantly, from nuclear wastes. This form of energy production has the unprecedented effect of threatening all living matter. Moreover, the toxic, high level radioactive wastes need to be *stored* ("disposal" is too comforting a word because of its connotation of an acceptable "solution" under existing technology) in isolation and in safety from all forms of life for extraordinary periods of time. As we have seen our modern human species (*homo sapiens*) is roughly 50,000 years old: certain man-made radioactive wastes need to be preserved in safety for over five times that period. In this century of mass violence it appears odd that the storage proposition can be contemplated. Hitler's "one thousand year Reich" lasted a little more than a violent decade. Who can guarantee sufficient human political stability for 250,000 years to enable the wastes to decay into harmless substances? In the short term, moreover, safe storage techniques — in the sense of effective physical containment — are not proven. As the British Flowers Report noted in 1976: "There are promising ideas for the disposal of these wastes but it may take 10 to 20 years to establish their feasibility" (p. 193).

The hazards of fission power are unique in the evolution of the human species. A few nations possess nuclear weapons that can kill all humans, destroy all forms of life on this planet. The acquisition of nuclear technology for civil purposes is a passport to a military option. In short, peaceful and warlike

*Jevons canvassed possible substitutes for coal: internal heat of the earth, tides, sun's rays, wind, water, timber fuel, and "any wholly new source of power (to) be some day discovered".

purposes of fission power cannot be distinguished. Within nuclear countries the example of terrorists in the 1970s has raised a new problem. What will be the effect of an increasingly nuclear-powered economy on existing civil liberties? In Britain the Flowers Commission faced this question. In the context of a possible increase in fission reactors powered by and generating additional plutonium, a man-made, highly toxic, and highly explosive metal, the Commission tried to assess the consequences emerging fifty years in the future. "What is most to be feared", they declared, "is an insidious growth in surveillance in response to a growing threat as the amount of plutonium in existence, and familiarity with its properties, increases; and the possibility that a single serious incident in the future might bring a realisation of the need to increase security measures and surveillance to a degree that would be regarded as wholly unacceptable, but which could not then be avoided because of the extent of our dependence on plutonium for energy supplies." (p. 128). The foreclosing of political, social, economic and energy options, the unique social implications of this form of energy production, have never been more clearly put. Indeed, decision-makers had never put them.

A global point of view of the industrial revolution over the past 200 years is sobering. Such a view is important for appreciating national objectives, pinpointing potential international conflicts, and gauging the scale of the problem of energy consumption. Equally, we have to make a beginning in seeing ourselves as a species. But in an age where nation-states form the largest unit for policy-making, we must, for practical purposes, focus on home ground. As Hannes Alfven has observed, we can tackle our inquiry into the problem of energy in two ways. First: what are the choices, and their social and economic implications, among various energy sources? Second: how much energy do we think we need? In Australia, because about 85 per cent of the people live in an urban environment, we have to ask our questions in the context of our urban way of life.

In the past it has been technologists, essentially, who have made decisions on the mode of energy use. Their monopoly role may have been appropriate. This is not to decry their real achievements in contributing to social welfare. But the message of these collected essays is that energy production and use has important social impacts. Moreover, there are no simple answers, let alone scope for narrowly-conceived policies. Accordingly, technologists cannot be left to make decisions by themselves, not least because the intractability of the political issues that we face may well be a choice between evils. It became clear, when the spotlight fell in the sixties on the environmental ill-effects of our urban-industrial civilisation, that these problems were, fundamentally, problems arising from human behaviour. Asking questions about energy policy not only involves other policy fields such as environment, urban planning, transport, housing and industry, but also implies an historical analysis of human behaviour, focussing in particular, on the high energy-using, "affluent" society as it emerged after 1945.

But, it might be asked, policy-making so conceived is so intricate that the makers are left bewildered. Where do we start? One way is to sharpen our ideas. Passmore's *Man's Responsibility for Nature* (1974), subtitled "Ecological Problems and Western Traditions", is both historical and analytical. So far as ideas are concerned, the chapters on pollution, conservation, the destruction of species and over-population are relevant here. A second way is to explore the obstacles confronting the making of policy conceived from a broad perspective. For example, the Botany Bay Project, an interdisciplinary group now disbanded, has explored the intricate institutional, technological, political, legal, economic and social obstacles in the making of an effective environmental policy in the context of the 1970's and beyond. These obstacles have been charted for Australia's largest city in the Project's first book, *Sydney's Environmental Amenity 1970-1975*. Its sub-title is "A study of the system of waste management and pollution control". A third way is to focus on a particular complex of problems, in conflict, requiring resolution. Another of the Project's books, *The Impact of Port Botany,* offers a rare example of what is, in effect, a comprehensive "environmental impact statement". *The Impact* can be regarded as a model for the sorts of things that ought to be considered by public authorities evaluating proposed physical changes in urban areas. In particular *The Impact* brings together first, the cumulative effect of several related projects in Port Botany; second, an evaluation of the total environmental effects; third, an assessment of social costs and benefits; fourth, a consideration of alternative solutions. There is no reason why the same mode of thought cannot in principle and in practice be applied to issues arising from the generation and the use of energy. It is possible to make sense out of seeming chaos.

In a relatively short space of time our human species, by harnessing on a large scale energy derived from ultimately exhaustible resources, has greatly extended our control over our physical environment; but we may now be putting at risk the basis of our continued existence. Thus some lucky nations, and some elites in poor nations, can achieve goals other than those related to biological existence. "Coal stands not beside but entirely above all other commodities", wrote Jevons in 1865. "It is the material energy of the country, the universal aid, the factor in everything we do. With coal almost any feat is possible or easy; without it we are thrown back into the laborious poverty of early times." Jevons was emphasising that Britain's economic growth was underpinned by coal-based energy. In turn, allied with a stream of technological innovations, the use of energy allowed a substantial increase in the productivity of labour. This sounds fine in general terms. In practice, what we now call "structural" economic change involved continuing, large scale displacement of people. This is a characteristic of the industrial revolution. It is present, in greater or lesser degrees of inhumanity, in industrialising countries of today like South Africa, Latin America and Asia. Or indeed, the experience of Australia's aborigines. The dispossession of tribes and their incorporation into modern nation states undergoing industrial expansion is often not an edifying story. In Britain, the losses for particular pioneering urban generations were substantial and, essentially, not ameliorated until the emergence and elaboration of programmes of the mid-twentieth century welfare state. The gains and losses of the nineteenth century revolution in Britain form a fiercely disputed battleground for historians. One reason for this is that, essentially, the debate has been a judgment on capitalism, about the social consequences of the operation of the free market economy. Without entering into this lively field, we can say that separating out the implications of the three contemporaneous and interpenetrating phenomena of industrialisation, urbanisation and industrial capitalism is a difficult task for social historians. But the evidence of social costs is there, whether one reads contemporary works like the *Poor Law Report* of 1834, Frederick Engels' *Condition of the Working Class in England* published in 1845, Henry Mayhew's *London Labour and London Poor* of 1849-51 or even Jevons' comments in 1865. "It is a melancholy fact", he writes, "which no Englishman dare deny or attempt to palliate, that the whole structure of our wealth and refined civilisation is built upon a basis of ignorance and pauperism and vice."

The heart of any substantial social change, induced in great part by the continual improvements in energy and other technologies, is the distribution of income. In the virtually income-tax free era of the nineteenth century, there existed no substantial means of redistributing income to groups of people — the sick, the poor, the under- and unemployed — in need of shelter from the storm. Perhaps we know better today. But urbanisation and our commercial civilisation has fostered an atomistic way of life: greater geographic mobility, greater anonymity and the decay of traditional kinship ties. A tax "revolt" by the well-off could substantially weaken the ability of industrial nation states to redistribute income to those hurt by changes (for example, in production and in the geographic structure of our sprawling cities), which will be partly shaped by rising costs of energy.

Collective policies, rationally speaking, are necessary. But how are we to perceive "economic growth" which is intimately related to energy policy? We need to distinguish the *purpose*, over time, of enabling, via the process of expanding economic cake, the redistribution (through employment) of income to those in poorer socio-economic groups. The fact that some prospects for material self-betterment exist enables this notion to flourish. Once upon a time, the idea of progress contained this comfortable vision of the future to some social groups. Of course, the view that things will continually become better has taken hard knocks in the twentieth century. But it survives in a residual form in "economic growth". If individuals continue to believe that things will improve for them at some vague future time, one effect, it has been argued, will be the weakening of the demand for necessary collective action. This is the view of Fred Hirsch in his book *Social Limits to Growth*. Accordingly, he argues, as expectations are not fulfilled, the outcome is likely to be politically destabilising and moreover, likely to increase the cost of distribution policies. In short, the promise of "economic growth" is extravagant, in some cases, illusory.

As we now realise, economic activity, particularly in industrialised nations, also yield "dis-economies". This fact is highlighted in the serious side-effects of environmental pollution and loss of amenity arising from the joint effects of industry, transport and households in cities. Ecologists and dissident economists revealed the impact of "the affluent society" in the late 'sixties. That level of material wealth conventionally depicted in terms of Gross National Product in part results from cheap energy prices. Over the past two decades or more we have put much store in the concept of "Gross National Product" as a measure of our national fortunes. Economists have questioned the adequacy of this concept in measuring improvements in human welfare. They are now devising non-economic indicators, for measuring changes in welfare, some of which are necessarily unquantifiable. Today we question whether high rates of growth have any close relation to social desiderata like reducing poverty or increasing leisure time. These things hinge rather on conscious policy arising from the economy, not on the "economy" which is a result of human activity. Equally, we might ask whether increased economic growth is dependent upon increased energy use. It is obvious that transport policy, fuel prices, energy conservation, the reduction of energy-intensive industries, changes in behaviour and so on will affect the demand for energy. Accordingly, we hit again the problem of values, raising the question, as said at the outset: what is to be the purpose of our industrial-urban society? To explore this question would require the writing of a book.

Bibliographical Note

This is a derivative essay. Acknowledgment in the text would mean a needless array of footnotes. Several ideas have been drawn from published works, particularly Carlo Cipolla's slim synthesis The Economic History of World Population *(7th ed., 1978). Readers wishing to explore further will find the following books, some of which have been mentioned in the text, helpful.*

References

(1) Macfarlane Burnet (1978). *Endurance of Life,* Melbourne University Press.

(2) N. G. Butlin (ed.) (1976). *Sydney's Environmental Amenity, 1970-1975,* Canberra, Australian National University Press.

(3) N. G. Butlin (ed.) (1976). *The Impact of Port Botany,* Canberra, Australian National University Press.

(4) Earl Cook, "The Flow of Energy in an Industrial Society", reprinted in Joseph G. Jorgensen (ed.) (1972). *Biology and Culture in Modern Perspective* (Readings from *Scientific American*), San Francisco.

(5) T. K. Derry and T. I. Williams (1960). *A Short History of Technology,* Oxford Univ. Press.

(6) Theodosius Dobzhansky (1972). *Mankind Evolving,* New Haven, Yale Univ. Press.

(7) Flowers Report (Royal Commission on Environmental Pollution) (1976) *Sixth Report: Nuclear Power and the Environment,* London, HMSO.

(8) W. K. Hancock (1973). *Today, Yesterday and Tomorrow* (1973 Boyer Lectures), Sydney, Australian Broadcasting Commission.

(9) Fred Hirsch (1977). *Social Limits to Growth,* London, Routledge and Kegan Paul.

(10) Gerald Leach (1975). *Energy and Food Production,* London, Int. Inst. for Environment & Development.

(11) E. J. Mishan (1969). *The Costs of Economic Growth,* Penguin.

(12) John Passmore (1974). *Man's Responsibility for Nature,* London, Duckworth.

(13) David Pilbeam (1972). *The Ascent of Man,* New York, Macmillan.

(14) William L. Thomas (ed.) (1956). *Man's Role in Changing the Face of the Earth* (2 vols), Univ. of Chicago Press.

Mobility and the Individual

R. B. Hamilton

Environmental Affairs Co-ordinator, Shell Australia

1. Introduction

The motor car has had a most significant impact on society's resources, particularly crude oil consumption. The world fleet of cars of some 200 million consumed in excess of 6% of world energy production and 12% of world oil production in 1970.

In Australia the motor car consumed 80% of all direct transportation energy in 1970/71 and is presently responsible for 38% of all crude oil consumption. It is clearly evident that the motor car is a large energy consumer and the very high level of mobility enjoyed by Australians is at a substantial energy cost.

In order that we may plan changes in the total transport system to optimise the use of scarce non-renewable resources it is important to know something of the underlying factors of mobility in Australia. It is proposed to examine the various determinants giving rise to vehicle ownership and use and the relationships between these and socio-economic factors.

This paper is not an attempt at an exhaustive study of personal mobility and the motor car but rather an attempt to provide a broad framework for the understanding of the use of the motor car in present day society.

2. The Process of Societal Decision Making

Ackoff a "father figure" of operations research in a paper Explaining and Understanding the Process of Societal Decision Making "Does the quality of life have to be quantified" — contends that thoughtful men would agree that considerable progress has been made in science and technology. Some would agree that progress has also been made in the domain of political economy and in ethics and morality. But one would be surprised to hear it argued that mankind has made aesthetic progress.

Decision making in the presence of doubt involves problem solving. Through science we have developed rich and useful models for decision making and problem solving.

Decision modes usually have four components:

1. the decision maker;
2. the alternative possible courses of action defined by controllable variables;
3. alternative possible outcomes; and
4. the environment defined by those uncontrolled variables that together with the controlled variables can affect the outcome.

The components are interrelated by three types of parameter:

1. the probability of choice, that is, the probability that the decision maker will select a particular course of action;
2. efficiency of choice, the probability that a course of action if selected will produce a particular outcome in the decision environment; and
3. the relative values of the outcomes to the decision maker in the decision environment.

The sum of the products of these measures overall the possible courses of action and outcomes is the expected relative value of the choice situation.

3. Mobility

Personal mobility intrudes into most areas of human activity.

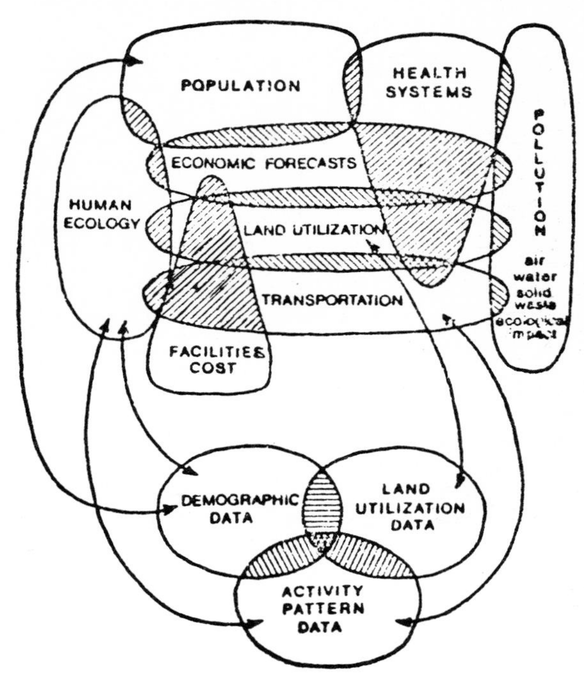

Figure 1.

There are many important elements of the environment in which we live — geographical, political, social, cultural. This presentation will focus on personal mobility and discuss energy usage in relation to personal mobility. This will be done mainly in the context of the use of the motor vehicle in contemporary Australian Society.

Long before the OPEC countries raised the price of world oil (the latter being the main source of energy for transportation) the OECD had expressed concern at the amount of the world's resources devoted to the motor car. It has been stated that (1):

> *The real problem was how governments and communities will act in the event of the world running short because of the growth in motor vehicles.*

Many critical social problems compete for the limited resources that are or can be made available to a society. Therefore, there is a need for some type of cost-benefit analysis of what is proposed for each problem area to enable society to do the most good with what resources it has available, and to determine how much of its resources should be allocated to each of its problem areas.

Oil is one such resource which considered inexhaustible (until a couple of decades ago) provided the basis for strong economic growth in most of the industrialised nations. The oil price rises implemented by OPEC in 1973 and 1974 brought into sharp

Diesendorf, M. (ed.) (1979). *Energy and People*. Canberra, Society for Social Responsibility in Science (A.C.T.)

focus the limited nature of the supply of this energy source. Consequently, we need to take a harder look at the utilisation of this diminishing resource and carefully weigh the various considerations which determine the nature and extent of its use by Society.

Energy consumed in the private motor car is of particular interest as some 38% of Australia's crude oil is consumed in the form of petrol. If we are to effectively conserve energy and ensure an equitable and rational allocation of this resource for the overall benefit of society then a thorough understanding of the factors underlying its consumption is required.

Lovins (2) recognises the challenge of modifying or satisfying the demand for mobility. He contends that:

> *The most important difficult and neglected questions on energy strategies are not mainly technical or economic but rather social and ethical.*

Clearly, we must examine the implications of continuing to increase personal mobility in the light of the social implications. Some of the side effects are clearly evident in the large cities of the world; high levels of pollution, congestion, increased personal injuries, and a disproportionately high usage of liquid transport fuels relative to non-urban transportation.

It is proposed here to examine the various aspects of motor vehicle usage and ownership with a view to highlighting some of the energy implications.

Mobility and energy consumption

As the degree of mobility increases (kilometres per person per year) so does the energy consumed. The increase in energy consumption is at a greater rate than the increase in mobility.

For Australia, the number of vehicles per capita has increased from 0.20 in 1961 to 0.37 in 1976, an increase of approximately 4.2% per annum on an exponential growth basis, while petrol consumption has grown at the rate of 5.8% per annum on the same basis over the same period (3).

The nexus between mobility and energy consumption is also illustrated by the growth in total mobility with increase in car ownership.

In Australia we are at the high end of the scale of vehicle ownership with 2.2 persons per vehicle or 2.7 persons per car.

The average person spends a relatively constant amount of time on travel around 500 hours/year and with increased disposable income, the desire for mobility is increased with a corresponding increase in energy consumed per unit distance travelled.

The relationship between mobility and energy efficiency is shown in Figure 2 where efficiency is defined as the ratio of

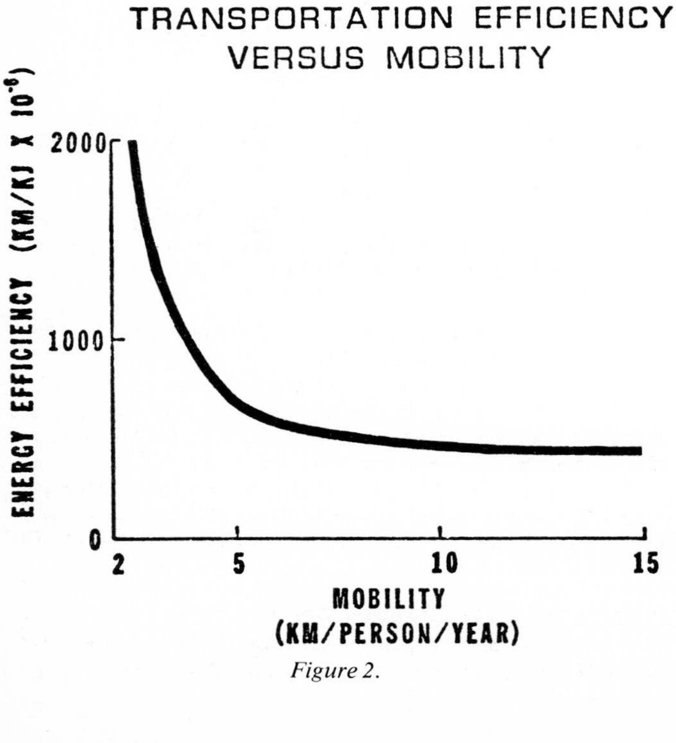

TRANSPORTATION EFFICIENCY VERSUS MOBILITY

ENERGY EFFICIENCY (KM/KJ X 10⁻⁶)

2000
1000
0
2 5 10 15

MOBILITY (KM/PERSON/YEAR)

Figure 2.

passenger kilometres of travel to the total transportation energy consumed.

The curve is based on the work of Clark (4) who has postulated the breakdown of travel modes at various standards of mobility and examined the relative energy usage for these levels of mobility for a constant travel time budget.

It is evident that increased mobility results in an exponential decline in the efficiency of energy utilisation on a per unit distance travelled basis as the result of a dramatic increase in energy consumed to travel the same aggregate number of kilometres per person.

Trip lengths in U.K. and U.S.A.

The predominant use of the motor vehicle is for relatively short trips with nearly 80% of journeys being less than 100 kilometres in length. The cumulative trip lengths in U.K. and U.S.A. are rather similar. Similar data for Australia is not available where city traffic survey data is mainly for urban week day travel. A comparison of data for the U.K. and U.S.A. is given in Figure 3 below.

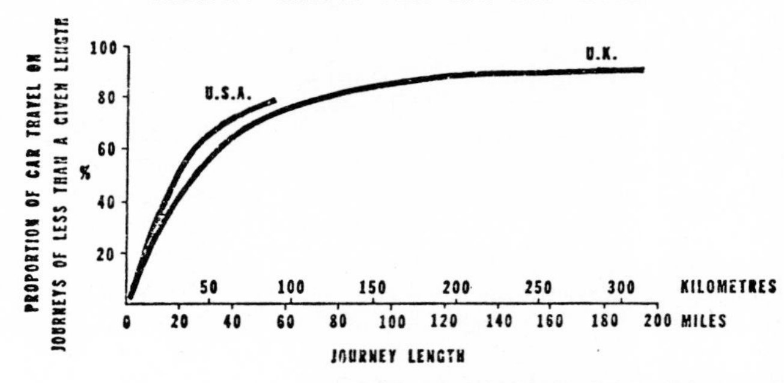

Figure 3.

Effects of population changes in large cities

There has been a distinct trend in recent years for people to move out from the inner suburbs and for continued population growth in the fringe urban areas.

An analysis of population movement trends for the Melbourne metropolitan area between 1971 and 1976 is shown in Figure 4. It is evident that the inner city municipalities are going through a decline in population with those suburbs outside of the 12 kilometre radius from the G.P.O. realising a net gain in population.

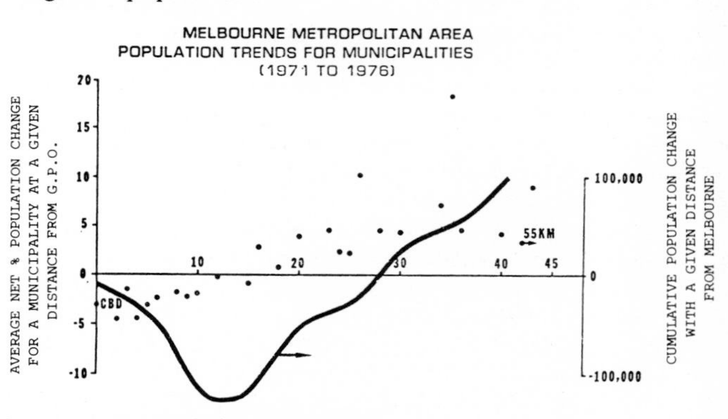

Source: ABS, population survey, 1976.

Figure 4.

In the older established inner and middle urban area of Melbourne the population fell by over 100,000 people between 1971 and 1976.

In the same period the population in the central inner Western and inner Northern Suburbs of Sydney fell by 53,860 persons.

In this time a population equal to a city about the size of

Ballarat left the established inner suburbs. This exceeded the total growth in Sydney during the same period of 40,176. What is relevant to energy usage for transportation is that there was a net outward migration from the areas where public transport availability was the highest.

This decline in population of the inner urban areas is also evident in the United States where the Central City areas saw a decline of 3.1% in the period 1970-1975 while the suburbs have undergone a reduction in growth rate (still positive growth but down 30% on that for the previous five year period) (5).

The implication of such population movement is that travel distances to place of work have increased as the density of work places is greatest nearest the Central Business District (CBD) (see Figure 6).

Effects of new housing development on energy consumption

The comparative energy consumption for different types of housing development in the U.S. is given in Figure 5 below (6).

Community Cost Analysis
Annual Energy Consumption

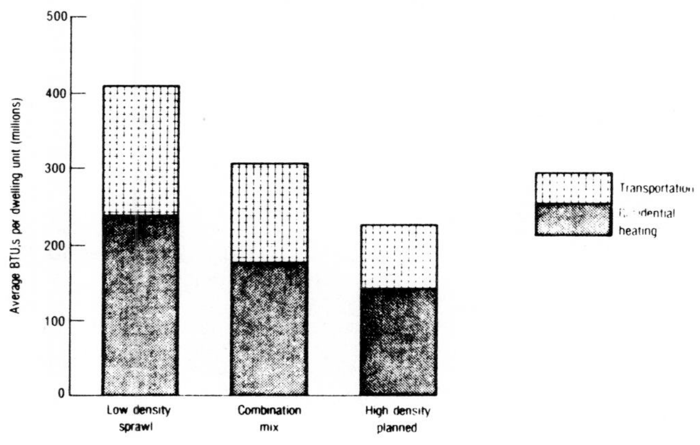

Figure 5.

Source: Council on Environmental Quality, U.S. 1974

At the fringe of our largest cities we still use the same basic planning assumptions on land use.

The community development pattern can also have significant impacts on energy consumption, affecting the degree of car use. The Australian CSIRO findings support those of the U.S. EPA (7).

Better planning, clustering, and higher density development can all significantly reduce reliance on automobile travel in terms of number of trips taken, number of kilometres driven, and amount of time spent in a car. These relationships hold true even when the amount of energy consumed in commuting to work is excluded, since commuting may not be directly affected by the development pattern of the residential community. Increased density also reduces the amount of transportation required for the delivery of urban goods and services.

The journey to work in Australia

The journey to work accounts for about 45% of the person trips for those between 20 and 60 years of age. At peak hour the traffic concentrations are highest and the average speed is the lowest for the day. A car is used by about 66% of the work force to travel to work (as driver or passenger) and is considered essential for 50% of the workforce to earn a living. At peak hour the fixed rail public transport systems serving the CBD are at or near capacity. Workers must travel across "saturated space" to marginal jobs in the CBD area thus increasing the direct and indirect energy consumption.

The time taken to travel to work has increased in the past 10 years. Some people travel long distances to work; professional people travel the furthest.

The data required to make a systematic study of changes in the length of the journey to work has not been obtained on an Australian wide basis since 1971 (8).

Work by Vaughan (9), using data from the Sydney Transportation Study, shows the relationship between the density of work places in Sydney local government areas, and the distance from the CBD in 1971 and is reproduced in Figure 6 below.

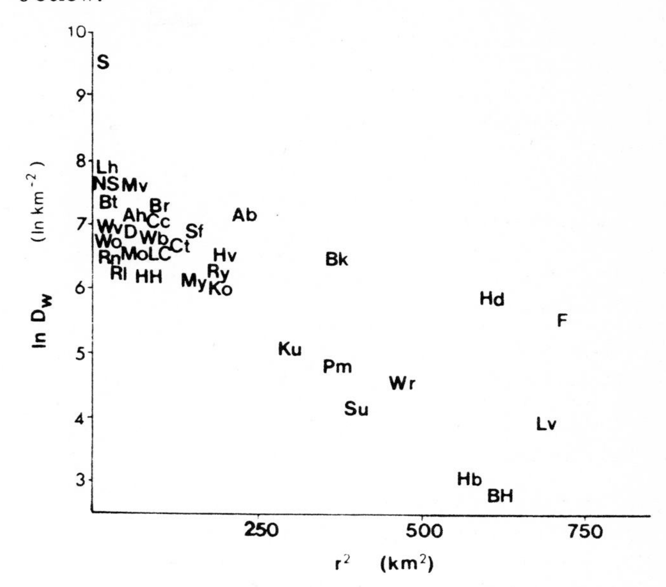

The density of work places D_w related to the distance from the city centre r

Figure 6.

Age and sex determinants of motor vehicle usage

The average annual distance driven by persons of different ages is almost constant for ages between 20 and 60 years in the U.S. (Note the peak at 30-35 years of age.)

Table 1

Average Annual Distance Driven (Kilometres)
U.S.A.

Age	*Male*	*Female*	*All*
16-19	8789	5771	7456
20-24	18387	8565	13293
25-29	22420	8914	15794
30-34	23329	9257	16534
35-39	20978	10029	15897
40-44	21136	9576	15825
45-49	20629	10092	15892
50-54	19867	8777	15203
55-59	18499	8753	14499
60-64	15627	8515	13055
65-69	11129	6716	9415
70 and over	8533	5123	7474
All ages	18269	8708	13977

Source: Based on unpublished data from National Personal Transportation Survey conducted by the Bureau of the Census for the Federal Highway Administration 1969-70.

Based on the above data, female drivers in the U.S. accounted for 32% of the total distance.

The percentages of total distance travelled in Australia by male and female drivers are not directly comparable with those in the U.S. as the survey dates are different. In Australia in 1971 women drove 20% of the distance covered.

The total number of trips per person per day is strongly related to the age of the traveller as shown in Figure 7 which is based on data from the Sydney Area Transportation Study (10) carried out in 1971.

Households and vehicle availability

The levels of car ownership per household in U.S. and Australia are comparable. In many parts of U.S. the level of

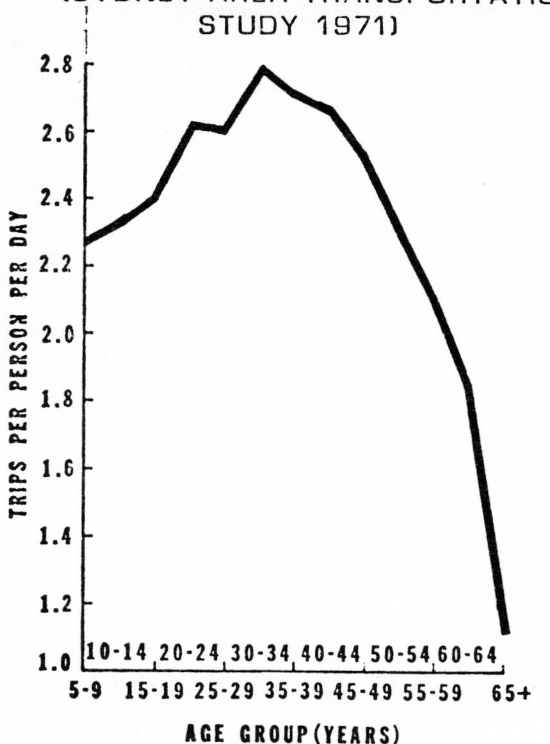

Figure 7.

ownership is below the average for all Australia which has about 5% of the vehicle population of the U.S.

Table 2

Car Ownership and Usage in Australia and the U.S. in 1972 (Per cent)

Cars per Household	*U.S.*	*Aust.*
0	20.5	21.0
1 or more	79.5	79.0
1 car	49.3	50.2
2 cars	24.6	22.3
3 or more	5.6	6.5

Source: "1973/74 Automobile Facts & Figures". U.S. Motor Vehicle Manufacturers Association, Detroit, Michigan.
Commonwealth Bureau of Roads, "Report on Roads in Australia", (1975).

Table 3

Ratios of Vehicles per Dwelling and Vehicles per person in Major Urban Areas* in Australia, 1966 Through 1971

	Year		*Increase Per cent*
	1966	*1971*	
Population (millions)			
•Total	7.124	8.227	+ 15.5
•Aged 17 and over	4.913	5.690	+ 15.8
Vehicles (millions)	1.910	2.642	+ 28.3
Dwellings (millions)	2.000	2.429	+ 21.5
Persons per dwelling	3.562	3.387	— 4.9
Persons aged 17 + per dwelling	2.457	2.343	— 4.6
Vehicles per person	0.268	0.321	+ 19.8
Vehicles per person aged 17 +	0.389	0.464	+ 19.3

*Centres with population over 100,000.
Source: Australian Bureau of Statistics, Census of the Commonwealth of Australia, 1966, 1971 (after L. Lawlor: Electric Transport Conference, Canberra, 1977).

Between 1966 and 1971 the Australian vehicle population increased by 28% with a 1.09 vehicles per dwelling ratio as compared with 0.96 vehicles per dwelling in 1966. (The results of the 1976 census relating vehicles to households were not available at the time of printing.)

Income and mobility

In both America and Australia where vehicle usage levels are among the highest in the world, the effect of income on ownership and usage is highly significant.

Data from the Melbourne (11) and Geelong (12) Transportation studies has been graphically represented in Figures 8 and 9 below and demonstrates the relationship between low household income (one of three categories used to classify all households) and vehicle ownership.

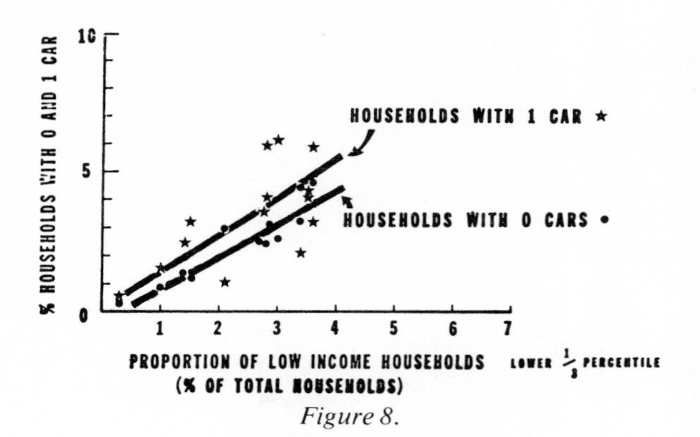

Figure 8.

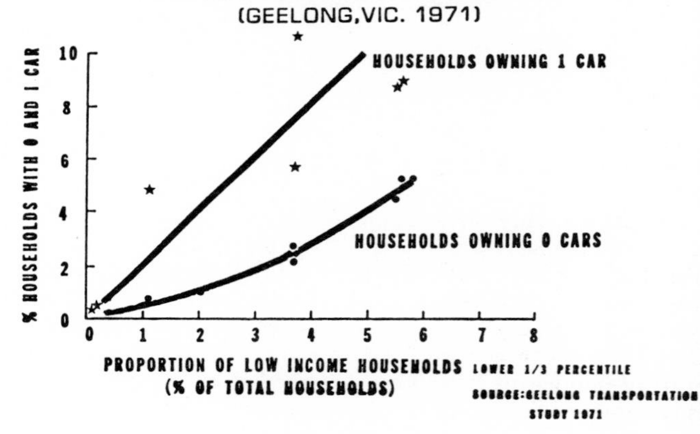

Figure 9.

Similarly data from the Sydney Area Transportation Study (1971) on income and the total number of trips per person and family per day has been plotted in Figures 10 and 11.

It is clear that the number of trips is strongly dependent on motor vehicle ownership and income level. The upper income households undertake three times as many trips as the lowest income level, on the basis of no vehicle ownership, and just under twice the number of trips where one vehicle is owned by each of these income groups.

In Figure 12 the average annual household income is plotted against the average annual miles of travel for the United States for the year 1975. The average distance travelled is relatively constant in the middle income range ($5000-$15000 in 1975) with a very wide gap between the lowest and highest income ranges. This middle income group of households represents 49% of the car owners and 66% of the total nation's car travel. Similar data for Australia is not available. However, the

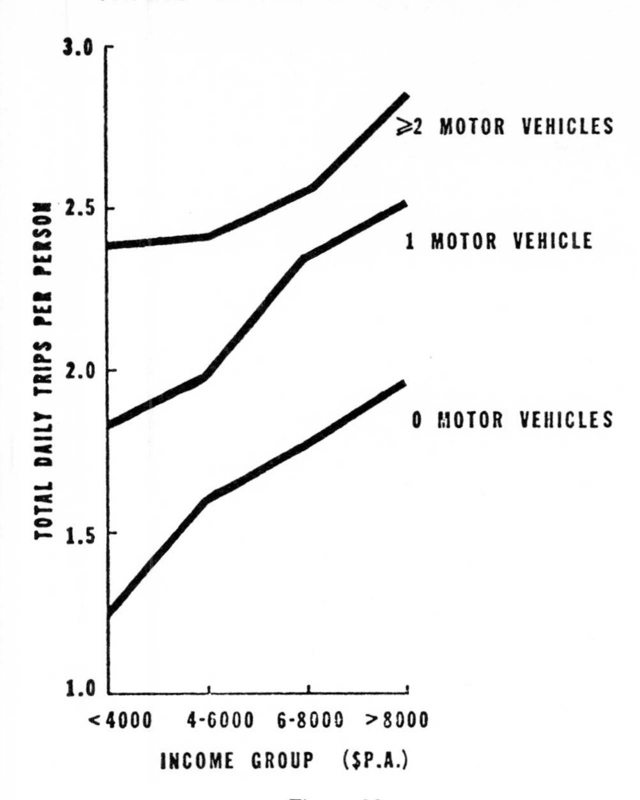

Figure 10.

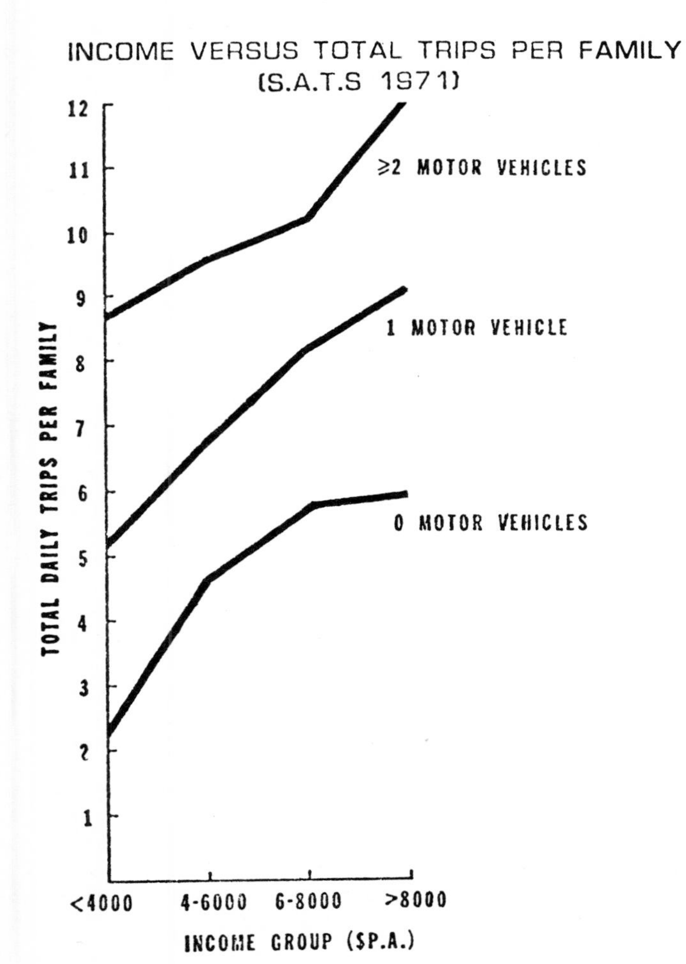

Figure 11.

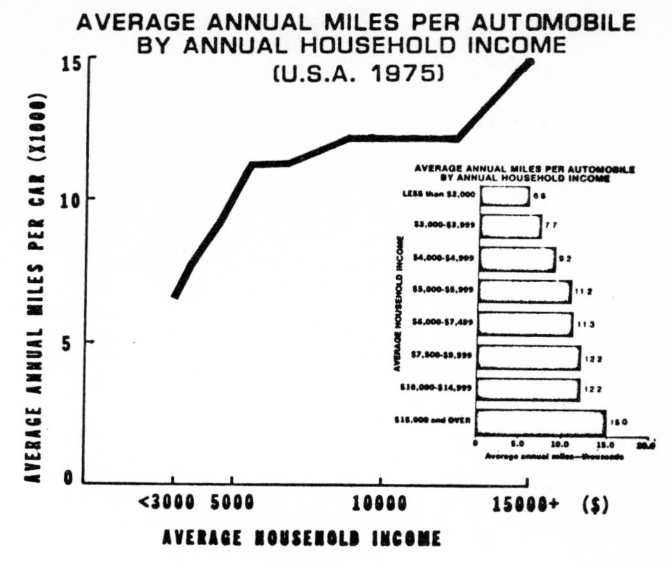

Figure 12.

average distance travelled by cars in the two countries is compared in Table 4 below.

Table 4

Annual Average Kilometres/per year

	U.S.A.	*Australia*
	16058 (1970)	15900 (1971)
	15504 (1975)	15700 (1976)
% Change	—3.4	—1.3

In the U.S.A. the longest trips are for social and recreational purposes with the next longest for earning a living.

The highest income earners travel the furthest to work. Trips for educational, civic and religious purposes are the shortest.

The growth in second car ownership may be reducing the average annual distance as reflected above.

Income and petrol consumption

It is instructive to examine the relationship between average household income and the proportion of that income which is spent on petrol.

In Australia the lowest income group (refer Table 5 below) in 1974/75 expended 2.09% of the average weekly household outgoings on petrol while the top income group expended 2.99%. While this difference appears relatively small, when compared to the income differential it may be inferred that the higher income groups spend at least double the amount on petrol and hence are high energy consumers in respect of motor vehicle usage. The more recent data for 1975/76 is somewhat different and suggests that further study is justified.

Table 5

Expenditure on Petrol as a Percentage of Total Expenditure by Weekly Household Income, Australia 1974-75

	Average Weekly Household Income						
	Under $80	*80-140*	*140-200*	*200-260*	*260-340*	*340+*	*All Households*
1974/75 (13)	2.09	2.90	3.20	3.38	3.22	2.99	3.09
1975/76 (14)	2.72	3.33	4.12	3.76	3.86	3.56	3.67

Note that the income group spending the highest proportion of all household expenditure has dropped from $200-260 to the $140-200 bracket.

Table 6

Levels of Car Ownership and Household Income in Australian Capital Cities, 1974-75

*Number of cars**	*Average weekly household (gross) income (A$)*
No car	$113
1 car	$197
2 cars	$298
3 cars	$384

*Includes utilities, station wagons, trucks and vans as well as cars owned by or available for continuous use by members of the household.
Source: Morris (15) October, 1977.

A similar situation is indicated by U.S. data as shown in Table 7 and Figure 13.

Table 7

Vehicle Ownership and Petrol Expenditures (U.S.A.) July 1973 — June 1974 (16)

Average Family Income (Gross) per Year	*Average Weekly Expenditure on Petrol (all Families)*	*Expenditure on Petrol as % of Family Income (all Families)*
$	$	%
1198	2.14	9.3
2957	3.18	5.6
4626	4.84	5.4
6447	6.24	5.0
8360	7.56	4.7
10307	8.81	4.4
12330	8.82	3.7
14916	10.20	3.6
18654	10.87	3.0
30079	12.37	2.1

The high level of petrol consumption by the upper income groups is illustrated by the fact that the top 27% of household income groups are responsible for around 44% of all expenditure on petrol (alternately the top 30% use half the petrol). In Australia it is evident that the most affluent families consume the greatest proportion of the nation's petrol.

PETROL CONSUMPTION AND HOUSEHOLD INCOME
U.S.A. 1973

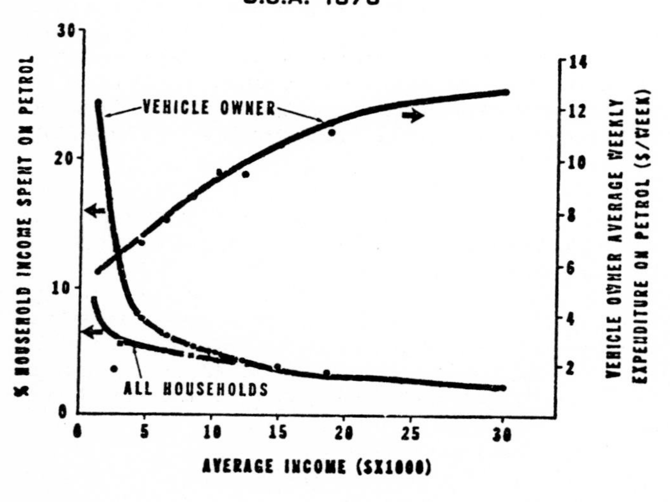

Figure 13.

4. Transport Energy in Australia

A breakdown of direct energy consumption by various transportation modes is given in the table below.

Table 8

Direct Energy Consumption by Passenger Transport in Australia, 1970-71

Vehicle/Mode	*Urban*		*Non-Urban*		*Total*	
	Energy 10^{12}kJ	*Per cent*	*Energy* 10^{12}kJ	*Per cent*	*Energy* 10^{12}kJ	*Per cent*
Car and station wagon	188.8	(56.2)	82.0	(24.4)	270.9	(80.6)
Motor cycle	1.0	(0.3)	0.5	(0.1)	1.5	(0.4)
Commercial vehicle	9.0		5.6	(1.7)	14.6	(4.3)
Bus	4.4	(1.3)	2.7	(0.8)	7.1	(2.1)
	203.2	(60.5)	90.8	(27.0)	294.2	(87.4)
Tram	0.7	(0.2)	—	—	0.7	(0.2)
Train	11.5	(3.4)	6.4	(1.9)	17.9	(5.3)
Air	—	—	23.1	(6.9)	23.1	(6.9)
Sea	—	—	—	—	—	—
Total	215.4	(64.1)	120.3	(35.8)	335.9	(100.0)

Source: Bureau of Transport Economics.

Both direct and indirect energy is used in transport. With transportation the largest amount of energy in Australia is used by the private car in the urban area (56% of all transport energy).

The most recent data as provided by the 1976 Motor Vehicle Usage Survey showed that 67% of private car travel is in urban areas.

The macro effects of urban mobility

Greater personal mobility, expressed in terms of car ownership, increases our demands for space and leads to new forms of urban life style and job availability.

Australian cities have many similar characteristics which may be summarised as:

1. Concentration of activities around a Central Business District (CBD) and a radial road system.
2. Energy consumption per unit area is highest in the CBD area.
3. Concentration of traffic at peak hours.
4. Average speed increases at increasing radial distance from the CBD.
5. Congestion is related to city size.

NORMALISED VEHICLE DISTANCE TRAVELLED VERSUS NORMALISED RADIUS FOR THREE CITIES

Figure 14.
Source: Hamilton, R. B. and Watson, H. C. (17).

For Australian cities, the relationship between the usage of motor vehicles and the amount of energy they consume is now more widely known. The influence of the Central Business District is very significant, for traffic densities per unit area are highest in this area. The traffic density, km/km²/day, decreases uniformly as the radial distance from the CBD increases. The author has calculated that 27% of the total national vehicle kilometres of travel (VKT) occurs in the inner urban areas of the mainland capital cities, i.e. the area equivalent to that within an 8 km radius of the Sydney GPO. Some 40% of Australia's petrol is consumed in this small inner area of our capital cities which is equivalent in aggregate to 550 km².

On a national basis, 67% of car travel is in urban areas and accounts for 74% of the total petrol used. This disproportionate fuel usage is because of the congestion which reduces the average speed of travel. At peak hour in Melbourne and Sydney 30% of the vehicle kilometres of travel is below 25 kph.

The following figure illustrates the distribution of energy consumption by the motor car for the Melbourne metropolitan area (18).

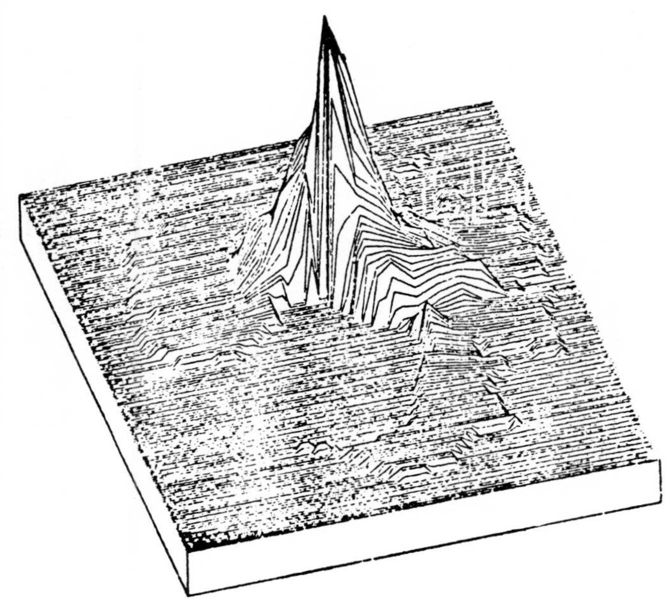

Figure 15.

The effect of proximity to the CBD, on average speed, is shown in Figure 16 together with the traffic density for selected major cities. The combination of increased VKT and reduced average speed in the urban areas has a considerable adverse effect on fuel consumption (see Section 4, Figure 17).

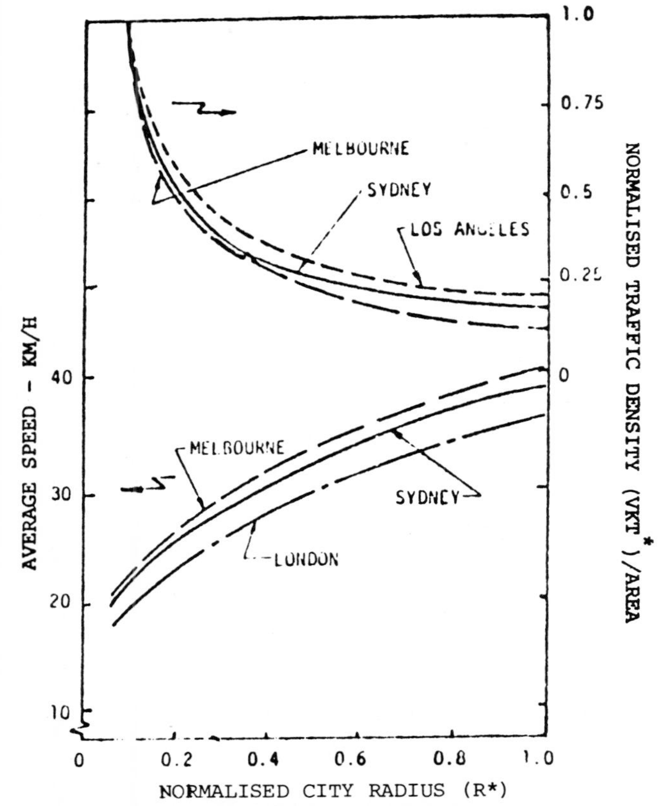

Figure 16: Variation in Normalised Traffic Density and Average Speed with Normalised City Radius

Source: Watson, H. C. (19)

VKT* = VKT/VKT_{max}. R = city radius/max city radius

Vehicle fuel consumption and average speed

Figure 17: Predicted fuel consumption variation with average speed for 1350 kg vehicle for individual links in Melbourne driving

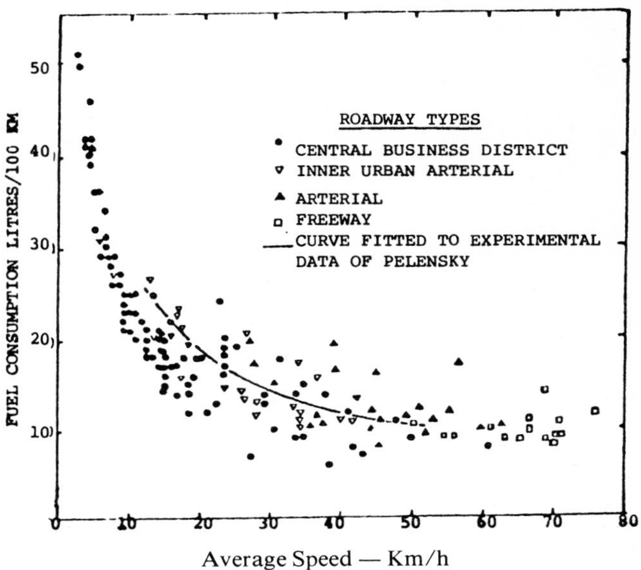

Average Speed — Km/h

Source: H. C. Watson (19)

Traffic in Australian cities is concentrated on the freeway/arterial road system, i.e. 50% of city VKT occurs on about 7% of the total road network system (Sydney has 16900 km of roads). The fuel consumption of a specific vehicle is highly dependent on the average speed and idle time during a journey. At peak hour when average speeds are low the average fuel consumption can exceed the optimum, which occurs at about 56 kph, by a factor of two. In Sydney about 30% of fuel is consumed at speed below 24 kph and 65% below 32 kph (20).

Traffic congestion and city size

The major Australian cities have many similar characteristics relating total vehicle travel to the spatial geography of the road network system.

There is a considerable theoretical background of knowledge and much practical evidence upon which to quantify the effects of congestion in our cities. At the present time some 10% of the petrol or an extra $90 million of crude oil per year is consumed in our capital cities, under congested condition when traffic speeds at peak hour are less than 25 kph.

With the existing land use patterns:

- The level of congestion C is related to the city size and the total Vehicle Kilometres of Travel (VKT) by the approximate relationship (21): $C = \frac{1}{3}(VKT)^{\frac{1}{2}}$.
- The indirect effects of city congestion in Sydney which in the past has determined future motor vehicle emission

limits has increased the energy consumed in private transport in Australia. The effect is felt in all parts of Australia and is not confined to the areas of highest pollutant density.

- The benefits of restructuring city development in Australia to conserve energy will also satisfy other desirable community goals by increasing leisure time and reducing pollution in our largest cities.

Figure 18 shows the relationship between congestion and city size.

RELATIONSHIP BETWEEN CONGESTION AND CITY SIZE FOR AUSTRALIAN CAPITAL CITIES (1975)

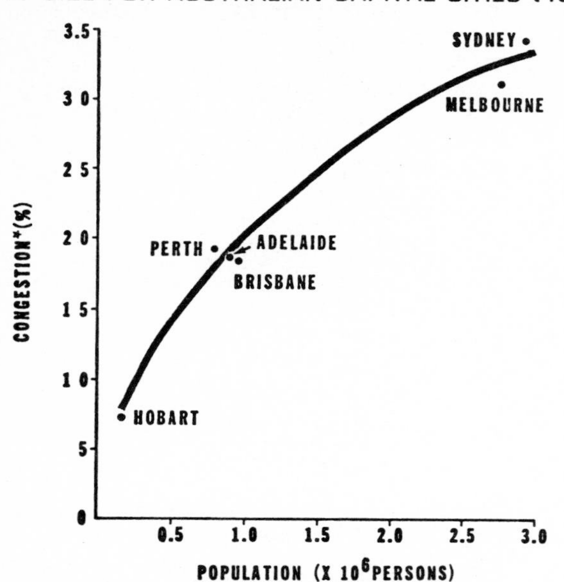

POPULATION (X 10^6 PERSONS)

*CONGESTION IS DEFINED BY THE EXPRESSION: $C = 0.294V^{0.482}$

WHERE V = ANNUAL VEHICLE KILOMETRES OF TRAVEL

SOURCE: R.B.HAMILTON I.E. AUST. TRANSPORTATION CONF. ORANGE 24-26 OCT. 1977

Figure 18.

Past petrol consumption and estimate of future demand

Table 9

Consumption of Petrol per Motor Vehicle in Australia, 1961-62 through 1975-76

Year	*Total Consumption (million kilo-*	*Motor Vehicles on Register* litres)*	*Unit Consumption (kilolitres per vehicle per year)*
1961-62	6.051	2.446	2.47
1962-63	6.467	2.604	2.48
1963-64	7.027	2.790	2.52
1964-65	7.558	2.976	2.54
1965-66	7.912	3.117	2.54
1966-67	8.274	3.267	2.53
1967-68	8.684	3.400	2.55
1968-69	9.314	3.573	2.61
1969-70	9.946	2.771	1.64
1970-71	10.374	3.966	2.62
1971-72	11.077	4.159	2.66
1972-73	11.519	4.393	2.62
1973-74	12.282	4.629	2.65
1974-75	12.766	4.904	2.60
1975-76	13.261	5.156	2.57

*Cars, station wagons and light commercial vehicles only.
Source: L. Lawlor.

The average amount of petrol consumed by vehicles has remained almost constant since 1960 and is equal to about 2,600 litres per year.

Official forecasts of petrol consumption for the future have been reduced significantly, e.g. in 1975 the Department of National Resources (DNR) forecast for 1984-85 was 20.1×10^9 litres. In April 1978 the Department of National Development (formerly DNR) estimated the demand for the same period to be 17.2×10^9 litres.

Thus while the demand for petrol is growing the rate of growth has declined and is expected to decline even further in the light of the recently announced goal of the motor vehicle manufacturers to reduce the National Average Fuel Consumption figure from 11.2 litres/100 km to about 8.96 litres/100 km (a 20% reduction) by 1987.

The dynamics of the light vehicle population

The vehicle population is continuously changing as new vehicles enter the population and old vehicles are scrapped. The mean age in Australia is about 5.5 years.

The annual distance driven by an individual vehicle declines with age. The scrappage rate curve of the individual vehicle falls steeply after the 10th year. The proportion of the total VKT is derived from the combination of the two and is not uniformly distributed with vehicle age with some 50% of the VKT being done by vehicles four years of age or less.

The relationship can be used to model changes in energy consumption, gaseous and particulate tailpipe emissions and octane requirements of the total vehicle population based on reasonable assumptions and proposed changes in legislation on vehicle design characteristics.

National average fuel consumption

Recently the motor industry announced a voluntary uniform code of practice for passenger car fuel consumption reduction.

The industry through the Federal Chamber of Automotive Industries has given an undertaking to the Government that the national average consumption of petrol will be decreased by some 20% by 1987 when an average of 8.96 litres/100 km will be achieved (22).

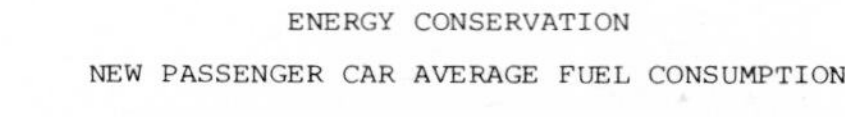

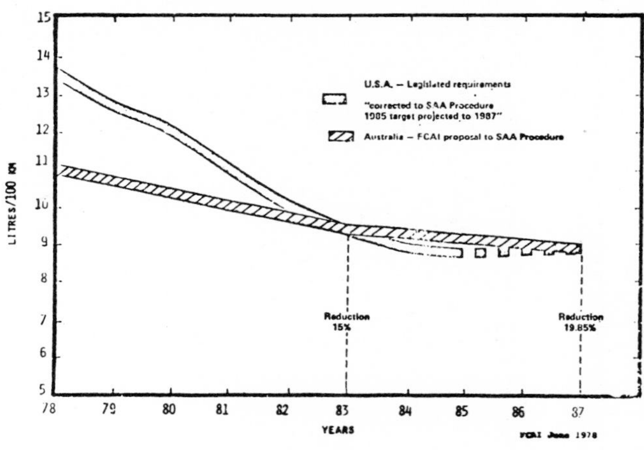

Figure 19.

The program of National Average Fuel Consumption (NAFC) reduction requires:

1. Capital investment in the provision of new engine designs, bodies and transmissions.
2. Continued stable economic conditions for a long time ahead i.e. no sudden change of rules relating to import quotas or local content plans.
3. Some reliance on the continuation of a reasonable level of local vehicle production with accompanying employment opportunities.

The motor industry is the largest secondary manufacturing industry in Australia. The effect of the planned reduction in NAFC may result in the demand for petrol for cars in 1987 falling below that for 1977. The lowest growth rate in demand would appear to occur around 1984.

The demand for petrol for cars is still subject to a wide variety of factors and the estimates must take into account the social/economic condition prevailing over the next 10-12 years. The actual demand in 1990 depends on the number of cars per capita and the average annual distance travelled, a range of scenarios is shown in Table 10 below.

Table 10

Potential Petrol Demand by Cars in 1990 (Litres × 10⁹)

Car Ownership Scenarios (Cars per Capita)	Fuel Consumption Rate (Litres/Veh.) 2500	2000	1500	1400	1200	1125
0.46	18.4	14.7	11.0	10.3	8.8	8.3
0.44	17.6	14.1	10.6	9.8	8.5	7.9
0.42	16.8	13.4	10.1	9.4	8.1	7.6
0.40	16.0	12.8	9.6	9.0	7.7	7.2
0.38	15.2	12.2	9.1	8.5	7.3	6.8
0.36	14.4	11.5	8.6	8.1	6.9	6.5
litres/100 km	16.67	13.33	10.00	9.33	8.00	7.50

Note: Population projected to 16 million by 1990 and an average VKT of 15000 km/vehicle/yr assumed. Present cars per capita value is 0.37 and average fuel consumption is around 2025 litres/car/yr.

Estimate of cost of emission control

To meet the present motor vehicle emission standards in the period 1976 to 1980 the increased cost to transportation will be some $500 m. This additional cost will be mainly to control pollution in the densely populated central business areas of the capital cities.

Indicative costs 1976-1980 for Australia are (23), (24), (25):

1. Increase cost of vehicle @ $100-150 each 2.3 m vehicles	$250-350 m
2. Increased petrol @ 7% (ATAC)	$ 40- 90 m
3. Premature Oil refinery investment	$ 30- 50 m
4. Extra refinery fuel @ 1% of petrol for lead reduction in N.S.W., Vic., Tas.	$ 15- 20 m
TOTAL:	$335-510 m

These costs are borne by the new car population in order to contain/reduce the emissions in our largest cities.

5. *Optimising Resource Allocation*

Both the oil industry and the motor vehicle manufacturers are studying the means and methods of meeting the challenge of the diminishing traditional fuel resource oil.

Shell utilises mathematical models to gain a better understanding of the important factors in optimising the allocation of resources for meeting society's needs for energy, especially relative to transport.

A diagramatic layout of the interrelationships between the main submodels which make up the present group is shown in Figure 20.

Vehicle engine and oil refinery as a single system

Resources and energy can be saved by considering the vehicle engine and the oil refinery as a single technical and economic system.

Important constraints in the system are:

Industry fuel consumption targets.

Motor vehicle emission laws; present and projected.

Fuel quality laws; allowable lead content of petrol.

Complete elimination of lead from petrol would increase consumption in the joint system by 9-11% overall.

The basic principles involved are:

1. The total investment in the vehicle engine manufacturing and oil refining facilities to minimise energy consumption should be based on sound economic criteria which recognise both continuing investment and operating costs.
2. If vehicle engines can make full use of the octane qualities of the available fuels, the higher the pool octane the greater will be the distance which can be travelled per unit of energy used in the combined refinery and engine system up to an optimum, largely determined by emission constraints.
3. Offsetting this, as the leaded pool research octane number increases so does the energy required for the production. Maximising lead additives minimises energy consumption in producing fuel of any specified octane number.
4. Minimising emission constraints on engine and petrol design by employing alternative pollution control strategies could have valuable energy conservation results at moderate cost.

The counteracting effects of energy savings in the refinery versus those on the road are illustrated in Figure 21 below (24).

The average fuel consumption saving derived by General Motors, DuPont and the European motor industry together with the energy consumption information from a study by Shell, has enabled the estimation of the net fuel saving for the leaded petrol option for Australia. At the optimum pool octane quality, the net fuel saved due to leaded fuel is 9 per cent greater than with unleaded fuel. The continued use of engines designed for high octane petrol will save energy in the future.

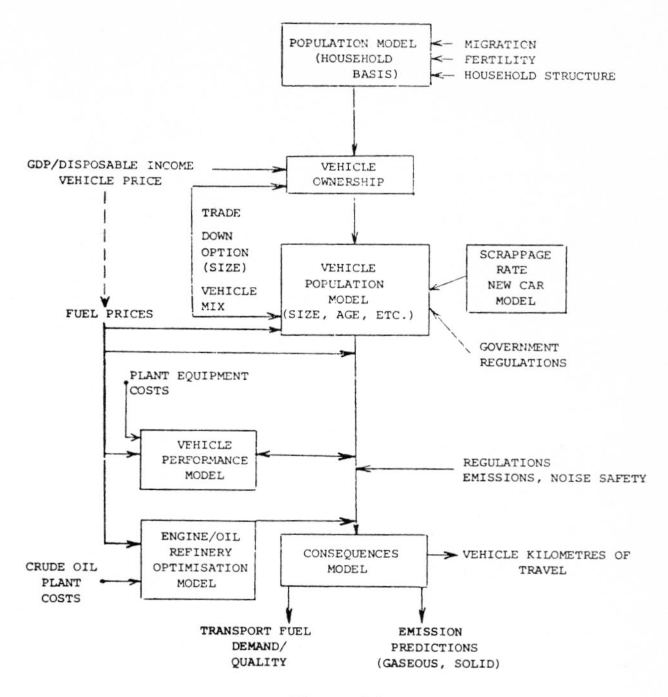

Figure 20

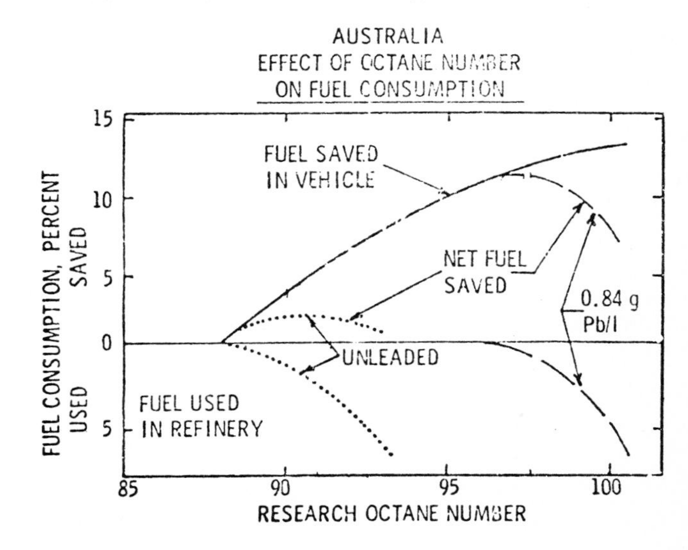

Figure 21.

Existing and new technology will be used in the oil refineries to improve the yield of petrol and other transport fuels to satisfy Australia's needs.

Efficient use of existing resources

It is not the purpose of this paper to explore in detail the numerous alternative courses but rather to examine some of those which are socially desirable and cost effective in terms of energy usage.

Some examples are:

(i) Traffic Management:
Computer traffic control, higher parking fees/tolls at peak hours. Staggered working hours, flexitime, and 9 day fortnight/4 day week.

(ii) Increase Utilisation:
Van pooling, multi-hiring of taxis, car pooling, matching vehicle to the transport task.

(iii) Federal and State Fiscal Policies:
Sales tax, customs and excise duty, taxation, registration. Predictable long term policies to achieve national objectives.

(iv) Education:
Proper maintenance, using fuel of appropriate quality, observing speed limits, reducing acceleration rates.

6. Summary and Conclusions

Mobility is a highly valued and sought after personal need of modern day man. The attainment and maintenance of this need, however, has certain profound implications for energy consumption.

As our personal mobility increases, the use of energy for transportation increases exponentially. Working man has a limited amount of time for travel and therefore tends to consume an increasing amount of energy. This is consistent with the urban sprawl and low vehicle occupancy as is evidenced in most of our cities.

Many critical social problems compete for the limited resources which are available to society and long before OPEC countries raised the price of crude oil the OECD expressed concern at the resources devoted to the car.

In Australia at present, some 50% of oil is used for transportation needs (38% of crude oil consumed is petrol). Some 67% of national car travel occurs in the cities and consumes 88% of the direct energy used in personal transportation in the urban area.

By world standards the Australian household population enjoys a level of car ownership equal to the highest in the world and Australians travel the second longest average annual distance.

A car is considered as essential for about 50% of workers to travel to work.

The present radial city structure based on a central business district leads to a significant increase in energy consumption in our major cities.

In Australia, the most urbanised nation in the world, 40% of the nation's petrol is consumed in an aggregate area of some 550 sq. kilometres comprising the inner urban districts of our capital cities (15 miles × 15 miles).

The decline in the population level of the established inner areas of our cities and the present land planning laws lead to new development on the fringes of our largest cities which are intrinsically highly energy intensive. This high energy consumption arises from the increased direct and indirect energy consumption in private transport. Public transport can only provide for a small fraction of the increasing community demands.

Energy consumption for private transportation is related to our level of personal disposable income; with half of the petrol being consumed by families in the top third of the income range.

Significant changes in transport energy usage which will halt if not reverse the growth in demand for transport fuels for the private car, are already taking place in Australia.

The relative importance to the community of both cleaner air and reduced energy consumption must be known by the facility planners as the two goals are to some extent opposed to one another.

The vehicle engine and oil refinery should be optimised as a single technological system.

A restructuring of our use of energy for private transport involves the complex questions of the individual's freedom of choice and how communities and governments will react to many improperly perceived problems.

To bring about changes requires a thorough reappraisal of community planning and the ability to articulate the constraints and the solutions to the community.

We must carefully examine the factors involved to gain a better understanding of the choices, implications and decisions which flow from our actions. The problem is to harmonise socio-economic and environmental goals by redefining patterns of resource use and the uses of growth.

In contemporary Australian society, this involves how we plan our cities, where we live in relation to our work place, and how we spend our leisure time.

There is not an energy supply crisis as such but a crisis of understanding. Few people understand the complex interacting factors involved and their effects on energy consumption in meeting personal demands for mobility in the urban areas of Australia.

References

(1) Leach, G. (1972). *Enquiry into the Impact of the Motor Vehicle on the Environment.* Report to OECD, Dec. (unpublished).

(2) Lovins, A. B. (1977). *Soft Energy Paths, Towards a Durable Peace.* Penguin.

(3) Aust. Inst. Petroleum (1977). *Oil and Australia.*

(4) Clark, N. (1975). Occasional Paper 2, *Transport and Energy in Aust.* Part I — Review, BTE, June.

(5) U.S. Government, Council on Env. Quality (1977). *Environmental Quality,* 8th Annual Report.

(6) U.S. Government, Council of Environmental Quality (1974). *Environmental Quality,* 5th Annual Report.

(7) United States Real Estate Research Corporation (1974). *The Costs of Sprawl,* prepared for U.S. Council of Environmental Quality, April.

(8) Australian Bureau of Statistics (1974). *Journey to Work and Journey to School.* August.

(9) Vaughan, R. M. (1977). Some Traffic Characteristics of Sydney. *Australian Road Research,* Vol. 7, No. 4, December.

(10) *Sydney Area Transportation Study (1971).* Prepared by the Departments of Transport and Local Government.

(11) *Melbourne Transportation Study (1964).* Prepared for the Met. Transport Committee, 1969.

(12) *Geelong Transportation Study* (1970/71). Prepared for the C.R.B., Vol. I, Nov. 1971.

(13) Morris, J. and Wigan, M. R. (1977). Family Expenditure Survey Data and Their Reference to Transport Planning. A.R.T.F. Conf. Proc.

(14) Australian Bureau of Statistics, Unpublished Data, 1975-76 Household Expenditure Survey.

(15) Morris, J. (1978). Social Implications of Urban Transport. Paper to Urban Land Inst. Publ. ARRB March.

(16) Motor Vehicle Facts and Figures 1977. Motor Vehicle Manufacturers Association of the U.S. Inc.

(17) Hamilton, R. B. and Watson, H. C. (1976). Int. Conf. on Photochemical Oxidant Pollution, Raleigh, U.S.A. Sept.

(18) Sharpe, R. (1977). The Effect of Urban Form on Transport Energy Patterns. *Urban Eco.* x:xx-xx.

(19) Watson, H. C. (1977). *Traffic-Vehicle-Energy Interactions.* A.A.A. Transport Energy Seminar, Hobart, Nov.

(20) Hamilton, R. B. and Hocking, D. M. (1976). *The influence of the CBD on Energy for Transport in the Urban Area.* Inst. Eng. Aust. Conf., Townsville, May.

(21) Hamilton, R. B. (1977). *The Consequences of the Large Congested City on Energy and Transport.* Transport Conf. Orange, Oct.

(22) Federal Chamber of Automotive Industries (June 1978). *A Uniform Code of Practice for Passenger Car Fuel Consumption Reduction.*

(23) Hamilton, R. B. (1978). Paper to Clean Air Society Australia and New Zealand, Hobart, 15th March.

(24) Court, J. W., and Coughlin, R. C. (1978). *An Analysis of Air Pollution Control Costs in N.S.W.* Proc. International Clean Air Conference Brisbane, May.

(25) Little, A. D. (1978). *The Lead Costing Studies of COMVE, The Cost of Reducing the Lead Content of Petrol in Australia. (Refinery Aspects).* Australian Transport Advisory Council Committee on Motor Vehicle Emissions, Department of Transport, July.

(26) Bettoney, W. E. and Cantwell, E. N. (1978). *Fuel Quality and Energy Conservation.* Petroleum Lab. E. I. du Pont de Nemours & Co. Inc. Pub. A.I.P. April.

Energy Costs of Commuting

Bryan Furnass

University Health Service, Australian National University, Canberra, A.C.T.

Energy for human movement between shelter and source of food supply is derived from two types of chemical reaction. First is the carbohydrate cycle, involving the step-wise transfer of electrons to molecular oxygen in skeletal muscle by a process which is the reverse of photosynthesis:

$$\underset{\text{glucose}}{C_6H_{12}O_6} + 6O_2 \underset{\text{photosynthesis}}{\overset{\text{respiration}}{\rightleftharpoons}} 6H_2O + 6CO_2 + 2.8\ MJ$$

This source of energy is renewable, its time relationship with the sun being limited to a season of plant growth. The second type of reaction is the explosive release of energy from hydrocarbons in fossil fuels, carried either as fuel in a transport vehicle or transmitted from some central point by electric power. This process is irreversible, its time relationship with solar energy exceeding 10^8 years.

The Carbohydrate Cycle

The muscular and cardio-respiratory systems of human beings had evolved long before the hunter-gatherer stage of development two million years ago. Personal transport then consisted of walking, running, climbing, jumping and swimming over a wide variety of rough terrain. Under the adaptive influence of different habitats the efficiency of human muscle power has been enhanced by a variety of simple tools and attachments, including canoes, sails, snowshoes, skis and skates. The taming of animals by nomads and by peasant farming communities during the neolithic development added another dimension to human transport. This involved the release of stored solar energy through the muscles of horses, donkeys, mules, camels, dogs, bullocks, elephants or lamas, the food for these animals being derived either from grazing in the natural environment or from specially designed cropping arrangements.

Despite its invention over 8000 years ago it is only during the past century that the wheel has been adapted as a direct adjunct to human muscle power, namely in the form of the bicycle. Basically the bicycle converts the vertical thrust of leg muscle contraction into horizontal propulsion, the development of freewheeling and gearing making it in energy terms the most economical type of personal transport yet devised.

Fossil Fuels for Transport

The discovery of coal, the invention of the steam engine and its adaptation for movement over rails during the industrial revolution provided the opportunity for relatively economical mass transportation of people which formed the basis of suburban commuter trains. Invention of the internal combustion engine and the discovery of vast reserves of oil at the beginning of the present century opened up new possibilities for independent personal movement in the journey to work. Unlike the relatively compact older cities of Europe which favoured pedestrian, bicycle, train, tram or bus transport between home and work, the growth of suburban living particularly in the towns of North America and Australia has been based on the widespread private ownership of motor cars. This difference is born out by contrasting the population density of Vienna (a city frequently cited as one of the fine examples of a truly "urbane" environment) which has 19,600 persons per square mile with Los Angeles-Long Beach (usually described as the epitome of the "motorised city" built, that is, for automobiles, not people) which houses 5,500 persons per square mile. By comparison the population density of the built-up area of Melbourne is 7,300 persons per square mile and of Canberra (in 1971) 2,600 persons per square mile (1).

Clarke (2) has calculated standards of mobility for different countries by postulating that there is a constant amount of time available for travel, of the order of 500 hours per year, the differences in distance travelled being due to the availability of modes of travel. Thus the greater the availability of fast modes of travel, the greater the standards of mobility and the shorter is the distance covered on foot or bicycle. Table 1 shows that while the overall standard of mobility for Australia is over six times as great as for Indonesia, we travel less than a third of the distance by foot or bicycle as our non-industrialised neighbours.

Table 1

Estimated Standard of Mobility for Selected Countries — Late 1960's

	Person-km/head, by mode					
Country	*Rail*	*Bus**	*Air*	*Car*	*Walk/cycle†*	*Total*
U.S.A.	80	200	1010	12,750	500	14,500
Australia	790	700	1102	10,150	500	12,742
Canada	160	(200)	1000	9,680	500	11,540
U.K.	640	900	370	6,100	750	8,850
France	800	390	270	5,560	750	7,700
Japan	2740	980	160	1,730	1000	6,010
Poland	1140	990	20	340	1500	3,990
India	210	190	10	20	2000	2,430
Brazil	120	(200)	50	490	1500	2,360
Thailand	110	(200)	30	80	1750	2,170
Indonesia	30	(200)	10	40	1750	2,030

*Values in brackets assumed.
†Walk/cycle is wholly estimated from a scale ranging from 2000-500 person-km per head depending on the level of motorised mobility.
Source: Clark (1973).

Direct Energy Costs of Commuting

Direct energy costs of carbohydrate transport may be computed from the energy equivalents of measured rates of personal oxygen consumption during walking and cycling (3). Similar costs of hydrocarbon transport may be calculated from the heat. of combustion of petroleum related to the average number of passengers carried per vehicle distance (2).

Comparative approximate figures are presented in Table 2, together with the range of 24 hour metabolic energy requirements of people in sedentary occupations. These figures indicate that a round trip of 5 km on foot would account for something in the region of 12-15% of daily energy expenditure and occupy an hour of time, or 12% of the working day. The same distance covered by car would require the burning of the equivalent of the total daily allowance of metabolic energy as fossil fuel and would "save" 50 minutes of time for non-physical activities such as drinking or television-watching. For the short journey to work the advantage of the bicycle becomes apparent since a 5 km round trip would account for only 4-6% of 24 hour energy expenditure and occupy the equivalent of only 4% of working time, or only 10 minutes more than when travelling by car.

Values for hydrocarbon transport are subject to considerable variation, depending on factors such as passenger occupancy per vehicle, traffic congestion, distance between bus stop or parking lot and place of work, and petroleum prices. Clark's

Diesendorf, M. (ed.) (1979). *Energy and People*. Canberra, Society for Social Responsibility in Science (A.C.T.)

figures are based on the Melbourne mass transport system which like many others in the Western world operates at a very low load factor, such that the energy consumption per seat kilometre could be as little as one third of the values shown for buses, trams and trains and even less if one allows for standing passengers. Similarly the energy cost of private car transport, based on 1.3 occupants per vehicle, could be less than a third of the value quoted during periods of car pooling such as have been observed during fuel shortages or public transportation strikes.

Table 2

Direct Energy and Time Costs of Commuter Transport (Australia)

Mode	*Approximate Journey Speed* (km/hr)	*Costs* (cents/ person/km)	*Energy* (MJ/person /km)	*Time to Travel 1 km* (min)
Walking	5	0	0.3	12
Cycling	10-15	0	0.1	4
Bus	10-15	3	0.8	4
Tram	10-15	3	1.3	4
Train	15-20	3	1.3	3
Motor Cycle	30-40	3	1.4	2
Car	30-40	5-15	2.3	2

(Average 24 hour metabolic energy requirements for people working in sedentary occupations = 8-12 M.J.)

Indirect Energy Costs of Carbohydrate Transport

Indirect energy costs of travel by foot or bicycle are generally calculated on the basis that extra food must be consumed and that the energy costs of production, distribution, packaging and cooking of this food must be included in the estimate. The Steinharts (4) have reviewed the historical influence of high energy technology on the U.S. food system. Thus, during the present century there has been a five-fold *increase* in the energy costs of food production in parallel to a five-fold *decrease* in the number of man-hours spent in farm work (Fig. 1). The use of high energy technology in the food industry has led to a situation where no less than ten units of fossil fuel energy are required for each unit of food energy consumed, in contrast with the situation in Third World countries where solar energy remains the main component of input to food production (Fig. 2).

Gifford (5) has argued that since each calorie of food used in exercise requires a commitment of 10 calories of petroleum to the food system, walking may be no less wasteful than a car as a means of locomotion. This argument presupposes that the modest increments of energy expenditure required for personal transport will lead to an increase in food consumption, for which there is no evidence — in fact, exercise often inhibits rather than stimulates appetite. Obesity, which is endemic in modern Western society, represents the storage as fat of energy consumed in excess of energy expended, being frequently

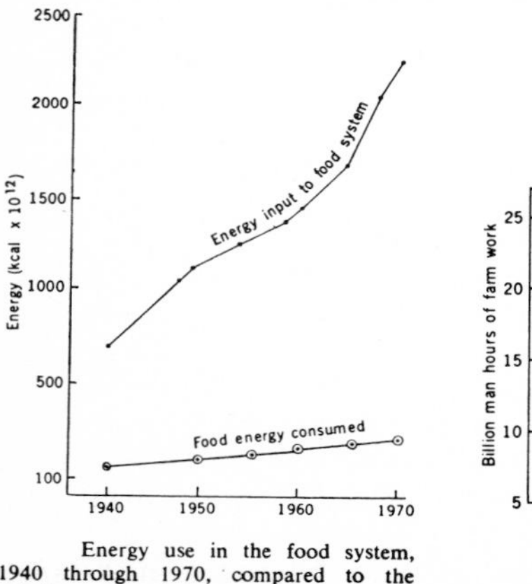

Energy use in the food system, 1940 through 1970, compared to the caloric content of food consumed.

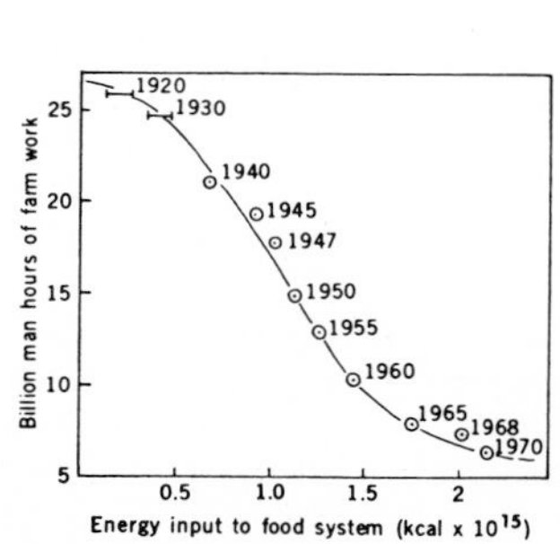

Labor use on farms as a function of energy use in the food system.

Figure 1.

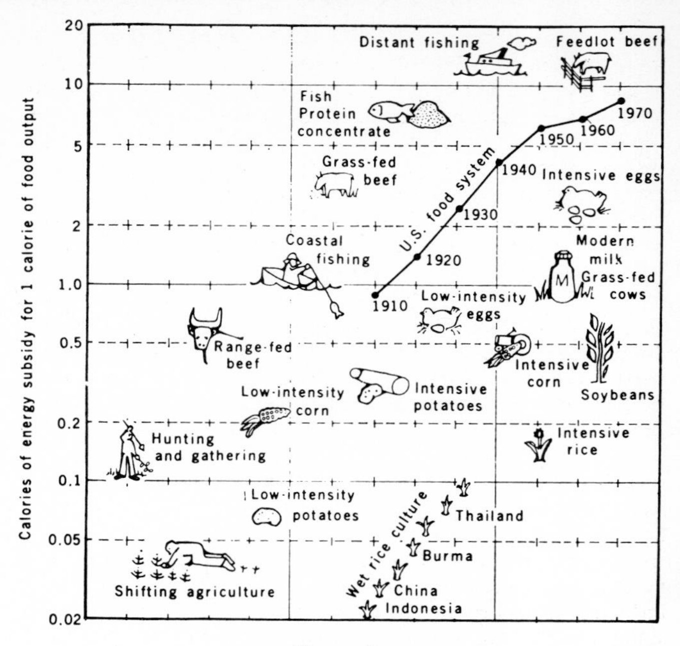

Figure 2.

associated with underactivity rather than with overeating. One therapeutic implication of this is that a 5 km commuter walk daily, if food intake remained unchanged, could lead to an energy deficit of about 6 MJ per week, representing the oxidation of 8 kg of stored excess fat in a year, and resulting in a lowered rate of energy cost per unit of distance travelled.

In an era of high technology medical (or *illth*) care the journey to work on foot or bicycle, in promoting the positive and preventive aspects of health and wellbeing, could stabilise and eventually reduce the demand for hospital beds and associated fossil fuel use. Thus the incidence of coronary heart disease, which accounts for more than a third of premature deaths in Western society, is much less common amongst physically active men than amongst sedentary controls (6). Further supportive evidence in favour of physical activity as a preventive measure comes from the finding of a significantly lower incidence of abnormalities in the electrocardiogram of men who walk at least part of the way to work than amongst those who remain seated throughout their journey (7). Motivation to include some physical activity in the daily routine comes from the fact that moderate exercise improves both physiological function and psychological wellbeing. An added bonus is the immeasurable aesthetic pleasure which accompanies walking or cycling to work through a parkland setting, which is possible in a few Australian cities, notably Canberra and Adelaide.

Indirect Energy Costs of Hydrocarbon Transport

The construction and repair of roadways and motor vehicles involves a fossil fuel consumption which is at least as great as that used in actual movement of traffic (2). Ivan Illich (8) has contrasted the impact of various modes of travel on the area of cities devoted to transport by pointing out that eighteen bicycles can be parked in the place of one car and thirty of them can move along the space devoured by a single automobile. It takes two lanes of a given size to move 40,000 people across a bridge in one hour by using modern trains, four to move them on buses, twelve to move them in their cars, and only one lane to pedal across on bicycles. Moreover, when viewed in economic terms the model American puts in 1600 hours of work a year to run his car over 12,000 km giving him a net speed of 7.5 km per hour or less than half the speed of a bicycle.

To the fuel costs of motorised transport must be added the energy and people costs of road crashes — not only the high technology burden of hospital treatment but also all the paraphernalia of insurance and litigation. Some idea of the extent of this problem in Australia may be seen from the fact that during 1976 there were 64,281 road traffic crashes involving casualties with 3,583 people killed and 87,807 injured at an estimated economic cost of $1000 million. This loss of life and limb if regularly sustained on the battlefield would be viewed with considerable dismay by military leaders. A less readily assessable but increasingly apparent by-product of motorised city centres is atmospheric pollution. Thus, figures published by the State Pollution Control Commission of New South Wales have calculated that Sydney's 1.2 million motor vehicles daily add to the atmosphere 290 tonnes of hydrocrabons, 90 tonnes of nitric oxide and 2000 tonnes of carbon monoxide. There is circumstantial and epidemiological evidence that air pollution contributes to many diseases of the respiratory system, particularly bronchitis and asthma, and to heart disease (9).

Confronting the Future

Costs of commuting must be seen in the context of total energy use by the human species as a whole. International disparities are such that the United States with only 6 per cent of the world's population consumes 40 per cent of the world's non-renewable energy supplies. An important component of the individual's contribution to our private affluence and public squalor is the illusion of power which comes from having a Tiger in the Tank. Kenneth Boulding (10) has described the emotional motivation to car commuting as follows:

The automobile, especially, is remarkably addictive. I have described it as a suit of armour with 200 horses inside, big enough to make love in. It is not surprising that it is popular. It turns the driver into a knight with the mobility of the aristocrat and perhaps some of his other vices. The pedestrian and the person who rides public transportation are, by comparison, peasants looking up with almost inevitable envy at the knights riding by in their mechanical steeds. Once having tasted the delights of a society in which almost everyone can be a knight, it is hard to go back to being peasants. I suspect, therefore, that there will be very strong technological pressures to preserve the automobile in some form, even if we have to go to nuclear fusion for the ultimate source of power and to liquid hydrogen for the gasoline substitute. The alternative would seem to be a society of contented peasants, each cultivating his own little garden and riding to work on the bus or even on an electric streetcar. Somehow this outcome seems less plausible than a desperate attempt to find new sources of energy to sustain our knightly mobility.

Unhappily the current political and economic attitude in the Western world is to regard fossil fuel as a source of renewable income rather than a rapidly diminishing capital resource. The race towards increasing entropy can only be slowed according to Schumacher (11) by urgent investment of human and material resources in intermediate-energy technology both for the developing countries of the Third World and for the materially over-developed societies of the West.

A reduction of energy costs for commuting will basically involve a greater investment in public transportation, construction of smaller powered private cars and provision of a network of cycle paths within cities, including the facility of carrying bicycles on buses and trains. Although such developments would exert a relatively small impact on overall energy economy their main value would be to create a greater public awareness of the finite nature of energy resources and, in an era of increasing unemployment, encourage a gradual return to more labour intensive activity such as mixed organic farming. Incentives to adaptation, as in other fields of human endeavour are more likely to depend on crises than on rational planning, and one suspects that rising oil prices will be the main arbiter of change.

References

(1) Day, L. H. (1977). Report of workshop on habitat and health. *The Impact of Environment and Lifestyle on Human Health,* 228-232. Eds M. Diesendorf and S. B. Furnass. Canberra, Society for Social Responsibility in Science (A.C.T.).

(2) Clark, N. (1973). Energy in transport, in *Energy and How We Live,* Australian UNESCO Committee for Man and the Biosphere, 226-237.

(3) Passmore, R., Durnin, J.V.G.A. (1955). Human Energy expenditure. *Physiol. Rev.* **35**, 801-840.

(4) Steinhart, J. S., and Steinhart, C. E. (1974). Energy use in the U.S. food system. *Science* **184**, 307-315.

(5) Gifford, R. M. (1978). Energy in modern agriculture and the rural-urban relationship: a one-way cul-de-sac? *Energy, Agriculture and the Built Environment: Towards an Integrative Perspective,* King R. Ed. Centre for Environmental Studies, University of Melbourne, 150 p.

(6) Fox, S. M. (1973). Relationship of activity habits to coronary heart disease. *Exercise Testing and Exercise Training in Coronary Heart Disease.* Academic Press, New York, 3-21.

(7) Rose, G. (1972). Epidemiology of ischaemic heart disease. *Br. J. Hosp. Med.* 285-288.

(8) Illich, I. D. (1974). *Energy and Equity.* Calder and Boyars, London.

(9) Gibson, J. B., Johansen, A. (1978). *The Quick and the Dead. A Biomedical Atlas of Sydney.* Australian National University Press, Canberra.

(10) Boulding, K. (1974). The social system and the energy crisis. *Science* **174**, 255-257.

(11) Schumacher, E. F. (1973). *Small is Beautiful.* (Harper and Rowe, New York.

Energy Prices, Urban Structure and Social Effects

Ross King

Centre for Environmental Studies
University of Melbourne

1. Introduction

Households are dependent on energy use in three main ways: first for the production and operation of their dwellings, secondly for their transport and consequent use of the opportunities of urban areas, and thirdly for their jobs, their manufactured goods and their food supply. The first is related to dwelling design, the other two to dwelling location and urban structure. It is on those latter two that this paper concentrates.

Dwelling location in relation to jobs, schools, recreation opportunities, the sources of our food supply and the sources of our manufactured goods is certainly important: location generates the demand for transport, and transport is second only to the manufacturing and utilities sectors as a user of fuels in Australia, it is the major consumer of petroleum, the least efficient sector in its energy use, and probably the most difficult to reform by conventional changes in technology (1).

It is common to hear calls for salvation through a technological fix — electric cars, more efficient petrol-driven vehicles, synthetic petroleum, liquid hydrogen — although the evidence is that such solutions have lead times that are too long, offering only limited relief in the short term of a decade or so (2). It is also common to hear calls for certain behavioural changes such as increased car pooling and increased use of public transport, despite the evidence of declining car sharing and declining use of public transport (3). It is far less common to hear calls for a better understanding of the causes of our transport dependence, even though it must be self-evident that ultimately the most effective path to energy conservation will be through a reduction in both the length and frequency of journeys, therefore through a relocation of households and of their activities. This paper is about their present locations, and about location-dependent energy use; it is also about some of the social effects of price rises or non-price rationing, and some of the social transformations involved in achieving a "conserver society".

2. Home, Work and the Suburbs

Four aspects of urban structure are particularly important to an understanding of transport dependence, therefore of oil dependence:

- With the outward growth of the cities, the proportion of the population in lower density suburbs has increased.
- Population densities have tended to decline in established suburbs, and particularly in the inner suburbs, as standards of housing and of dwelling occupancy have risen.
- The population has tended to become clearly segregated in terms of socio-economic status or class.
- Jobs, particularly in the blue-collar sectors, have increasingly dispersed to suburban locations.

None of these are new trends — the first three in particular have persisted over several generations — although each has assumed new characteristics in recent decades following the increased availability of automobiles. Each is discussed following.

Suburbanisation

The first trend — the growth of lower density suburbs — is well described by Neutze (4), where its relationship to the development of public and private transport is stressed. He observes that, with the expansion of train and tram services up to the 1920s in Australian cities and with the growth of bus services from the mid-1920s, the suburbs that developed from the 1890s to the Second World War were well served by public transport, and the public transport was well used. Although car ownership grew very rapidly in the 1920s (though not in the 1930s or 1940s), it seems to have had little effect in permitting residential development beyond walking distance of public transport.

Immediate post-War residential development seems to have largely followed the public transport routes, so that during that period of high employment and petrol rationing, "greater use was made of public transport than ever before or since"(5). Increasingly from that time however, new development occurred beyond walking distance of radial public transport (trains, trams and some buses to the city centre, ferries in Sydney), although it was frequently served by local or "feeder" buses to the radial system.

The new suburbs of the 1960s and 1970s were almost entirely beyond the radial public transport system, and dependent on feeder buses and the private automobile. As Neutze observes:

> *Since the Second World War car ownership in Australia has increased rapidly. Although less rapid than in the 1920s, the absolute growth, from 80 cars per thousand in 1950 to 330 in 1973 is much greater. The 1950s and 1960s were the periods when car ownership spread to the majority of Australian families. At the 1971 Census 74 per cent of private dwellings in Sydney and 77 per cent in Melbourne had one or more cars.* (6).

The results of this expansion can be observed by comparing growth (and decline) of population in the various "rings" of the Sydney urban area as a case study (Table 1) (7).

Despite substantial flat development in the City of Sydney and the inner ring in the period 1947 to 1971, population declined in both. The shift to the more distant and lower density suburbs was thus both relative and absolute.

Standards of housing and of dwelling occupancy

The population decline of the inner city and inner suburbs reflects the second trend that is significant to population distribution and transport dependence: the rise in standards of housing and of dwelling occupancy. The trend is easily summarised:

- Occupied dwellings per thousand persons rose from 229 in 1947 to 280 in 1971, mainly because of the increasing tendency of the young to establish separate households earlier in their lives.
- Persons per room in private houses fell from 0.76 in 1947 to 0.66 in 1971, and in flats from 0.70 in 1947 to 0.62 in 1954 (although the occupancy rate has risen again since then, to 0.65 in 1971). As the average number of rooms per dwelling has not changed substantially over that time, the overall fall in occupancy has been accounted for principally by a fall in the size of households.

Diesendorf, M. (ed.) (1979). *Energy and People*. Canberra, Society for Social Responsibility in Science (A.C.T.)

Table 1

Distribution of Population, Sydney, 1861-1971

	City of[1] Sydney	*Inner[2] Ring*	*Middle[3] Ring*	*Outer[4] Ring*	*Total*
1000 persons					
1861	79	14	11	24	128
1891	214	130	60	43	447
1921	244	410	291	107	1052
1947	214	582	635	262	1694
1971	145	567	960	1110	2781
Percent of total					
1861	60.1	11.6	28.3		100
1891	47.9	29.2	22.9		100
1921	23.2	39.0	27.7	10.2	100
1947	12.6	34.4	37.5	15.5	100
1971	5.2	20.4	34.5	39.9	100
Density: persons per hectare					
1861	25.6	1.1	0.09		0.3
1891	73.3	9.8	0.27		1.1
1921	83.6	31.0	7.4	0.3	2.7
1947	73.3	44.0	16.2	0.8	4.3
1971	49.6	42.8	24.5	3.3	7.1

Note: 1. Defined by 1949-1967 boundaries.
2. Mosman, North Sydney, Leichhardt, Drummoyne, Ashfield, Marrickville, Botany, Woollahra, Waverley, Randwick.
3. Manly, Willoughby, Lane Cove, Hunters Hill, Ryde, Parramatta, Auburn, Concord, Strathfield, Burwood, Bankstown, Canterbury, Rockdale, Kogarah, Hurstville.
4. Warringah, Ku-ring-gai, Hornsby, Baulkham Hills, Blacktown, Windsor, Penrith, Holroyd, Fairfield, Liverpool, Camden, Campbelltown, Sutherland.

- Overcrowding of dwellings has fallen generally since 1954 (although there has been some increase in overcrowding of flats), while under-utilisation of dwellings has increased, particularly since 1961 (8).

There may be evidence that these changes are reversing, or at least slowing, as unemployment or its threat keeps young people in their parental homes a while longer. Nevertheless the overall effect is still for population densities of suburbs to fall generally, and to fall even more dramatically where the population is ageing: in the suburbs of the 1930s, 1940s and even 1950s, that mainly followed the radial public transport system.

Segregation of population

Social differentiation — the tendency of populations to segregate themselves spatially in the city in terms of socio-economic status, ethnic background, lifestyle, etc. — has been much explored, principally through studies in the "factorial ecology" of urban areas. Census tract (or CCD) data, summarising characteristics of individuals, households and dwellings, are typically manipulated by computers using various multi-variate analysis programs in order to explore the ways in which those relatively small tracts differ from each other (9).

Most factorial ecological models are heuristic: they look for patterns in data, but assume neither characteristics nor causes of those patterns. Some however are based on *theories* of social differentiation, and of these the best known are those categorised as "social area analysis". Shevky and Bell suggested that "the necessities of economic expansion" are the underlying "prime mover in the recent transformation of Western Society" in the direction of increasing *societal scale* (a scale of society representing the number of people in relation and the intensity of these relations). This increasing societal scale is associated with (i) changes in the distribution of skills, (ii) changes in the structure of productive activity, and (iii) changes in the composition of population. They suggested that the first of these leads to changes in the spatial distribution of occupations in the city (measurable in terms of a construct called *social rank* or *economic status*); the second leads to changes in the way of living, to the movement of women into the workforce, and to the spread of alternative family patterns (measurable by a construct *urbanisation* or *family status*, to do with fertility, women at work, single family dwelling units); and the third is measurable in terms of *population segregation* or *ethnic status*. Within any community, differentiation in terms of each of these will be increasing; and the increase may be examined through multi-variate analysis of areal data (10).

Because this differentiation has been a consequence of, among other things, the increasing scale of industrialisation and of institutions and a consequence of the transport technologies (and prices) that have enabled households to remove their residences from the areas of their workplaces, it may be that some reduction in such differentiation — or at least some modification of it — is a necessary condition for a greater localisation of economic activity and for reduced dependence on private transport.

We will return to this question of social differentiation later in the paper.

Suburbanisation of jobs

Job opportunities, and particularly those for blue collar workers, are dispersing towards the perimeter of the city. Data for Sydney can illustrate these trends (Table 2) (11).

Table 2

Major Occupation Groups in each Major Ring. Sydney 1961 and 1971.

Occupation Group		*% of Metropolitan Total* CBD	*Remainder of city centre*	*Inner Ring*	*Middle Ring*	*Outer Ring*	*Metropolitan total* 1000's	*% increase*
Professional, technical	M 1961	33	16	17	23	11	55	
	1971	28	14	17	25	15	80	45.5
	F 1961	15	21	19	29	17	35	
	1971	10	15	20	30	26	54	54.3
Administrative, managerial	M 1961	30	14	17	26	12	63	
	1971	20 d	13	19	30	18	80	27.0
	F 1961	25	14	22	24	15	10	
	1971	18 d	11	21	28	22	10	0
Clerical	M 1961	55	15	11	15	5	70	
	1971	48	14	13	18	8	84	20.0
	F 1961	50	15	12	17	6	97	
	1971	36	12	15	24	13	159	63.9
Sales	M 1961	31	16	17	26	10	47	
	1971	19 d	13	19	32	17	53	12.8
	F 1961	27	8	17	32	16	30	
	1971	16 d	6	16	34	28	44	46.7
Transport, communication	M 1961	23	19	19	27	12	52	
	1971	11 d	21	20	30	18	57	9.6
	F 1961	45	17	12	17	8	5	
	1971	32	17	15	23	13	10	100.0
Craftsmen, process workers, labourers	M 1961	12	24	21	33	10	329	
	1971	7 d	18 d	19 d	37	20	344	4.6
	F 1961	14	25	20	31	10	60	
	1971	7 d	17 d	19	37	21	69	15.0
Service, sports, recreation	M 1961	30	17	19	24	10	34	
	1971	20 d	16	22	26	17	38	11.8
	F 2962	20	18	25	25	13	38	
	1971	13 d	12 d	21	30	23	49	28.9
Other, not stated	M 1961	8	12	12	16	52	25	
	1971	7	13	14	17	50	53	112.0
	F 1961	19	19	13	19	30	4	
	1971	13	15	17	22	33	20	400.0
Total	M 1961	24	19	18	27	12	674	
	1971	17 d	16 d	18	30	18	790	17.2
	F 1961	30	17	17	25	11	279	
	1971	22	13	17	29	20	415	48.7
Total	1961	26	19	18	26	12	953	
	1971	19 d	15	18	30	19	1205	26.4

Note: Spatial categories of Local Government Areas as for Table 1.
d: decline in absolute number of jobs from 1961 to 1971.

There is a strong proportional shift away from blue collar employment (transport and communication, craftsmen, process workers and labourers) towards white collar employment. Within the blue collar categories, job opportunities are moving from the city centre and inner ring to the middle ring and outer ring.

Similar shifts are observable in Melbourne and in other Australian cities. A study of manufacturing relocation in Melbourne between 1968-69 and 1971-72 found that 80% of all gross declines in manufacturing employment in the Melbourne Statistical Division occurred in the inner areas of Collingwood, Fitzroy, Melbourne, Port Melbourne, Richmond and South Melbourne, while nearly half of the increases occurred in the seven neighbouring south-eastern areas of Moorabbin, Springvale, Oakleigh, Waverley, Knox, Nunawading and Box Hill (12). The blue collar jobs are dispersing to the suburbs, but *not* to the same suburbs that the low socio-economic status households are going to.

Meanwhile white collar employment remains firmly concentrated around the old city centre, although it may be dispersing *within* the centre and to a few surrounding suburbs (to North Sydney and Surry Hills in Sydney, St Kilda Road, South Melbourne and Parkville in Melbourne).

O'Connor and Maher, observing these patterns of changes in Melbourne, have attempted to monitor the changing work-residence relationships that have accompanied a shift in employment. They concluded:

> *Throughout a band of middle and outer suburban areas, cross regional journey to work patterns are very typical and increasingly more important than the trip to the central city. This analysis shows that Melbourne has shifted towards "a fairly dispersed multi-nodal metropolis", though that needs to be qualified on two counts. First, the central city is still a major employer of labour and — probably more fundamentally — the suburban growth has been diffused among a large number of adjoining LGAs in a semi-continuous fashion which has mitigated against the development of suburban nodes.* (13)

The consequences of these shifts in Sydney and Melbourne — and by the implication in the other major cities — include the following:

(i) The dispersal of blue collar jobs to the suburbs compels increasing numbers of workers to travel by car. There is generally no effective circumferential or cross-regional public transport.

(ii) Although accessibility to employment for outer suburban residents has generally increased, for those without use of cars it has declined seriously.

(iii) White collar employment, still concentrated in the city centre and nearby areas, remains well served by the radial public transport systems.

(iv) Newly formed households are likely to be distant from public transport of any sort however, and so they are likely to be compelled to car use regardless of whether they are white collar or blue collar workers. This is of some concern: if the husband must use the family car to drive to work or even to a railway station, the wife is without use of that car and therefore isolated, or else the household must obtain and operate a second car.

3. Journeys and Preferences

In this increasingly dispersed urban structure, it is useful to look at the journeys that people make, both non-discretionary (to work, to school, to shopping) and discretionary (to a better and more preferred shopping centre than the one nearest to hand, to recreation, to visit friends). Preferences and choices are important to both groups of journeys: they underlie mode choice in the case of non-discretionary travel (i.e. the choice between car travel, bus, train, walking or whatever), and the decision whether or not to make the journey as well as mode choice in the case of discretionary travel.

The value of time: the disincentive to travel

Perhaps the easiest way to understand preferences and mode choice is to take a "micro-scale" approach, and to consider the behaviours of hypothetical "typical" individuals in a major city such as Melbourne or Sydney. There is now a substantial literature on the way that different individuals value their time savings. It is a literature particularly relating to studies of choices between transport modes (walking *versus* driving *versus* public transport . . .) and of the implicit trade-offs involved in people's choices of transport mode (14).

Although derived from only one city (Leeds) at one time (1967), the average values of time savings (or "modal disutilities") found by Quarmby are fairly representative of the sorts of conclusions from this area of work (Table 3) (15).

Table 3

Average Values of Time Savings

Time spent in vehicles (cars)	⅓ hourly earnings
Time spent in buses	⅔ hourly earnings
Time spent walking and waiting	⅔-1 hourly earnings

Assuming that something like these time values apply in a city such as Melbourne or Sydney, and that the disutility or disincentive associated above with buses applies fairly generally to the whole public transport system, estimates have been made for three "typical" individuals' valuations of their time losses relating to their journeys-to-work. The three are assumed to have net weekly incomes of $100, $200 and $300, and to drive to work taking 30 minutes door-to-door. The valuations of time losses by each have then been estimated for three alternative public transport situations that could confront him: (i) a 30 minute bus ride door-to-door . . . improbable but possible; (ii) a 45 minute bus ride with 15 minutes walking and/or waiting; and (iii) two or more buses taking 60 minutes overall, with 30 minutes walking and/or waiting. To each valuation was then added an appropriate estimate of motoring costs or public transport fares; and the resulting implicit costs are displayed in Table 4 (16).

Even though these estimates are based on assumed *average* time valuations and make no allowance for individual variations, the conclusion seems clear: if a direct door-to-door public transport system could be devised, then foreseeable increases in motoring costs *could* induce the majority of "average" individuals out of their cars, regardless of income . . . although the lower income would be the first. But if such a public transport system cannot be devised, then only the low income can really be affected. To induce the rest out of their cars by *pricing*, the price increases will need to be massive indeed.

Trends in prices of travel

It should be noted however that historic changes in travel prices have been the opposite of those foreseen as likely in the future: the prices associated with private motoring have *fallen* in real terms (i.e. relative to incomes and to other prices), while public transport fares have *risen.* Relative to the Consumer Price Index, public transport fares have risen by 50% since 1960, and motoring costs have declined by 25% (17).

The principal effect of these trends, particularly since 1945, has been to make private motoring the feasible and normal transport mode of the lower income. Within the geographical restrictions imposed by their housing markets, they have chosen their dwellings and their workplaces accordingly; and the dispersal and the structure of cities described previously reflects their choices . . . is a direct function of them, in fact. They are choices that have *compelled* them to a particular use of the automobile, and which will be examined below. If the price trends are now suddenly reversed — and foreseeable changes in energy prices seem to threaten such a reversal — then it will be a very dramatic and traumatic process.

Income and preferences: discretionary travel

With *non*-discretionary travel, the assumption is that the trips cannot be avoided, that the location of the dwelling *vis-a-vis*

Table 4

COMMUTERS, MODES OF TRAVEL, AND IMPLICIT COSTS

Individual	Income net per week	Likely occupation	Car travel (implicit cost per day)	Possible public transport alternatives (extra disincentive per day)		
			30 minutes door-to-door	1 bus, 30 minutes door-to-door	1 bus, 45 minutes with 15 minutes walking/waiting	2 or more buses, 60 minutes, with 30 minutes walking/waiting
(a)	$100	blue-collar unskilled	(a) ($1.83)	(a)1 ($0.83)	(a)2 ($2.29)	(a)3 ($3.75)
(b)	$200	blue-collar skilled, white-collar	(b) ($2.92)	(b)1 ($1.42)	(b)2 ($4.33)	(b)3 ($7.25)
(c)	$300	white-collar, professional	(c) ($4.00)	(c)1 ($2.00)	(c)2 ($6.38)	(c)3 (10.75)

Table 5

EXPENDITURE ON COMMODITY GROUPS AS A PERCENTAGE OF TOTAL EXPENDITURE, BY WEEKLY HOUSEHOLD INCOME GROUP, 1974-1975

	Weekly household income						
Commodity or service	Under $80	$80 and under $140	$140 and under $200	$200 and under $260	$260 and under $340	$340 or more	*All households*
average income	$47.36	$114.31	$168.72	$228.34	$293.77	$461.20	$205.92
average expenditure	$61.69	$113.97	$143.68	$169.94	$203.20	$285.66	$157.00
	%	%	%	%	%	%	%
Current housing costs	16.0	16.2	15.8	15.4	13.7	11.7	14.5
Fuel and power	4.0	2.7	2.4	2.1	2.1	1.7	2.3
Food	26.0	22.4	21.6	20.5	19.6	18.2	20.6
Alcohol and tobacco	5.1	5.9	5.7	6.0	6.0	6.9	5.9
Clothing and footwear	8.0	7.9	8.2	8.6	8.9	10.4	8.8
Household equipment and operation	10.0	9.0	9.5	8.8	9.3	9.7	9.3
Medical care and health expenses	3.6	4.2	4.0	3.8	3.6	3.2	3.7
Transport and communication	12.5	15.9	16.3	17.7	17.9	16.8	16.7
(Transport)	10.1	14.5	15.1	16.5	16.8	15.8	15.5
Recreation and education	5.8	7.5	7.8	8.2	9.4	11.3	8.8
Miscellaneous goods and services	9.0	8.3	8.6	9.0	9.6	10.9	9.4

Note: Maximum expenditures for each commodity are underlined.

Table 6

EXPENDITURE ON TRANSPORT AND COMMUNICATION AS A PERCENTAGE OF TOTAL EXPENDITURE, BY WEEKLY HOUSEHOLD INCOME GROUP, 1974-1975

	Weekly household income						
Commodity group	Under $80	$80 and under $140	$140 and under $200	$200 and under $260	$260 and under $340	$340 or more	*All households*
Average expenditure: Transport and comm.	$7.70	$18.17	$23.43	$29.97	$36.33	$48.17	$26.26
Transport	$6.21	$16.43	$21.58	$28.01	$34.00	$45.15	$24.27
	%	%	%	%	%	%	%
Transport and communication	12.48	15.92	16.35	17.68	17.90	16.82	16.73
Travel Independent	4.65	6.17	6.58	7.03	7.14	6.85	6.67
Car purchase	2.67	3.93	4.46	4.81	4.90	4.70	4.46
Other vehicle purchase	0.39	0.50	0.40	0.36	0.30	0.40	0.36
Motor cycle	0.05	0.23	0.08	0.15	0.16	0.15	0.14
Caravan	0.26	0.17	0.10	0.12	0.09	0.18	0.13
Trailer	0.03	0.00	0.01	0.02	0.01	0.04	0.02
Bicycle	0.05	0.10	0.06	0.08	0.03	0.04	0.06
Vehicle registration and insurance.	1.57	1.74	1.88	1.86	1.94	1.75	1.82
Travel Dependent	4.10	6.30	7.03	7.67	7.95	7.24	7.14
Petrol	2.09	2.90	3.20	3.38	3.22	2.99	3.09
Other running expenses	1.99	3.40	3.83	4.29	4.73	4.25	4.05
Public Transport	1.34	2.01	1.45	1.71	2.28	1.68	1.66
Rail fares	0.18	0.38	0.39	0.47	0.53	0.41	0.43
Bus/tram	0.60	0.70	0.52	0.58	0.59	0.47	0.56
Other public transport and freight	0.57	0.93	0.54	0.66	0.54	0.80	0.67
Transport	10.08	14.48	15.06	16.53	16.75	15.77	15.47
Postal/Telephone	2.42	1.45	1.29	1.15	1.15	1.05	1.27

Note: Maximum expenditures for each commodity are underlined.

that of jobs, schools, shopping, health services etc. determines what those trips will be, and that the costs of making them must simply be borne. With discretionary travel, it is assumed that preferences are more important; the physical constraints of location and distance are still significant, but also significant are constraints of income and the individual's preferences for the expenditure of both his money and his time.

The ways that different income groups allocate their expenditure can be explored through Household Expedition Survey data published by the Australian Bureau of Statistics. The following comments are based principally on results of the 1974-75 survey which was restricted to State capital cities. (The 1975-76 survey was extended to cover other regions, and although data are available for observing capital city expenditures, the sample size is smaller.)

Housing costs fall in relation to rising income, as do expenditures on other "necessities": fuel and power, food, medical care and health expenses. But transport, like alcohol and tobacco, clothing and footwear, and recreation and education, is a "luxury" to which spare or discretionary income is directed (Table 5) (18).

If this expenditure on transport is disaggregated into its components, the dependence of discretionary travel on the automobile becomes clearer: virtually 90% of all transport costs incurred by households are related to private transport; buses, the one public transport mode that is not entirely radial, are relatively more important for those on lower incomes, as are motorcycles and bikes; car use is relatively more important for the upper income group (Table 6) (19).

Morris and Wigan have calculated income elasticities of demand for these expenditure items (i.e. the rates at which expenditure appears to rise relative to rises in income, so that an income elasticity of 1.40 would signify that, for an increase in income of 10%, expenditure on that item would increase by 14%). These elasticities are listed in Table 7 (20).

Table 7

Estimated Income Elasticities of Demand for Commodities 1974-75

Expenditure category	*Income elasticity*
Current housing costs	0.81
Fuel and power	0.45
Food	0.77
Alcohol and tobacco	1.11
Clothing and footwear	1.16
Household equipment and operation	0.98
Medical care and health expenses	0.93
Recreation and education	1.42
Transport and communication (excluding holiday transport).	1.22
Holiday transport	1.25
Miscellaneous goods and services other than holiday transport.	1.32
TRANSPORT AND COMMUNICATION	
Travel Independent	1.28
Car purchase	1.40
Other vehicle purchase	0.92
Vehicle registration and insurance	1.10
Travel Dependent	1.40
Petrol (including holiday petrol)	1.25
Other running expenses	1.54
Non-Holiday Public Transport	1.11
Rail fares	1.61
Bus and tram fares	0.85
Other public transport and freight	1.08
Holiday fares	1.26
Total Public Transport	1.20
Total Transport	1.31
Communication	0.45

It is interesting to observe that fuel and power is the least income elastic of all commodity groups, while recreation and education is the most elastic. Most components of transport expenditure are also very income elastic, and require some comment.

- Expenditure on bus and tram fares is relatively *in*elastic, and this may be the one substantial element of transport expenditure that is a "necessity".
- Expenditure on rail fares is revealed to be particularly income elastic. However as Morris and Wigan suggest, "This reflects the importance of rail transport for the journey to work, especially to the CBD. The CBD draws workers from all sectors of the city, but the largest number by far are white collar workers, including a greater proportion on high incomes" (21).
- Also particularly elastic is expenditure on the purchase and running of cars. Here the explanation may be a genuine elasticity: people on higher incomes buy more expensive cars that use more petrol, and they may use them more.

It is possible that increased bus and tram fares would lead to little reduction in use, but would be socially regressive (in affecting the lower income inordinately); that increased train fares might also have very little effect on use, and particularly on the journey to work, but might be socially progressive; and that increased motoring costs could lead to reduced use in the medium term, but would be socially regressive in view of some blue-collar and other lower-income dependence on private motoring for the journey to work.

It should be noted, however, that low income may restrict travel *generally* rather than force people from one mode to another. Data from the Sydney Area Transportation Study (SATS) reveal that, with rising incomes, the proportion of travel by various modes (car, bus or train) does not vary substantially (22). *Amount* of travel has certainly risen with income; but *mode* seems to be determined more by urban structure (and presumably by age and health of individuals).

A few points should be made here on the high elasticity for education and recreation. First, households on high incomes clearly direct comparatively large proportions of their incomes to education, principally to the higher fees of "better"schools. There is also strong evidence that they seek those suburbs where such schools are seen to be concentrated — Melbourne's Hawthorn, Kew, Camberwell, for instance — as access to children's education opportunities seems generally to be far more significant to households than is access to parents' employment. Secondly, the increasing expenditure on recreation with rising income levels should be seen in relation to the similarly increasing expenditure on private motoring and on holiday fares. Recreation is particularly dependent on transport, and probably accounts for the principal component of discretionary travel. It seems reasonable to suggest that restrictions on travel, either through pricing or through other arrangements, will especially hit recreation and leisure, but mainly (or at least initially) for the lower income.

Restrictions on transport and lower real incomes will be reflected in greater social segregation in urban areas, reduced education opportunities for the lower income, and reduced opportunities for the enjoyment of leisure in lower income groups.

4. *Where People Live: Locations and Preferences*

It is useful to try to relate the various income groups discussed above, and the inferable differences in their behaviour, to the urban structure discussed previously. The question to be asked is: what are the most fruitful directions of reform for energy conservation.

Insofar as various population groups segregate themselves (or are segregated by housing markets), their concentrations can be explored by means of factor analysis of census data, as described earlier. Such an exploration of Melbourne at Census Collectors District (CCD) scale has indicated that, in that city, the population seems to be differentiated principally in terms of four dimensions, to which meanings can be inferred as follows (23):

(i) *Educational and professional status,* where high scores are associated with high levels of education, university and other tertiary graduates, concentrations of doctors, dentists, lawyers, workers in the finance industry, employers and the self-employed, and concentrations of larger dwellings. The dimension was equated with *command of resources*.

(ii) *Life-style orientation: material possessions,* suburbia, where high positive scores are associated with families with children, in one-family dwellings, detached houses and an absence of flats, high levels of home ownership, high vehicle ownership, and televisions. Negative scores are associated with the overseas born and particularly with more recently arrived migrants, and with "disrupted" households of the permanently separated and divorced.

(iii) *Life cycle: young households,* with pre-school and school children, industrial employment, working women (also often in manufacturing industry). Negative scores are associated with an aged, widowed population, one-person dwellings, and an absence of vehicles.

(iv) *Ethnic background: Southern European,* with few long-established migrants, uneducated population, women employed in manufacturing industry, multiple-family dwellings, and apparently limited use of private vehicles.

CCDs have been grouped in terms of three classifications: (a) command of resources (dimension (i) above, with CCDs categorised in equal thirds . . . high third, middle third, low third); (b) stage in life cycle (dimension (iii) above, again with CCDs categorised in equal thirds); and (c) access to public transport. For this third dimension, road distances were measured from maps, from the mid-point of each block on each residential street, to the nearest railway station or tram stop. From these distances the average for each CCD was calculated, and CCDs then categorised in thirds . . . closest third (less than 0.55 km, i.e. within immediate walking distance), middle third (0.55 km to 1.35 km, i,e. within normal walking distance), and most distant third (greater than 1.35 km, i.e. beyond normal walking distance).

The categories that result from this classification, together with the numbers of CCDs (out of a total sample of 318 CCDs) in each category, are displayed in Table 8.

Table 8

Classification of Sample CCDs. Melbourne 1971
(Total Sample: 318)

(a) *Command of resources*		(b) *Stage in life cycle*		(c) *Access to public transport*	
Low	(106)	Late *(Map 1)*	(20)	distant	(2)
				medium	(5)
				close	(13)
		Middle *(Map 2)*	(39)	distant	(13)
				medium	(14)
				close	(12)
		Early *(Map 3)*	(47)	distant	(29)
				medium	(11)
				close	(7)
Medium	(106)	Late *(Map 4)*	(33)	distant	(9)
				medium	(9)
				close	(15)
		Middle *(Map 5)*	(43)	distant	(3)
				medium	(23)
				close	(17)
		Early *(Map 6)*	(30)	distant	(20)
				medium	(9)
				close	(1)
High	(106)	Late *(Map 7)*	(53)	distant	(3)
				medium	(15)
				close	(35)
		Middle *(Unmapped)*	(24)	distant	(8)
				medium	(9)
				close	(7)
		Early *(Unmapped)*	(29)	distant	(22)
				medium	(5)
				close	(2)

While making no claims of correspondence between *command of resources* as measured above, *income groups* described in relation to time savings and transport mode choice (Table 4), and *income groups* used in classifying income and expenditure patterns (Tables 5 and 6), nevertheless some comparisons are possible, and useful for the further questions that they raise. Observations on such comparisons follow.

Low command of resources, late in the life cycle (Map 1)

This has one of the clearest spatial patterns of any of the categories of Table 8: insofar as the social group is concentrated in Melbourne, it tends to occupy a zone of inner suburbs fairly evenly ranged around the city centre. South Melbourne, Northcote, Brunswick and Footscray are particularly represented. Access to public transport is generally very good, as these areas tend to fall within the cordon of Melbourne's better public transport provision . . . within the area served by trams, and where the railway network is still relatively dense.

The remainder of this category is represented by just a few outer CCDs in Flinders, Hastings and the market gardens area of Moorabbin.

(The other categories late in the life cycle — with *medium* and *high* command of resources — generally follow the same pattern: inner and middle distance suburbs with good public transport provisions, together with some far distant "retirement" areas in the hills, etc.)

Those who are not retired or otherwise unemployed very likely come into one of a small number of categories.

(i) They may still follow the 19th century pattern: live near their work, in an inner suburb, and either walk to work or take the tram or bus. (Transport options (a), (a)1 or (a)2 of Table 4 are most likely to apply.)

(ii) Their jobs are frequently dispersed to an outer suburb as their firm "decentralises". They may commute out to it, "backloading" on the public transport system (taking up options (a)2 or (a)3), or more likely being forced to drive (option (a)).
Alternatively they may elect not to follow their old jobs but to seek another in the dwindling unskilled labour market of the inner city. If they succeed in finding such a job, then (i) applies; if they do not, they become part of the structural unemployment and forced early retirement among the unskilled.

(iii) They may of course live away from these areas of concentration, dispersed in areas that are more predominantly characterised by other social groups. In that case they are most likely in the middle-distance and outer suburbs.

Only for some of those in (i) and (ii) is there likely to be a feasible public transport alternative to current private car use. For most however, any increase in motoring costs will merely have the effect of inordinately transfering income away from them.

Low command of resources, middle of the life cycle (Map 2)

Here there is some overlap with the previous category. Additionally however there is a sprinkling of more distant areas: in Sunshine, Keilor, Whittlesea, Waverley (Chadstone), Dandenong and Frankston. In these, access to public transport is likely to be poor (though not as poor as for comparable areas that are *early* in the life cycle, as indicated in Map 3).

Beyond these are some areas in the outer east and south: Lillydale, Sherbrooke, Cranbourne and Mornington.

(The other categories in the middle of the life cycle, but with *medium* and *high* command of resources, tend to follow a similar pattern: mainly along the major railway routes, and within walking distance of the stations).

Again transport categories can be suggested.

(iv) They may live in inner suburbs, following a pattern very similar to that of (i) above.

(v) If in the middle distance suburbs, their work is likely to

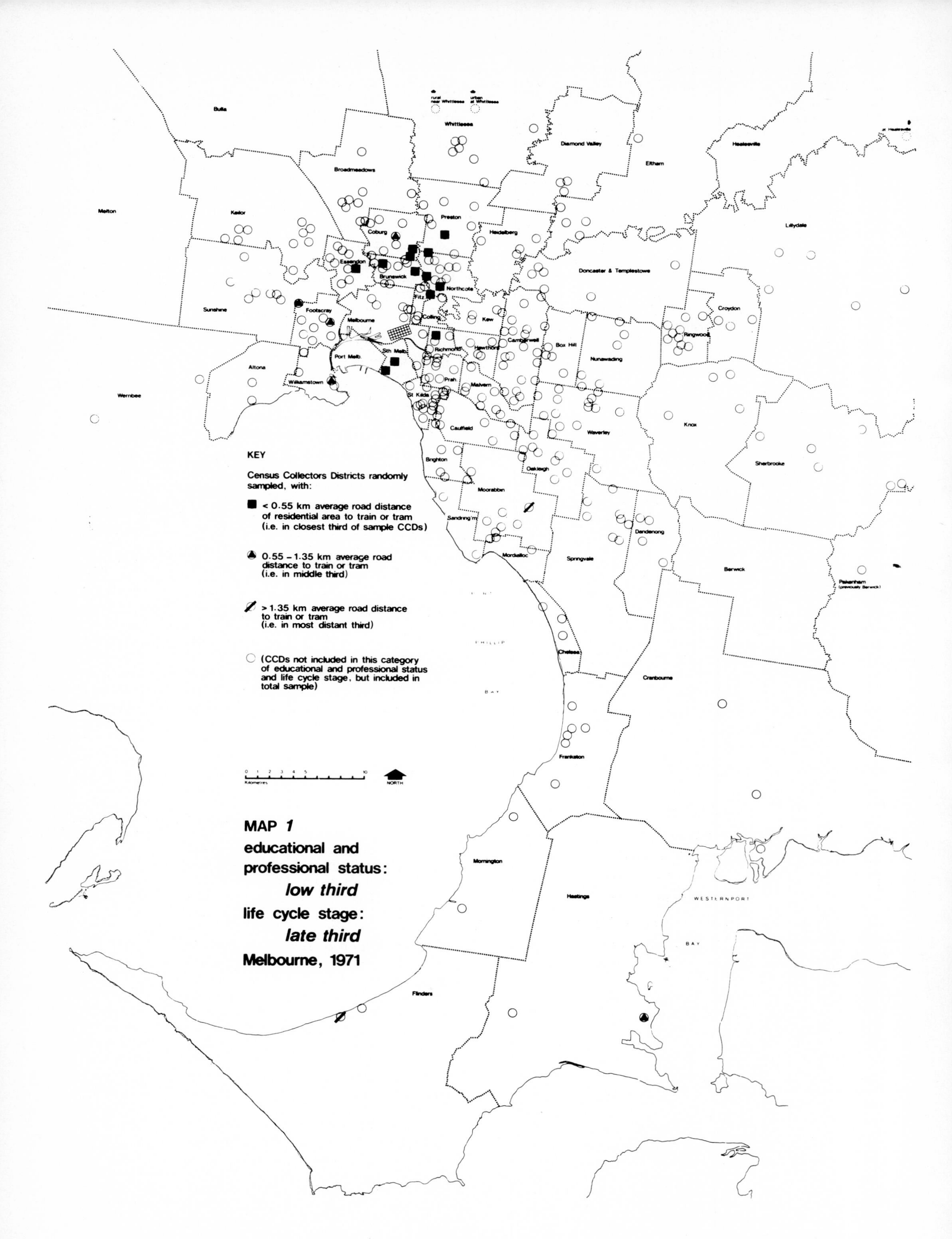
KEY
Census Collectors Districts randomly sampled, with:
< 0.55 km average road distance of residential area to train or tram (i.e. in closest third of sample CCDs)
0.55 – 1.35 km average road distance to train or tram (i.e. in middle third)
> 1.35 km average road distance to train or tram (i.e. in most distant third)
(CCDs not included in this category of educational and professional status and life cycle stage, but included in total sample)
Kilometres
NORTH
MAP 1
educational and professional status: low third
life cycle stage: late third
Melbourne, 1971
Bulla
Whittlesea
Broadmeadows
Diamond Valley
Eltham
Healesville
Melton
Keilor
Preston
Coburg
Heidelberg
Lillydale
Essendon
Brunswick
Northcote
Doncaster & Templestowe
Sunshine
Footscray
Fitz
Melbourne
Colling
Kew
Croydon
Ringwood
Camberwell
Box Hill
Nunawading
Richmond
Hawthorn
Port Melb.
Sth Melb.
Altona
Williamstown
Prah.
Malvern
Wernbee
St Kilda
Caulfield
Waverley
Knox
Brighton
Oakleigh
Sherbrooke
Moorabbin
Sandring'm
Dandenong
Mordialloc
Springvale
Berwick
Chelsea
Cranbourne
Frankston
Mornington
Hastings
Flinders
WESTERNPORT
BAY

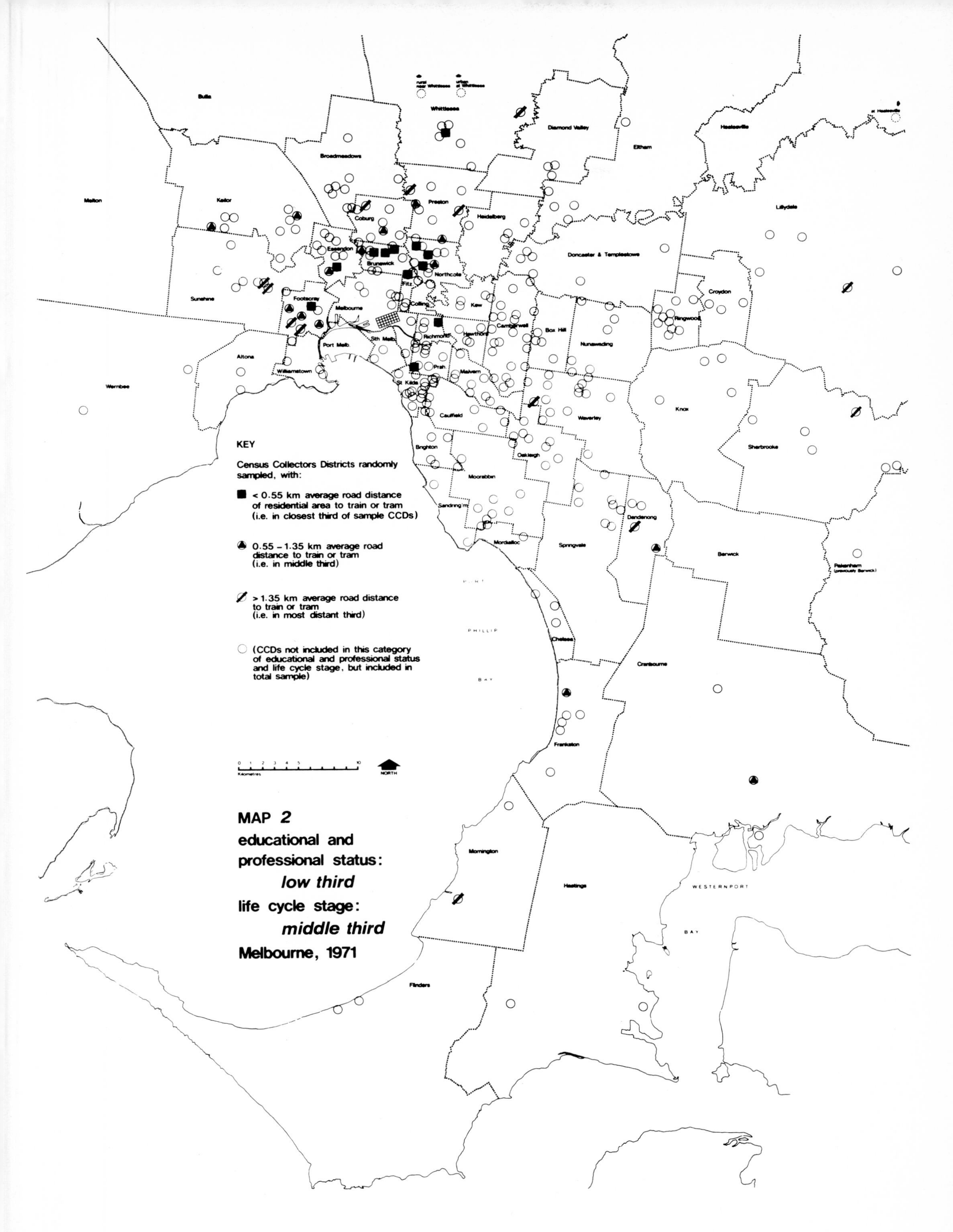

MAP 2
educational and professional status: *low third*
life cycle stage: *middle third*
Melbourne, 1971

be fairly dispersed, and a substantial proportion will be forced to use private cars. (Transport options (a), (a)1, (a)2 and (a)3 are all likely to apply to different individuals.)

(vi) Those in outer suburbs are generally distant from public transport. Their work will also generally be in an outer suburb, and generally distant from where they live; there is rarely any feasible alternative to driving (option (a)). Even if there is a "dispersed" or "circumferential" public transport option, it is likely to be by multiple buses (option (a)3 or worse).

Some of those in (iv) and (v) who currently drive to work will have a feasible public transport option. Most however will not.

Low command of resources, early in the life cycle (Map 3)

The pattern here is in some contrast to that of the immediately preceding classification (*middle* of the life cycle). Insofar as this social group is concentrated, it tends to be found in the middle and outer north and west: Heidelberg, Preston, Whittlesea, Broadmeadows, Keilor, Sunshine, Footscray, Williamstown, Altona and Werribee. Representative CCDs are generally distant from trains and trams.

Another concentration is in the south east, around Oakleigh, Springvale and Dandenong; access to public transport may be marginally better than for the group in the north west.

Other representative CCDs are to be found in the outer east and south: Sherbrooke, Cranbourne, Chelsea, Frankston. They are very distant from public transport.

Breaking this pattern of middle and outer suburbs is an important group around the city centre, in City of Melbourne, Fitzroy, Collingwood and Richmond. These areas have good access to public transport, and generally have resulted from Victorian Housing Commission activity.

(The categories early in the life cycle with *medium* and *high* command of resources are even more clearly concentrated in the corridors between the radial transport routes, and distant from those routes. They are generally "infill" areas between the earlier corridors that developed *along* the radial routes.)

Likely transport categories appear to be as follows:

(vii) Those in inner suburbs usually have very good access to public transport and to a variety of job opportunities; the transport options of (i) above would generally apply. It is significant however that these households are usually in flats, and so are paying for their lower access costs in the currency of certain personal costs imposed by characteristics of their dwelling type.

(viii) If they live in the suburbs — and they generally do — it is most likely in the outer suburbs and it is most likely distant from radial public transport, given the structure of housing markets and the geographic distribution of the lowest income group. Their work is also likely to be dispersed, and they will face the same transport options as (vi) above.

The majority will have no feasible alternative to private motoring for the journey to work.

There is a worrying aspect to this pattern. Whereas of the category *low* command of resources and *late* in the life cycle, only two representatives CCDs (0.6% of the sample) are distant from public transport, for the categories in the *middle* and *late* in the life cycle there are 13 and 29 distant CCDs (4.1% and 9.1%) respectively. As the population ages, more and more people will therefore be isolated by the coincidence of increasing physical disabilities, rising costs of private motoring, and a normal reluctance to drive on the part of the aged. The more affluent will be able to compete for more convenient locations — some of those represented in Map 2, for example — but the lower income will be left to their isolation. If the distribution of the population generally is somehow reflected in these distributions of its concentrations, then the problem of the isolated aged is today easily resolved . . . subsidised taxis would do the trick. But in two or three decades time, when an imputed 9.1% of the population may be old and poor and isolated, such solutions may not be feasible.

Medium command of resources, late in the life cycle (Map 4)

As with the comparable life cycle group with *low* command of resources (Map 1), this group tends to concentrate in some inner and middle suburbs, though more to the south-east than to the north-east: St Kilda, Caulfield, Brighton, Moorabbin, but also Richmond, Hawthorn, Malvern and Camberwell. Less typically there are representatives CCDs to the north (in City of Melbourne and Northcote) and in Essendon. Generally access to public transport is good, although there are some areas beyond the cordon of Melbourne's good transport provision (in Moorabbin and Essendon).

In sharp contrast to the pattern of this zone, and to the even greater concentration of the group with *low* command of resources and late in the life cycle (Map 1), are the CCDs at some considerable distance from the city: in Lillydale, Sherbrooke, Cranbourne, Frankston. Some such areas may have reasonable access to public transport because they are part of older settlements around railway stations (e.g. Cranbourne or Frankston), but many would be very distant indeed.

Those who are in the workforce very likely come into one of a small number of categories.

(ix) Those in inner and middle suburbs, if in blue-collar employment, will most likely work in other middle or inner suburbs. They will probably drive to work (option (b)), although a train-and-bus or a bus-and-train-and-bus option may be available (option (b)2 or (b)3).

(x) Some will travel out to decentralised jobs, possibly "backloading" on the public transport system (options (b)2 or (b)3), but more likely being forced to drive (option (b)).

(xi) If in white-collar employment, their jobs will most likely be in the city centre or an inner suburb. If the former, they will probably use public transport (options (b)2 or (b)3 or their equivalents) . . . the public transport system in each Australian city is, after all, basically a radial system. If the latter, they will probably drive (option (b)).

(xii) Those in far-flung outer areas — in Lillydale, Sherbrooke, Cranbourne, Frankston of the present sample — will almost certainly drive to work (option (b)).

Of those who currently drive to work, only some in (ix) and (xi) are likely to have feasible public transport options available to them.

It is important to note that those areas of this and the following category in the middle eastern and south-eastern suburbs (Caulfield, Brighton, Hawthorn, Malvern and Camberwell) are in the zone where Melbourne's better private schools are concentrated. As location preferences seem to be related more to access to schools than to access to work for the more affluent groups in the community, it is clear that these areas will be increasingly subject to "gentrification" or housing succession by better educated groups.

Medium command of resources middle of the life cycle (Map 5)

This is the only social group where the majority of representative CCDs are at *medium distance* from trains and trams. It is also relatively unconcentrated: typical areas are to be found along the public transport routes to the north, east and south east, sometimes at considerable distance from the city centre (areas at Ringwood, for example), and almost invariably within one kilometre of railway stations.

The work and transport options will most likely be those of (ix) to (xii) above. As this is the social group principally concentrated along the radial public transport routes but beyond the cordon of the trams and *dense* rail network, public transport options are likely to be available if workplace and residence are close to the same radial link.

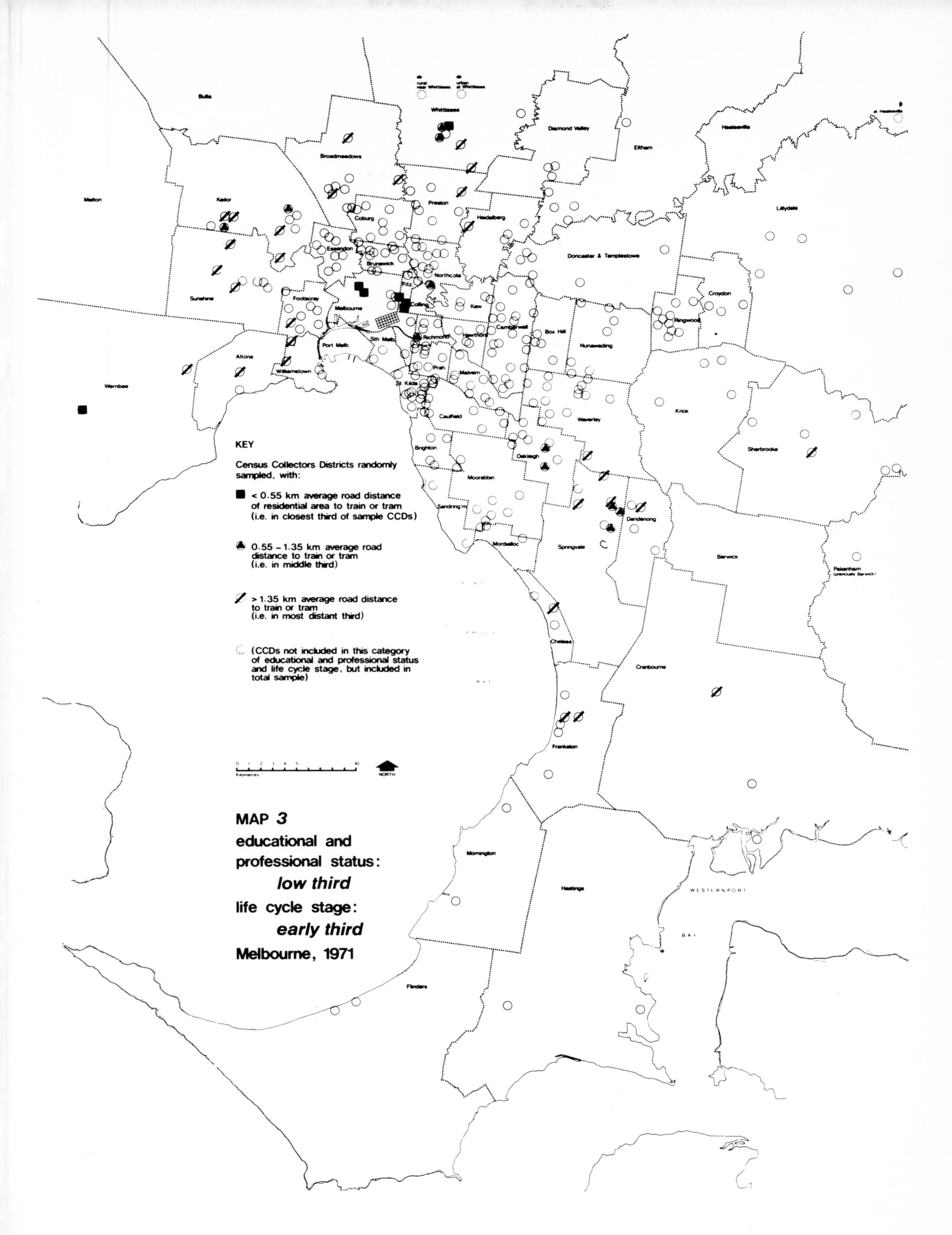
KEY
Census Collectors Districts randomly sampled, with:
< 0.55 km average road distance of residential area to train or tram (i.e. in closest third of sample CCDs)
0.55 – 1.35 km average road distance to train or tram (i.e. in middle third)
> 1.35 km average road distance to train or tram (i.e. in most distant third)
(CCDs not included in this category of educational and professional status and life cycle stage, but included in total sample)
MAP 3
educational and professional status:
low third
life cycle stage:
early third
Melbourne, 1971
NORTH
Kilometres
Bulla
Whittlesea
Diamond Valley
Eltham
Healesville
Broadmeadows
Melton
Keilor
Preston
Coburg
Heidelberg
Lillydale
Essendon
Brunswick
Doncaster & Templestowe
Northcote
Sunshine
Footscray
Melbourne
Fitz
Colling
Kew
Croydon
Ringwood
Camberwell
Box Hill
Nunawading
Hawthorn
Richmond
Port Melb
Sth Melb
Altona
Williamstown
Prah
St Kilda
Malvern
Werribee
Knox
Caulfield
Waverley
Brighton
Oakleigh
Sherbrooke
Moorabbin
Sandringham
Dandenong
Mordialloc
Springvale
Berwick
Pakenham (previously Berwick)
Chelsea
Cranbourne
Frankston
Mornington
Hastings
WESTERNPORT
BAY
Flinders

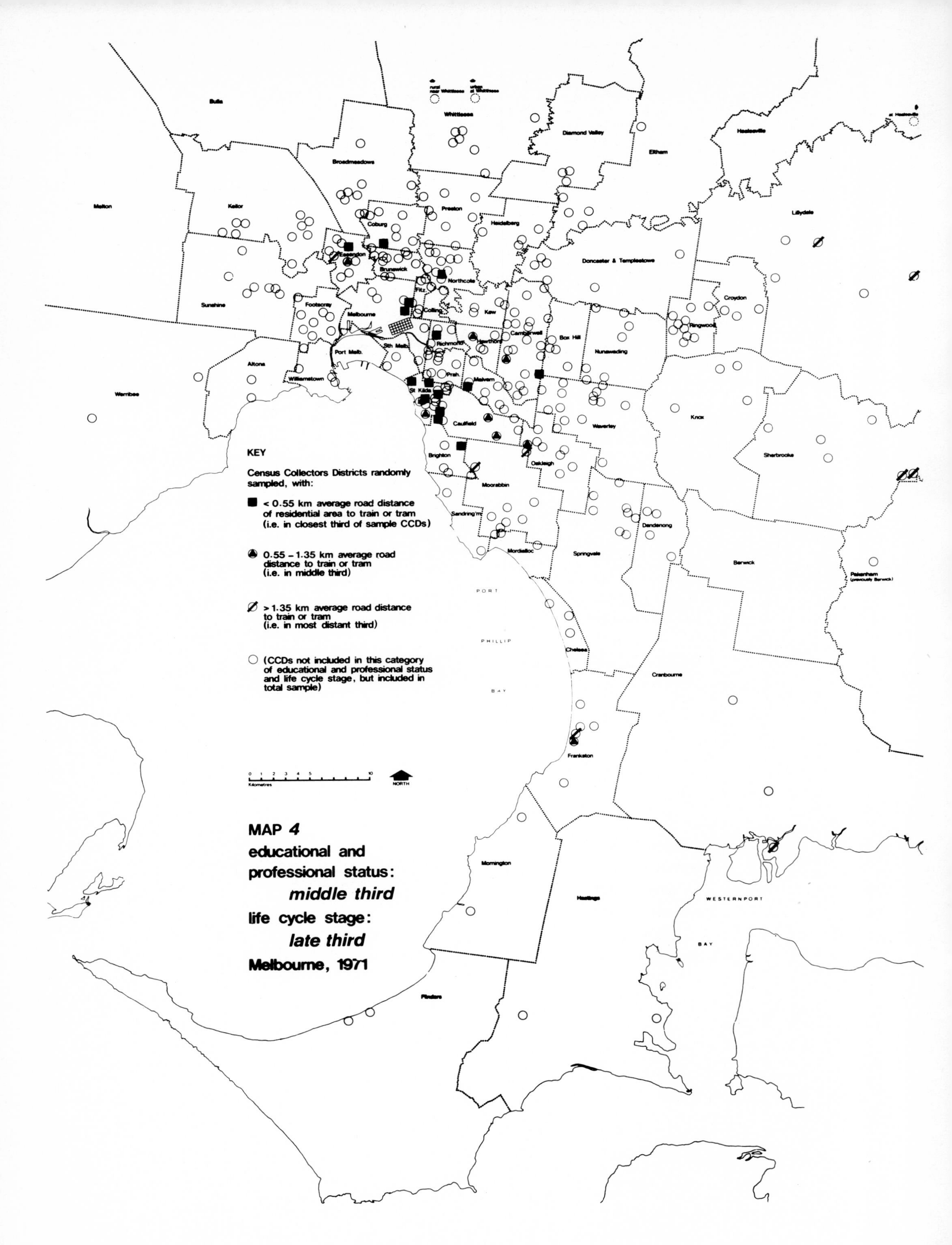

MAP 4
educational and professional status: *middle third*
life cycle stage: *late third*
Melbourne, 1971

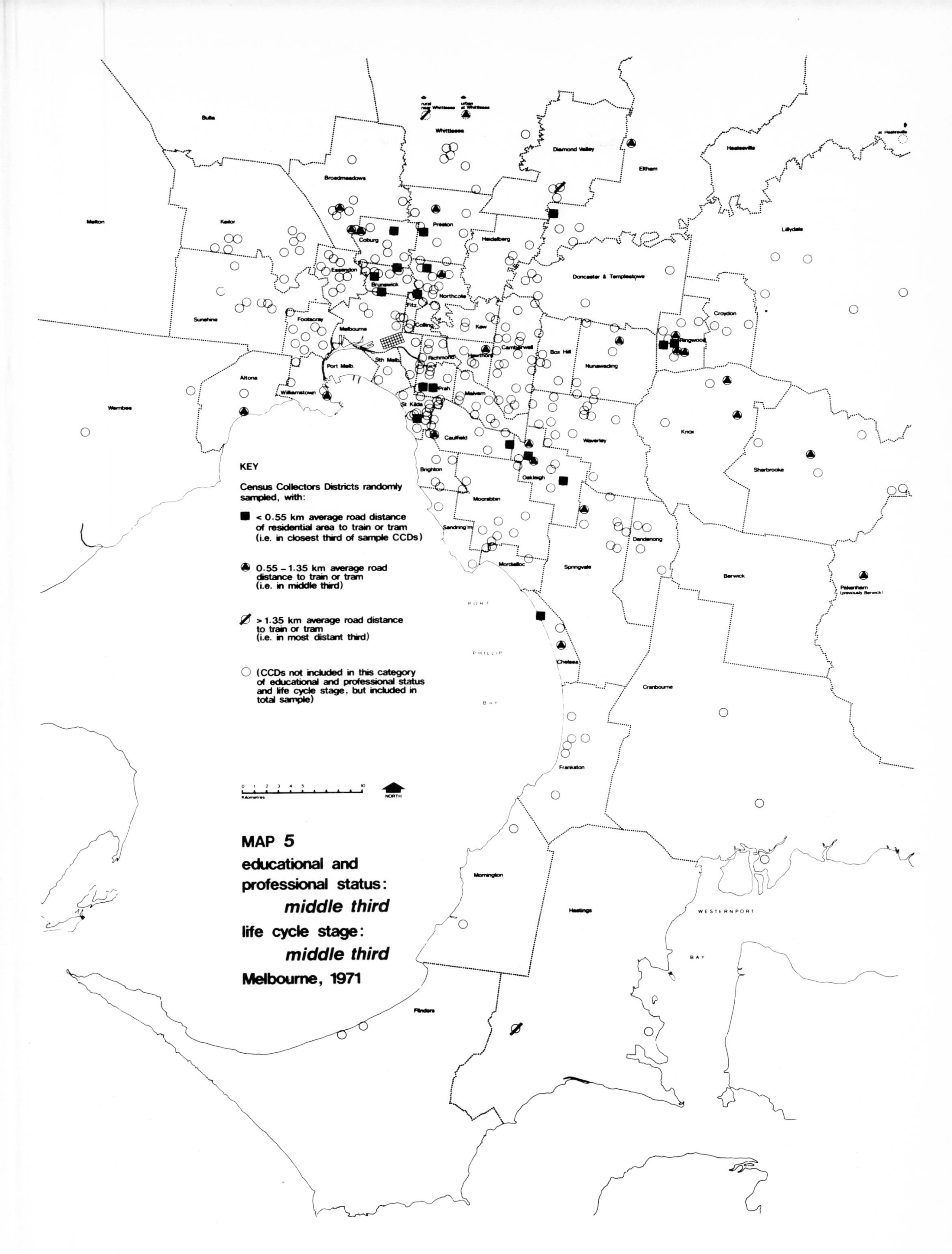

Bulla
rural near Whittlesea
urban at Whittlesea
Whittlesea
Diamond Valley
Eltham
Healesville
at Healesville
Broadmeadows
Melton
Keilor
Preston
Coburg
Heidelberg
Lilydale
Essendon
Brunswick
Northcote
Doncaster & Templestowe
Fitz.
Croydon
Sunshine
Footscray
Melbourne
Colling.
Kew
Ringwood
Camberwell
Box Hill
Hawthorn
Richmond
Sth Melb.
Port Melb.
Nunawading
Altona
Williamstown
Prah.
Malvern
Werribee
St Kilda
Knox
Caulfield
Waverley
Sherbrooke
Brighton
Oakleigh
Moorabbin
Sandring'm
Dandenong
Mordialloc
Springvale
Berwick
Pakenham
(previously Berwick)
PORT
PHILLIP
BAY
Chelsea
Cranbourne
Frankston
Mornington
Hastings
WESTERNPORT
BAY
Flinders
KEY
Census Collectors Districts randomly sampled, with:
< 0.55 km average road distance of residential area to train or tram (i.e. in closest third of sample CCDs)
0.55 – 1.35 km average road distance to train or tram (i.e. in middle third)
> 1.35 km average road distance to train or tram (i.e. in most distant third)
(CCDs not included in this category of educational and professional status and life cycle stage, but included in total sample)
0 1 2 3 4 5 10
Kilometres
NORTH
MAP 5
educational and professional status: *middle third*
life cycle stage: *middle third*
Melbourne, 1971

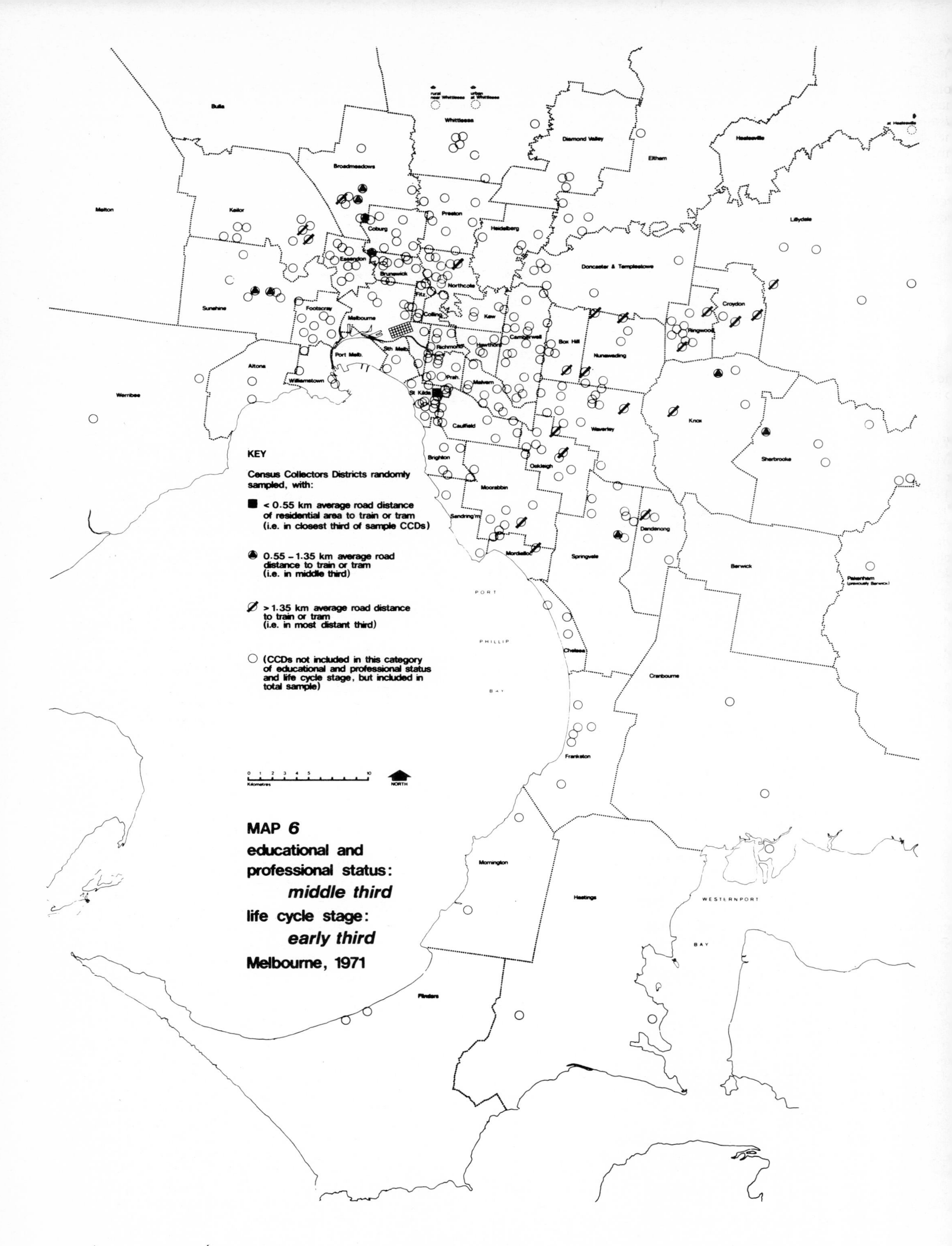
KEY
Census Collectors Districts randomly sampled, with:
< 0.55 km average road distance of residential area to train or tram (i.e. in closest third of sample CCDs)
0.55 – 1.35 km average road distance to train or tram (i.e. in middle third)
> 1.35 km average road distance to train or tram (i.e. in most distant third)
(CCDs not included in this category of educational and professional status and life cycle stage, but included in total sample)
0 1 2 3 4 5 10
Kilometres
NORTH
MAP 6
educational and professional status:
middle third
life cycle stage:
early third
Melbourne, 1971
PORT
PHILLIP
BAY
WESTERNPORT
BAY
Bulla
Whittlesea
Diamond Valley
Eltham
Healesville
Broadmeadows
Melton
Keilor
Preston
Heidelberg
Lillydale
Coburg
Essendon
Brunswick
Northcote
Doncaster & Templestowe
Sunshine
Footscray
Fitz
Collingwood
Kew
Croydon
Melbourne
Camberwell
Box Hill
Ringwood
Sth Melb
Richmond
Hawthorn
Port Melb.
Nunawading
Altona
Williamstown
Prahran
Malvern
Werribee
St Kilda
Knox
Caulfield
Waverley
Brighton
Oakleigh
Sherbrooke
Moorabbin
Sandringham
Dandenong
Mordialloc
Springvale
Berwick
Pakenham
Chelsea
Cranbourne
Frankston
Mornington
Hastings
Flinders

Medium command of resources, early in the life cycle (Map 6)

The pattern resembles that immediately preceding, for medium command of resources but *middle* of the life cycle (Map 5). The main difference is that areas that are early in the life cycle are *distant* from the railway stations and other public transport . . . they tend to be the middle-income areas that were developed during the 1960s and 1970s. Again the north, east and south-east predominate, but particularly the *outer* east.

The only groups with reasonable access to public transport would seem to be found along the railway lines to the north and west. Work and transport situations are easily summarised:

(xiii) If in blue-collar employment, it is probable that they drive to work (option (b)). The most likely public transport option is bus-and-train-and-bus ((b)3 or worse).

(xiv) If in white collar employment, the radial public transport system may be used provided that local buses are available to a railway station (option (b)2); otherwise, driving will be compelled (option (b)).

These, like other groups early in the life cycle, are mainly in the city's least convenient locations, and they are the least likely to have reasonable public transport options.

High command of resources, late in the life cycle (Map 7)

In this classification is the strongest single concentration found in the metropolitan area, namely in the middle eastern suburbs: Prahran, Malvern, Kew, Camberwell, Box Hill. It is a zone with invariably good access to a range of public transport and with the greatest concentration of Melbourne's better private schools. Beyond it are isolated representative CCDs in Nunawading and Ringwood.

Additionally there are lesser concentrations along the bayside suburbs, in St Kilda, Brighton, Sandringham and inland to Moorabbin. Here access to public transport is more varied.

Finally there is a group of representative CCDs in the City of Melbourne, and another in Essendon.

Those in the workforce will generally be professional, managerial or other white-collar workers. Their likely travel patterns can be categorised:

(xv) It is likely that they commute from their convenient suburbs to the central city or inner suburbs, possibly using the excellent public transport services available to them (options (c)1 or (c)2).

(xvi) There is a small but growing proportion of this group (and of those in the *middle* of and *early* in the life cycle) who are seeking inner suburban residences, usually to enjoy the better services of the inner suburbs and to reduce travel times to central city and inner suburb employment. They may drive to work (option (c)), or they may use the better public transport of the inner areas (adopting options (c)1 or (c)2).

This is the principal group for whom public transport is feasible for most journeys. But it is also the group least likely to be *priced* out of their cars, as explored earlier (e.g. Table 4).

High command of resources, middle of the life cycle (Unmapped)

This is a surprisingly dispersed group; representative CCDs are to be found in bayside areas (St Kilda, Brighton, Mordialloc, and further afield at Frankston and Mornington), in Moorabbin, in the mid and outer east (Camberwell, Waverley, Box Hill, Ringwood and Croydon), in some parts of the north-east (Heidelberg, Eltham), and in some parts of the mid north (Brunswick and Essendon).

This dispersal seems to suggest that the population group may itself be rather dispersed; that is, the group may not be particularly concentrated into specific CCDs and therefore not revealed by the method of analysis employed here (factor analysis of areal data).

The areas are evenly distributed in terms of access to public transport. The likely travel patterns will include (xi) and (xvi) above, but also one further:

(xvii) Those distant from radial public transport will almost certainly drive either to work (option (c)) or else to a railway station, thence taking the train to work (effectively a variant of options (c)2 or (c)3).

High command of resources, early in the life cycle (Unmapped)

As with other groups early in the life cycle, this group is generally isolated from radial public transport. It is strongly concentrated in Doncaster and Templestowe and in Waverley. Beyond is a less concentrated zone covering Nunawading, Ringwood and Lillydale. A further zone is to be found in the north east: parts of Heidelberg, Diamond Valley and Eltham.

The predominant pattern of work and travel will be (xvii) above.

5. *A Summary of the Problem*

It is certain that energy prices will rise relative to other prices and to incomes; indeed they may be vastly higher. Energy supplies may also be rationed through non-market restrictions. What is not certain is the rate at which these rises and restrictions might be applied. From the foregoing analysis and from the literature on energy use, housing use and urban structure, it seems that five inter-related problems will confront us following such rises and restrictions:

(i) Reduced consumption due to higher prices, therefore reduced production, exacerbates the falling demand for labour in communities such as Australia. And given the fundamental flaw in our society's processes for the division of labour, this reduced demand will be manifested in increased unemployment rather than in increased and equitably distributed leisure.

(ii) Because low energy prices have enabled the development of cities based on private car-dependence in virtually all income groups, we are now firmly locked in to the use of the automobile. Energy-related price rises in private motoring will therefore affect the lowest income groups most severely. Indeed, all energy price rises present serious equity problems, because a large part of the individual's energy use is non-discretionary.

(iii) Energy-related price rises in private motoring will only really affect discretionary travel — and particularly leisure trips — and then mainly for the low income. The poor may have more and more time on their hands, but comparatively few choices in how to use it.

(iv) Rising energy prices will increase the demand for convenient locations (which are generally occupied by households that are later in the life cycle), and house prices of these will rise relative to other house prices, so that the increasing vacancies occurring in those areas will tend to be taken up by more affluent households. Lower-income households will therefore tend to be allocated to less and less convenient areas of the city. Intervention to allocate convenient locations to the disadvantaged (but without putting the disadvantaged into flats or other dwelling types inappropriate to them) will prove particularly difficult.

A rationalised use of the current dwelling stock holds the key to resolving many problems associated with rising energy prices, urban congestion, quality of leisure and urban self-sufficiency. But effective tactics for rationalising housing use are generally elusive.

(v) The urban population is already highly segregated, as the previous analysis has revealed, and this segregation will increase due to the relegation of the less affluent to the least convenient locations ((iv) above). However any localisation of employment and of activities generally — therefore local self-sufficiency and a reduction in the *need* to travel — is dependent on a reduction in that segregation.

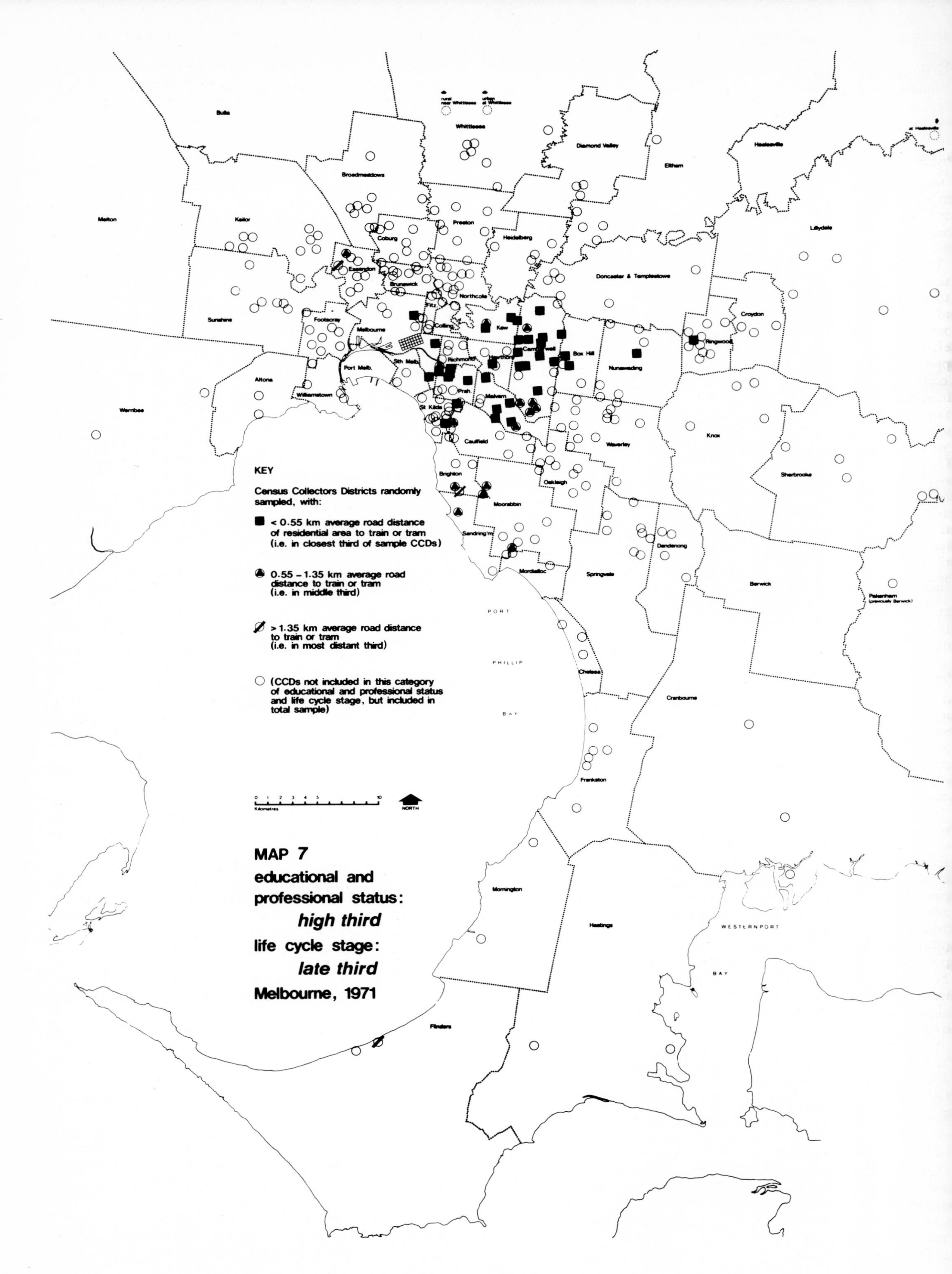

KEY
Census Collectors Districts randomly sampled, with:
< 0.55 km average road distance of residential area to train or tram (i.e. in closest third of sample CCDs)
0.55 – 1.35 km average road distance to train or tram (i.e. in middle third)
> 1.35 km average road distance to train or tram (i.e. in most distant third)
(CCDs not included in this category of educational and professional status and life cycle stage, but included in total sample)
0 1 2 3 4 5 10
Kilometres
NORTH
MAP 7
educational and
professional status:
high third
life cycle stage:
late third
Melbourne, 1971
Bulla
Whittlesea
Diamond Valley
Eltham
Healesville
Broadmeadows
Melton
Keilor
Preston
Coburg
Heidelberg
Lillydale
Essendon
Brunswick
Northcote
Doncaster & Templestowe
Sunshine
Footscray
Melbourne
Fitz.
Colling.
Kew
Camberwell
Box Hill
Croydon
Ringwood
Nunawading
Port Melb.
Sth Melb.
Richmond
Hawthorn
Altona
Williamstown
Werribee
Prah.
Malvern
St Kilda
Caulfield
Waverley
Knox
Brighton
Oakleigh
Sherbrooke
Moorabbin
Sandringham
Dandenong
Mordialloc
Springvale
Berwick
PORT
PHILLIP
BAY
Chelsea
Cranbourne
Frankston
Mornington
Hastings
WESTERNPORT
BAY
Flinders

6. *Some Paths for Reform*

'o tackle these problems, we need to do three things:

(i) Alter pricing;
(ii) Improve local services in each locality;
(iii) Relocate households.

Various aspects of these strategies are discussed following.

'ricing of energy

: seems incontrovertible that, in the medium term of say five to
:n years, all energy prices must rise — say to the replacement
osts of the fuels, where replacement is in terms of renewable
nergy sources — and that prices of transport and of other
nergy-dependent services must rise with them.

The real question for immediate policy is how prices will be
alculated, and how rises will be implemented. The question
an be illustrated by reference to the rail system. It may be
idged that the existing system has few social costs that require
ompensation, so that fares merely have to rise to remove the
eficit (thereby the subsidy from taxes) and to cover the
pportunity costs of primary fuels used to generate the
ectricity or provide the diesel oil to operate the system.
[owever it may then be judged that off-peak services cost more
o operate than do peak-hour services, and so should be priced
nore highly. Alternatively it may be considered that the overall
ost of the system would fall if peak-hour use was reduced and
ff-peak use increased, and so off-peak fares may be made
ower. Alternatively again it may be judged that the existing
ystem does indeed impose social costs through its centralising
ndency, and that *outward* movement should be encouraged in
e morning peak-hours, and *inward* in the evening peak.
mply, there is no value-free way of calculating and allocating
ail fares nor, for that matter, bus fares nor road-use charges
or motorists.

Transport pricing strategies will, in the medium term, partly
etermine house prices and land prices; and, in the longer term,
ey will partly determine the distribution of the population
nd use of the dwelling stock.

'rices of energy relative to prices of housing

s more energy conserving dwelling design and dwelling
cation will require more building inputs (in the form of
conditioning or "retrofitting" and in the form of new
onstruction) in order to reduce energy inputs for household
:tivities, reform will require different institutional constraints
n pricing of land, building and energy inputs, and hence
anges in public policy.

The *energy supply authority* could cover the capital cost of
e dwelling energy system (incremental cost of solar design
ver what is currently conventional design, insulation cost,
olar water heating system, etc), and then price its use on a basis
(i) lease of the system and (ii) energy inputs at their
pportunity cost. Alternatively, the *housing finance institution*
ould cost the dwelling in terms of life-time inputs
onstruction, maintenance, capital cost of energy provisions
oth on and off site in terms of their contribution to dwelling
peration), rather than in terms of construction inputs alone.
he housing finance institution would then provide an
surance cover for maintenance, and fund the dwelling's
share" of the capital cost of off-site energy plant (electricity
ower station, transmission lines, etc).

The effect of either of these measures, when allied with
ansport price reform to reflect replacement costs of primary
els (oil, coal etc.), would be to reduce substantially the real
ice of dwelling construction to the user, and to increase the
al price of energy inputs.

lleviation of income distribution effects of price rises

has been observed earlier that the effect of fuel price rises will
: a significant disadvantage to lower income groups, and
articularly to those who are comparatively early in the life
cle. Currently prevailing ideas are that income distribution
problems" should be resolved by income *re*distribution and
ot by subsidies attached to specific services.

In the short term however, specific subsidies may be necessary: fare concessions for the young, aged, invalid and otherwise immobile; subsidised taxis for the relatively immobile who do not have easy access to convenient public transport; etc. In the medium to long term, other forms of intervention may be necessary, for example in housing markets to retain convenient dwelling stock for the lower income.

Improved local services

The first reform must be improved local child care arrangements.

Secondly, we must concentrate on improving the *quality* of local schools. There is much evidence that people locate not so much as to limit their own journeys-to-work as to limit their children's journey-to-the-right-school. "Right" schools have to be made more ubiquitous.

The localisation of recreation

A restructuring of recreation opportunities is principally dependent on local action. Therefore it will be done well in areas of more affluent and better educated populations, but less well elsewhere. The needed changes include (i) increased and more diverse community education, addressed to all age groups, (ii) local arts and crafts clubs and workshops, (iii) improved local libraries and galleries, (iv) expanded local opportunities for passive recreation, with walking paths, bicycle tracks, etc, (v) a maintenance of present opportunities for group and team sports, and an expansion of them in areas currently poorly provided, (vi) hobby centres, particularly for the young, (vii) encouragement to community responsibility for landscape management, historical conservation, etc. These activities need to be catalysed by financial assistance and by the provision of trained personnel who can promote, teach and if necessary lead (24).

Relocation of employment

Regardless of how resources such as energy are priced, there is a case for reducing both goods movement and journeys-to-work in order to reduce the social costs of transport. This requires relocation of establishments. But how is that to be achieved?

Various approaches can be suggested: price labour and other inputs to production at their true cost; include part of individuals' travel time with work time; use differential payroll taxes. Most such methods are however open to abuse, and virtually all would be inflationary and aggravate unemployment. But there may ultimately be little alternative.

It is worth observing that with a reduced employment base but a more equitable distribution of activities and therefore of leisure (through job-sharing, for example), there would be substantial new demands for child care, recreation, opportunities in the arts . . . sectors that are *not* energy intensive, that would themselves create massive new demands for labour, but which would need to be supplied in a society of reduced personal transport. Due to generally reduced money incomes and to the higher price or reduced availability of many consumer goods (because of higher resource costs), there would be increased demands for recycled goods — reconditioned cars, furniture, household appliances, etc. — and recycling would also lead to new demands for labour. Effectively there would be a substitution of labour for energy and other inputs. All of these activities would be more localised than the activities they replace.

A low energy economy would be greatly different from the present one, with quite different sorts of activities and uses of time, which might be far more localised; but it would *not* necessarily be a society of reduced well-being . . . unless we fail to overcome the current weakness in the division of labour and the division of leisure discussed previously, whereby reduced demands for labour lead, through social competition to unemployment rather than to shared work and leisure.

Division and/or relocation of specific institutions

Institutions such as large manufacturing plants, hospitals, universities and colleges are often proud of their size, and work hard to achieve it and to maintain it. Some concentration and centralisation is inevitable, for example of highly specialised medical units, specialised research and teaching units, etc; and it is also inevitable that people will sometimes need to have access to such units from many areas of a city or region. But others can be split up and distributed through a community far more than at present . . . much tertiary education, specialised hospital clinics, and so on. Were they more evenly distributed, trip lengths and other access costs could be reduced. There may also be a case for relocating some industrial establishments over the medium to long term in order to facilitate "district heating" using waste heat from industrial processes: low-grade, waste heat can be used for hot water and space heating in buildings, for example.

The division and/or relocation of specific employment centres and institutions on to major public transport routes would confer certain benefits on users; there may also be costs incurred in the relocation itself, and due to losses in economy of scale. Conflict over these benefits and costs is to be expected, and will not be resolved simply by conventional economic analysis.

Relocation of housing

It is increasingly common to hear calls for the relocation of poorer households in the most inconvenient locations, to flats or "town houses" or "cluster housing" that is nearer to transport. I would however issue a few warnings.

(i) It is very difficult to achieve substantial increases in gross densities;
(ii) The numbers of households currently in inconvenient locations is large, as the previous analysis has made clear.
(iii) The cost of a substantial change to the dwelling stock itself will be substantial.

There is an argument for the retention of low densities in housing. The detached house on its own allotment offers a number of significant advantages: (i) it facilitates home-based leisure activities, (ii) it is more likely to permit home-based work, (iii) it can permit some home-based food production, and (iv) it is more easily adapted and altered than other dwelling types, as circumstances and demands on the dwelling change.

Lower income households appear to be more dependent on dwelling space and private open space than are upper income households: their leisure activities are generally more demanding of space (repairing automobiles or gardening takes more space than dinner parties or collecting stamps, records or books), and they are more likely to use home-based activities to supplement their incomes. With structural unemployment, reduced working hours and early retirement affecting the lower income more than the upper, the greater space needs of the lower income are likely to increase.

It may be sounder to aim for dwelling heterogeneity rather than for higher densities as such. A greater mixture of dwelling types in virtually every part of the city is to be favoured on a number of grounds: it permits a less constrained exercise of preferences, it allows households of different characteristics to locate near their work while still finding housing that is appropriate to their needs and preferences, and it permits households to change dwelling type but not their neighbourhood as they age or as circumstances otherwise alter, hence leading to a more efficient utilisation of the dwelling stock. The main arguments for retention of a high proportion of single-family houses in most areas are that (i) dwelling design must allow for increased home-based leisure activities, and probably for more home-based food production and work; and (ii) flats, while frequently suitable for households where both husband and wife work, will be far less suitable if both have increased leisure or if one ceases to be employed.

There will certainly need to be significant restructuring of the dwelling stock, and thereby an effective relocation of housing opportunities. But the greatest effects must be directed towards a relocation of households *within* the dwelling stock.

Relocation of households

The inappropriateness and inefficiency of households locations have been demonstrated previously. Households tha are early in the life cycle, in all socio-economic groups, tend t be distant from radial public transport routes and to be il served by any other public transport. If they have limited acces to private cars, they are seriously isolated and disadvantaged Rising energy costs will increase the demand for convenier locations (which are generally occupied by households that ar later in the life cycle), and house prices of these will rise relativ to other house prices, so that the increasing vacancies occurrin in those areas will tend to be taken up by more affluer households. Lower-income households will therefore tend to b allocated to less and less convenient areas of the city Intervention to allocate convenient locations to th disadvantaged (but without putting the disadvantaged into fla or other inappropriate dwelling types) will be very difficult, bu necessary.

It seems clear that households and workplaces will need t relocate to bring people closer to their employment: people wi increasingly need to live on the same transit line as the employment, or within cycling distance of it. But given th income distribution (restricting lower income households t lower priced housing) and given the long delay in househol relocating themselves in response to changing house pric and/or transport costs, this new pattern remains unlikely in th short term. For reasons of energy conservation allied wit reasons of minimising the social costs of the present patter (congestion, noise, danger, pollution and, most important c all, the increasing loss of increasingly valued leisure time), seems likely that there will be developing pressures for that ne pattern to be accepted. Housing relocation and employme relocation, it is suggested, will increasingly be advocated as way of reducing (i) energy consumption and (ii) the social cos of traffic and of automobile dependence.

There would seem to be three main consequences for urba planning in the short term:

- New housing production should be planned in relation t employment centres (which should themselves be mor diverse in the opportunities that they offer) and to trans lines serving those employment centres.
- There should already be a system of encouragements, even c incentives, for people to relocate both to new and to existin housing so as to minimise their dependence on automobile particularly for their journeys-to-work.

  Clearly there will be major problems in such a system. I all of the cities there are sectors of residential area sometimes along major transit lines or other communication arteries, but with no substantial employment base other tha at the city centre . . . there is simply no alternative t circumferential or dispersed travel to work. The neede restructuring in physical terms will be massive.
- Location of State Housing Authority dwelling stock mu particularly be determined relative to likely employme opportunities, and households should be encouraged t locate (and to relocate) to minimise their journeys-to-work.

  Because of cheap motoring, the previous practices of th State Housing Authorities in disregarding housing locatio relative to employment location have often seeme reasonable. However they are practices that can no longer b afforded in terms of community cost and in terms of th costs to low income households (25).

It is important to observe that income levels that are mo equal would also lead to reduced journeys-to-work. Where competition for "better" schools may largely account fo residential segregation, as discussed previously, competitic for "better" jobs seems to account in part for long journeys-to work. Thus Howe and O'Connor have observed the far great localisation of women's employment in Melbourne and the shorter journeys-to-work compared with those of me suggesting that the narrower wage differentials in women's jo reduce the incentive to travel further for a "better" job (26 With structural unemployment and currently widening wag differentials, competition for jobs and the incentive to travel are both increasing however.

7. Conclusion

The problem of energy, urban structure, housing location and transport is a *social* problem, not a *technical* problem. It relates to (i) effects of fuel shortages or price rises on real incomes and on lifestyles, and (ii) the social transformations involved in getting to a more equitable distribution of labour and leisure, to an improved spatial distribution of households' residences, employment and recreation opportunities, and to the ability of individuals to cope with increased leisure and reduced *material* well-being.

The income distribution effect

The principal conclusion to emerge from the foregoing is simply stated: there are many *possible* effects of an energy crisis, but the only one of which we can be certain is that the results will be socially regressive. Energy use in the home seems to be relatively inelastic, so that price rises will simply have to be borne; because we are locked in to private automobile use by the structure of our urban areas, much of our motoring is non-discretionary, particularly for lower-income households that are relatively early in the life cycle, so that increases in private motoring costs will merely erode real incomes; only discretionary or leisure travel for the lower income is likely to be significantly reduced, so that poorer households will have more and more leisure on their hands (following structural unemployment and under-employment), but fewer and fewer opportunities for its enjoyment; and the relative convenience of residential locations will be reflected in their house and land prices, so that the lower income will be increasingly forced to the least convenient locations where their isolation and transport costs will be severest.

The social transformations

It is not too difficult to imagine an alternative and better future than the very unequal one that confronts us. It might be characterised by a re-substitution of labour for energy in many production processes (with increased recycling and reuse of consumer durables, intensive urban-fringe agriculture, etc.); a more equitable sharing of work and of leisure; more local and even home-based work and recreation; a greater variety of leisure opportunities in every social group; a relocation of households and improved use of the dwelling stock, so that each household can maximise its convenience and the efficiency of its activities commensurate with a maximum convenience and efficiency for others; in the longer term a redistribution of population and of urban settlements so that people are living in a closer relationship with their food production and supply; etc. Few would disagree with the general outlines of such a future, particularly if it is also characterised by some sort of international equity in shares of resources. The problem is that we do not know how to get to it in our own community, let alone internationally. We do not know how to achieve the social transformations.

We are dealing here with conflicts of preferences . . . with the clash identified by Harsanyi between *personal preferences* (our selfish preferences related to our consumption activities and underlying social competition) and *ethical preferences* (to do with our altruism, ideals and visions of future states) (27). To suppress the former in favour of the latter requires community education to convince us of the logical consequences of our personal preferences, pricing changes to sheet home to us the real costs of the exercise of our personal preferences, and ultimately restrictions which we accept provided that they are applied uniformly.

In practical terms the issue is one of (i) assistance to lower income and other disadvantaged households who will be particularly affected by energy price rises and restrictions, but at a time of reduced economic growth when the political climate may be against such assistance; (ii) an intervention in housing and/or housing finance markets to achieve a better distribution of the dwelling stock and of the convenience that it delivers; (iii) an intervention in the decision making processes of firms and other institutions, to assist a closer proximity of population to jobs and other opportunities; (iv) job sharing; (v) voluntary energy conservation; and (vi) a willingness to support financially programs and opportunities for improved community education and for the creative use of leisure.

Notes and References

(1) The evidence for these generalisations is reviewed in King, R. J. (1978), *Review of housing research: housing location and energy*. DR 4. Melbourne, Centre for Environmental Studies, 83p.

(2) King, K. (1978). Energy for transport. *Search* **9**, 5.

(3) Australian Bureau of Statistics (1975), *Journey to work and journey to school, August 1974*. Canberra.

(4) Neutze, M. (1977), *Urban development in Australia*. Sydney, Allan and Unwin, 258p.

(5) *Ibid*, p. 26.

(6) *Ibid*, p. 27.

(7) *Ibid*, p. 31.

(8) King, R. J. (1978), *Review of housing research: studies of the housing problem and of access to housing*. PR 1. Melbourne, Centre for Environmental Studies.

(9) For a discussion of theory and review of studies, see Timms, D. W. G. (1971), *The urban mosaic*. Cambridge, Cambridge University Press. Among Australian studies are Jones, F. L. (1969), *Dimensions of urban social structure*. Canberra, ANU (looking at social differentiation and segregation in Melbourne); King, R. J. (1978), *Social differentiation as a key to assessing the social effects of transport related changes*. ARR 83. Melbourne, Australian Road Research Board (also looking at Melbourne); Stimson, R. J. (1974), The social structure of large cities, in Burnley, I. H. ed. (1974), *Urbanisation in Australia: the post-war experience*. Cambridge, Cambridge University Press, pp. 131-146 (looking at Adelaide), and Timms, D. W. G. (1971), *op cit*. (looking at Brisbane).

(10) For a more complete discussion of this theory and of its sources, see King, R. J. (1978), *op cit*.

(11) Derived from Neutze, M. (1977), *op cit*, p. 107.

(12) Cities Commission (1975), *Melbourne inner areas manufacturing industry relocation study*. Report prepared by Plant Location International. Canberra.

(13) O'Connor, K. and Maher, C. A. (1977), *Changing work-residence relationships in Melbourne, 1961-1971*. Paper presented to Section 21, 48th ANZAAS Congress, Melbourne.

(14) e.g. Beasley, M. E. (1965), The value of time spent in travelling; some new evidence. *Economica*, **32**, pp. 174-185; Quarmby, D. A. (1967), Choice of travel mode for the journey to work: some findings. *J. Transport Econ. Policy*. **1**, 3, pp. 273-314; Harrison, A. J. and Quarmby, D. A. (1969), The value of time, in Layard, R. ed. *Cost benefit analysis*. Penguin Modern Economics Readings, 1972; Beasley, M. E. (1977), Values of time, modal split and forecasting. Resource Paper for Third International Conference on Behavioural Travel Modelling, Tanunda, April.

(15) Quarmby, D. A. (1967), *op cit*.

(16) King, R. J. (1978), Energy, equity and lifestyle. in King, R. J. ed. *Energy, agriculture and the built environment: towards an integrative perspective*. Melbourne, Centre for Environmental Studies, pp. 123-148.

(17) Bureau of Transport Economics (1975), Urban passenger transport outlook. Transport Outlook Conference, July.

(18) Australian Bureau of Statistics (1977), *Household expenditure survey 1974-75, Expenditure classified by income of household*. Bulletin 4. Canberra.

(19) *Ibid*.

(20) Morris, J. and Wigan, M. R. (1978), *Transport planning: a family expenditure perspective*. ARR 71. Melbourne, Australian Road Research Board.

(21) *Ibid*, p. 24n.

(22) Commonwealth Bureau of Roads (1975), *Report on roads*. Melbourne.

(23) King, R. J. (1978), *Social impact of transport related changes: a classification of residential areas*. AIR 268-1. Melbourne, Australian Road Research Board.

(24) An application of these approaches to an outer local government area in Melbourne is to be found in King, R. J. and McGregor, A. (1977), *Youth, leisure and public policy: the case of Diamond Valley*. Melbourne, Centre for Environmental Studies.

(25) The enforced automobile dependence of this group, and the costs it imposes, are examined in particular depth in Commonwealth Bureau of Roads (1976), *An approach to developing transport improvement proposals*. Occasional Paper 2. Melbourne.

(26) Howe, A. and O'Connor, K. (1977). A working woman's place is near the home: journey to work of men and women, Melbourne 1971. Paper presented to Section 21, ANZAAS Congress, Melbourne.

(27) Harsanyi, J. C. (1955). Cardinal welfare, individualistic ethics and interpersonal comparisons of utility. *Journal of Political Economy*. August.

Designing Buildings for Minimum Energy Use

Tone Wheeler

School of Environmental Design
Canberra College of Advanced Education

During the last few years there has been a fast growing interest in energy conservation in buildings. Due in part to the energy "hiccup" in 1973, in part to the rising fuel prices and in part in reaction to the unnecessarily energy wasteful designs of recent modern architecture, this interest has spawned a plethora of books and articles, new magazines, many perfunctory conferences and the now obligatory Government inquiries. But, out of all this, very few energy conservation oriented buildings have emerged in the last few years. This is not, as one might at first presume, a result of the complexity of the technical issues involved. For although there is debate raging about design factors it is about detail; the basic issues are clear and by now well known. Further the principles of energy conservation in buildings are simple, both to establish and to incorporate into buildings if only the technical issues were to be coped with. However this has not been the case and many simple refinements in building design which could bring dramatic improvements in energy efficiency remain only on the drawing board. This lack of implementation can be traced to "non-technical" issues, three of which seem to bear particular importance: problems of economics in constructing and operating buildings, problems in town planning and zoning to allow for solar energy utilisation, and problems of acceptance by owners or users of these energy reducing features.

Before examining these three issues in detail, it is necessary to quickly review the technical issues involved, not only to establish that the principles are indeed simple and possible, but also because many of the "non-technical" issues arise from technical considerations and should be seen in that context.

Energy in Buildings — Technical Issues

Energy is consumed in buildings for a number of purposes, some of which are directly related to the building design (heating, cooling and lighting) and some which are only minimally related (hot water, cooking, power outlets, etc.). Of those related to building design, heating and cooling are the ones most highly dependent upon the building's external fabric. Although lighting consumes considerable energy in larger buildings its energy usage is not highly dependent on the building's design and how much natural light is available, but rather on the levels of lighting adopted and its distribution. In practice lighting energy can best be reduced by a reduction in lights per se and a lowering of unnecessarily high light levels, rather than by changes in the external "fabric" of the building. Indeed extra windows or roof lights will usually have greater consequences in raising heating and cooling energy use than they will in reducing the lighting energy required.

So it is the energy used for heating and cooling that is most determined by the design of the building itself and it is these areas which are primarily addressed in this paper.

1. *Domestic scale buildings* (houses, child care centres, small schools, etc.): the buildings are small in plan area and height, and have relatively low thermal requirements (wide variations in temperature and humidity can be tolerated, with occupants moving to more comfortable areas, adjusting clothing or leaving the building!).
2. *Other buildings:* Commercial, retail, educational, industrial and public meeting buildings which are usually larger and have more stringent requirements for temperature and humidity, either for the occupants who cannot adjust by altering their position or clothing, or for machinery (e.g. computers).

Opportunities to lower energy demand in both building types occur in the design of the building fabric (which modifies the "weather" to some extent) and in the mechanical devices which are used to make up the difference between this "modified weather" and the desired thermal comfort levels. It is desirable to therefore achieve as much of the building fabric (often called passive design) before energy-using mechanical devices are employed. It is only after these passive requirements are met that low energy "active systems" should be used, e.g. active solar space heating, solar air conditioning, etc.

Domestic scale buildings

In small buildings particularly houses, two design aims are important: firstly, to insulate the building from the extremes of the climate and secondly, to allow the house to benefit from times of benign weather conditions (e.g. sun on cold winter days, breezes on hot summer days).

The first aim can be met with the use of insulation on all surfaces of the house: bulk and reflective insulation in the roof space and walls, the use of concrete floors with edge insulation under timber floors, weathersealing of windows and doors to prevent cold air infiltration, reduction of unnecessary ventilation to prevent excess heat loss in winter, and good insulation on windows with tight fitting layered curtains or shutters and appropriate shading to all windows in summer. These factors, summarised in Fig. 1. are covered in greater detail elsewhere (1).

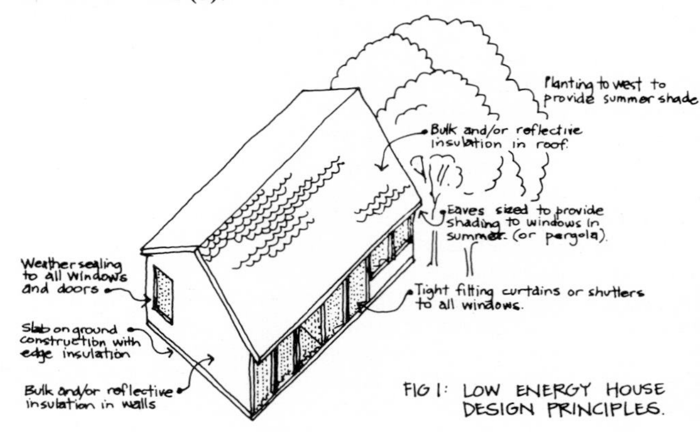

FIG 1: LOW ENERGY HOUSE DESIGN PRINCIPLES.

The second aim is met in winter by orienting windows to pick up the winter sun and by storing this "passive solar energy" in the thermal mass of the house. Studies (2) have shown that the optimum orientations for winter sunshine is north, and that as much glass as possible should be oriented between 30° E and 30° W of north, where maximum benefit can be gained. The sun's energy is stored in "thermal mass", for instance, brickwork or concrete floors. In this way the daytime sun can be held in the house to offset the cold conditions of night. For summer it is important to prevent sun penetration into the house, particularly from the west, to avoid overheating of the house. Since a cost penalty is involved in providing adjustable vertical shading needed to shade windows facing north east to south east and north west to south west, it is usually beneficial to avoid windows in these orientations. Further benefit can be gained in summer by storing night-time "coolth" in the thermal mass in order to cool the house in the day. This can be achieved by using fans or good cross ventilation at night and when the air temperature outside falls below the internal. The benefits of good cross ventilation in hot humid weather should not be forgotten. These design factors are illustrated in Fig. 2.

Diesendorf, M. (ed.) (1979). *Energy and People*. Canberra, Society for Social Responsibility in Science (A.C.T.)

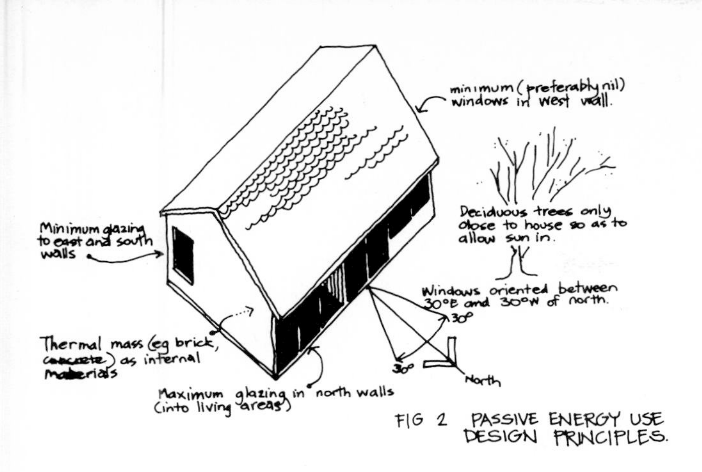

FIG 2 PASSIVE ENERGY USE DESIGN PRINCIPLES.

Active solar systems for heating or cooling of small buildings are rarely adopted in Australia, given the relatively moderate climate in most cities and the small amounts of energy required if the passive design factors outlined above are followed.

Other buildings

Like small buildings, larger buildings should be well insulated against climatic extremes and should seek to benefit from benign weather. Insulation techniques are similar to those outlined above; indeed higher levels of insulation are appropriate with more demanding levels of thermal comfort. One important design consideration, that is frequently overlooked is the problem of solar gain if air conditioning cooling is used. Given the unnecessary extra loading on the equipment, particularly from low angle western sun (and the consequent increase in energy use to overcome it) the days of the building with the same facade offering little sun protection for all orientations are hopefully numbered.

Opportunities to utilise passive energy for heating and cooling are usually more limited however: solar heat gain can present glare and local overheating problems in schools, office buildings and factories and natural ventilation cooling for buildings like theatres, cinemas and hospitals can present difficulties in maintaining comfort as well as acoustic, health and other conditions.

It is in larger buildings that active solar systems, mounted on the roof or at ground level are likely to replace the use of passive design measures in the building itself, yet the costs and relative mildness of climates will probably delay their immediate application.

It has been shown that the greatest area of energy reduction in existing large buildings is in alterations and maintenance to mechanical equipment, which has often become energy wasteful over time (3). One likely area of decreased energy usage is in the design and particularly the operation and maintenance of more efficient heating, ventilation and air-conditioning equipment.

Non Technical Issues

It must be remembered that thermal comfort is not the only aim in any building design, and that other factors such as views, site (boundaries, slope and orientation), internal building planning, health and fire regulations often conflict with optimum energy design. Nevertheless the technical issues are well known and relatively simple to apply and it has not been the design principles alone or building constraints that have thwarted their introduction. It has been the influence of other non technical issues of economics, planning and social attitudes that have impeded the introduction of "low energy" buildings.

The economics of low energy building design

The proportions of energy used for various purposes varies from building to building. In houses, heating costs may take up to 70% of the total energy used and figures of 40-60% are common for houses in the temperate coastal cities (4). Studies of office buildings have shown that air conditioning energy costs are often 40-45% of the total energy costs, with lighting taking a much larger share of the budget (5). These studies have also indicated that energy-reducing measures such as those outlined above can bring savings of up to 70% in heating costs in houses and 40% in air conditioning bills in larger buildings.

These savings are for the running costs of the building, whereas the costs of energy conservation are incurred during the construction of the building or during renovations and so are capital costs. Nearly all building projects (including houses) are only judged on the initial capital costs. Since lower initial costs are usually sought, often such items as conservation measures and more capital expensive mechanical equipment which will reduce energy consumption are not incorporated. The myopia of this approach can be seen if *life cycle costs* are considered. In this case the initial capital costs of establishing the building and the ongoing costs of operating the building (for a chosen "life" of the building) are considered together to give an overall cost for the building. It is in this way that the true cost to the owner or to society over time can be evaluated.

Williamson (6) has shown that when factors such as future fuel costs, discount rates and inflation rates, maintenance and replacement costs are considered for a conventional house and a low energy house, the latter has clear savings in life cycle costs.

The major problem however, is not in the proof of life cycle costings (although the investigation is fraught with such imponderables as future fuel costs and inflation rates) but to convince housing financiers that it is in their interests to consider life cycle costing, particularly from the point of view of the borrower's capacity to repay. It can be shown that, even assuming conservative figures, the owner of a low energy house (where fuel costs and maintenance are about one third of the average) has the capacity to repay a slightly larger mortgage, which would more than defray the initial extra expense in building in the energy conservation features.

For larger buildings Brown (7) has shown that it is currently cost effective to undertake an "energy audit" of larger air conditioned buildings. That is to check the air conditioning equipment for efficiency to examine the building fabric for potential savings and to check the lighting for possible reductions. Savings in the costs of fuel and maintenance in one or two years will often completely offset the cost of the energy audit (the engineering advice the contract works). Indeed in one case an investment of $8,300 in a 12 year old building brought energy savings of $17,000 per year!

As regards the design of new non-residential buildings, Wooldridge (8) cites a number of design factors and building services factors where costs can be recouped in energy savings in a short time (four years or less). Indeed some areas of design involving reduced areas of glazing, reduced lighting levels and reduced fresh air requirements can bring a decrease in capital costs as well as in running costs without a significant change in thermal, visual or lighting standards.

There are several other factors which should be considered at this stage although their direct economic consequences may be difficult to judge. These include the increased thermal comfort that often results from incorporating low energy design features (particularly in houses), the increased life expectancy of buildings and mechanical services within low energy buildings, and the benefits that may accrue to the community as a whole (including Governments in an Energy Conservation Program).

With regard to the last item many countries have thought it advantageous to introduce compulsory insulation levels and/or subsidies to building owners. In Belgium, Denmark, West Germany and Holland mandatory insulation levels have been established for all new buildings together with subsidies of 25-35% for installation of insulation. In the U.S.A. there are systems of grants, tax credits or tax exemptions in order to defray the initial capital costs of insulation, while in Sweden and the U.K. mandatory insulation levels for all buildings have been introduced and loans and grants can be made to building owners to assist the installation of insulation. In New Zealand a system of interest free loans totalling $NZ6 million was available in 1976 (9).

Planning and energy conservation

Obviously both passive and active solar energy use is dependent on direct access to the sun, but current urban planning does not often allow for this. Miller (10) has detailed several studies that show that for suburban developments this is unlikely to prove problematic as most houses are distanced far enough apart to avoid being shaded by neighbouring trees and buildings. However problems can arise, particularly at higher building densities and, as yet, few countries have taken decisions on the right to solar access. This issue is made particularly complex when it is considered that not only is access to existing houses or devices in question but also future access for devices not yet installed. To date few cases have been tested in court anywhere in the world, although several states in the U.S.A. have legislated for rights to solar access for all property.

Miller (10) points out that complex litigation can be avoided if attention to solar access is paid during the urban planning process, but this is unlikely to solve the major problems of existing cities' layout and subdivision. As solar energy becomes more popular a solution will have to be found for these existing situations, particularly where a change in building density occurs (Fig. 3).

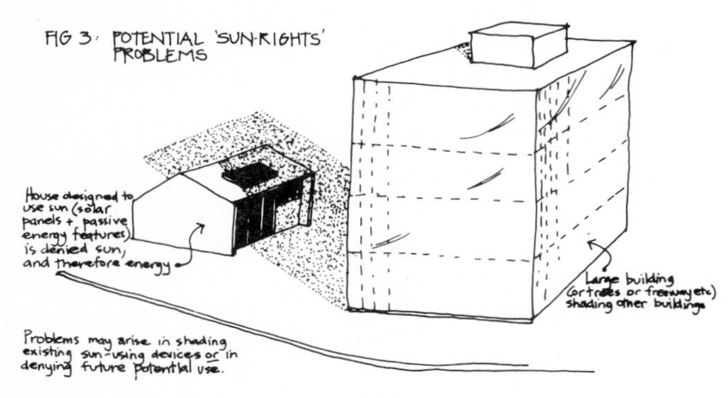

The solution to this problem will probably involve the establishment of rights, or at least an assessment to ensure solar access is gained. It would also be beneficial if future building and planning proposals were examined in this regard to determine whether any potential shadowing of adjacent buildings occurred (see Fig. 4).

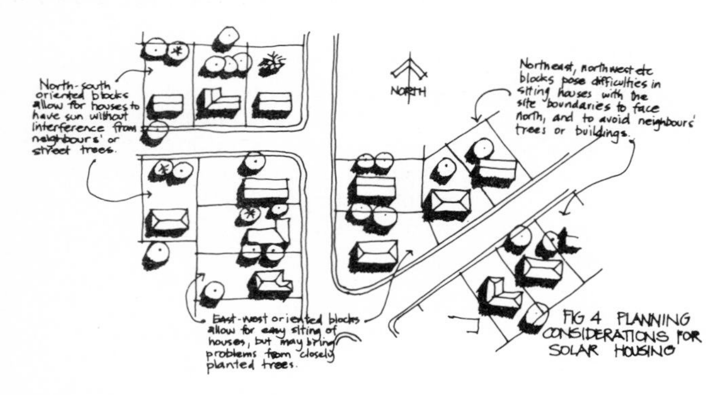

One further problem is the use of reflective film on windows to reduce heat gain (and hence air conditioning load) in the building. This is done fairly effectively, if only to divert it elsewhere, often onto adjacent buildings where glare problems, local overheating and additional air conditioning energy is required. This has led to some expensive litigation in the U.S.A. and is one energy conservation measure that must be handled with care.

Social attitudes to energy conservation measures

The measures outlined in the first section of this paper require a change in the building fabric and hence in the appearance of the buildings. They also require changes, to some extent, in the way in which occupants use the building. Occupants are likely to resist these changes for a number of reasons.

Firstly, Australian home owners have been notoriously conservative in the "style" of house that they desire. Or rather, the housing industry builds basically only one type of house, the "windows-to-the-street, multiple-fronted-brick-box-with-coloured-tile-roof", whether house buyers want that or not. Insulation measures are unlikely to change the style as such: facadism can continue with Corinthian columns and Spanish arch decorating a well insulated box, but in respect of passive solar energy use some changes are needed. The windows will no longer necessarily face the street (unless it is to the north!) and houses will be oriented to the sun rather than in the grid iron pattern. This has implications for privacy, both from house to house and particularly where north windows do face the street and curtains have usually remained closed day and night to achieve privacy. The alternatives are to provide screening by means of walls (to create courtyards perhaps) or the use of planting (hedges, trees) at a suitable distance from the house. Secondly, the occupants of a "passive" energy home will have to become slightly more "active" in opening and closing windows and curtains, and even though it means paying a little more attention to the weather, some resistance to change is likely!

In larger buildings the changes in attitudes may need to be more profound, not so much in accepting changes in building appearance (we are somewhat accustomed to that and in any event it is not so dramatic as for houses) but mainly in the standards that are expected in buildings. Thermal comfort levels are usually unrealistically narrow in large buildings and it has been estimated that if conditions in office buildings were more flexible say 22°C ± 5°C rather than 21°C ± 1°C, savings of at least 5% to 10% in energy would result. Furthermore lighting, which is one of the major loads on an air conditioning system, as pointed out earlier, is usually over designed to give unnecessarily high lighting levels.

Reductions in the numbers of lights to provide "adequate" conditions could save a further 6% in energy budgets.

A reduction in glazing in a building will usually affect the perception of internal space and views more than the external appearance but nevertheless it is something building owners and occupants will probably resist, seeing it as a lowering of standards.

More important than all of this, may well be the questioning of whether society actually *needs*, or at least whether it can afford, such expensive energy wasteful buildings. The ultimate (and most effective) energy conservation measure may be to not build at all, but to reuse another building.

Thus these "non technical" issues of economics, planning and user's attitudes may go some way to explain why simple, almost self-evident, technical solutions to energy conservation in buildings are finding such difficulty in implementation.

References

(1) D. Drysdale, *Designing Houses for Australian Climates* EBS Bulletin No. 8, *Low Energy House Design for Temperate Climates* NCDC 1977; V. Olgayay, *Design with Climate.*

(2) National Capital Development Commission, *Low Energy House Design for Temperate Climates*, NCDC 1977.

(3) Brown, A. Opportunities in existing buildings, in *Energy Conservation in the Built Environment,* Dept. of Environment, Housing and Community Development, 1978.

(4) *Thermal Economy in Buildings*, Australian Task Force on Energy, Aust. Inst. of Engineers, 1977 and in *Low Energy House Design for Temperate Climates*, Ref. (2).

(5) Op. cit. and A. Brown, Ref. (3).

(6) T. Williamson, Cost effectiveness of energy conservation measures, in *Energy Conservation in the Built Environment*, EHCD 1978, Ref. (3).

(7) See ref. (5).

(8) M. Wooldridge Energy consumption and conservation in building, *Energy Conservation in the Built Environment* EHCD 1978.

(9) *Thermal Insulation*, Vol. 1, No. 4, August 1976 p. 24 and *Domestic insulation standards, improved comfort levels, potential energy savings, mandatory home insulation*, Fibreglass Insulation Manufacturers Association, Oct. 1975.

(10) A Miller et al, *Solar Access and Land Use: State of the Law 1977.* Environmental Law Institute, Washington DC, U.S.A. 1977.

Energy and the City

Deborah White

Department of Architecture and Building
University of Melbourne,
Parkville 3052.

Abstract

The energy question is now recognised as central to many of the economic, social, political and environmental issues now requiring our attention. In an urbanised society such as Australia a significant proportion of the national energy budget is directly related to life in the cities, even disregarding their function as centres of industrial and commercial activity.

The urban living patterns and infrastructure we have developed have been in many ways directed and facilitated by the easy availability of energy. As access to convenient and cheap energy becomes more and more problematic, we must consider seriously whether the urban systems we have constructed are appropriate for the future. In fact the energy question offers us an opportunity to examine whether our cities in their present form were ever successful in providing for the social and environmental needs of their inhabitants.

Domestic consumption accounts for 25% of Victoria's primary energy. However, strategies for urban energy conservation should not be directed solely at the design and construction of buildings, and the encouragement of fuel-saving devices such as solar hot water services, although improvements to the existing building stock and standards for new buildings both offer the possibility of considerable and immediate energy savings.

From the long term point of view two other aspects are at least as important, but require more drastic measures for implementation; first, consideration of the variety, density and location of housing needed to facilitate *access* (to employment, to creative and social activities, to recreation) and *communication* (between people); and second, planning mechanisms which encourage intelligent *relationships* between the various functions of the city, the land they occupy, and access between them. Melbourne in the 70's appears almost to be the result of a conscious choice of the most energy-intensive options of building type, residential density, land-use strategy, and transport mode.

However, obscured as it may be by latter day random development, Melbourne provides an underlying structure upon which a 'low-energy city' could gradually be developed. The existing network of public transport routes could be upgraded, and the radial tram and train routes extended. Mixed-use areas could concentrate people-intensive residential, community, commercial, and some light industrial development around a hierarchy of local 'foci' and district centres along public transport routes, replacing our present car-based random mobility with functional access and an encouragement to community.

Large-scale demolition and reconstruction would be neither necessary nor desirable. Significant improvements in thermal performances are practicable in most existing buildings, particularly the older houses of the inner and middle suburbs best served by public transport. Housing infill involving local increases in density around transport nodes would allow not only for a halt to the spread of the suburbs, but for eventual withdrawal of built-up areas from the creek and river beds which provide the city's breathing space, in both physical and the recreational sense.

The problem and its solution are complex. But the main obstacles to positive action are not technical, but social and institutional; the difficulty of changing attitudes and practices developed over time and based in tradition, habit, and prejudice, and of revising the legislative structures built upon those attitudes and prejudices. All must be changed to become more responsive to the real needs, and more appropriate to the finite resources, of the community.

Diesendorf, M. (ed.) (1979). *Energy and People*. Canberra, Society for Social Responsibility in Science (A.C.T.)

Energy Policy and People

David J. Crossley

School of Australian Environmental Studies
Griffith University
Nathan, Brisbane.

Introduction

One of the most significant effects of the 1973-1974 OPEC-inspired "energy crisis" was the sudden advent of government energy policy as a matter for public discussion and government action. Government energy policy can be defined as the sum total of actions by government to influence energy production, distribution and use. In those countries which found themselves without sufficient supplies of oil in early 1974, the OPEC action effectively exposed the then-existing general lack of consideration of energy matters at the government level.

The energy crisis had two main effects on energy policy formulation. First, it forced the recognition of energy as a unitary concept. Previous to the crisis, government energy policy had been concerned with the individual forms of energy, so that separate strategies were formulated for coal, oil, gas and so on. The energy crisis brought home the fact that it was essential to conceive of energy as a unitary concept, and to plan to meet society's energy needs through an appropriate mix of different energy forms.

The second effect of the energy crisis was to achieve the speedy acknowledgment in government energy policy of a development in thinking about the role of energy in human affairs which first began to appear in the 1950's. Through studies of energy flows in animal and plant communities, and in some traditional human societies, energy began to be seen as a possible limiting resource in the development, or even the continuation, of human society. This idea was strongly supported, and popularised, by the publication of the Club of Rome's first report *The Limits to Growth* (1) and the 1973-74 energy crisis was its first practical demonstration.

Since the immediate crisis was one of energy shortages, these were seen as the major problem. However, as Lynton Caldwell points out, the comprehensive problem of energy in modern society has been developing for decades, and it is not exclusively a problem of shortages, prices, or allocation of supplies. In fact supply-demand problems are only one aspect of the limiting role which energy will play in human society. The basic problem is concerned with:

> *. . . choice and value in a world of finite capabilities. It is therefore also a moral and political problem, and for this reason will not yield to a purely technical solution* (2).

The moral and political aspects of energy policy are the areas which are central to the discussion in this paper. Moral questions arise when energy policy initiatives can be seen to have definite social implications which affect people's lives; the questions are concerned with the justification for choosing one policy option over several others. Political questions arise out of the actual decision making process by which a particular policy option is chosen. These questions are concerned with who actually makes the decision; on what information and advice; what procedures and channels are available for this decision to be reviewed, and who actually carries out this review. The two central concerns of this paper will therefore be possible social implications of energy policy, and procedures for energy policy formulation.

Social Consequences of Energy Use

The structure of modern post-industrial societies, and the lifestyles enjoyed by their members have been made possible by the ready availability of extrasomatic energy*. Indeed, there is good reason to believe that modern societies have evolved the way they have simply because of this ready availability of energy. The anthropologist, Leslie White has suggested that:

> *Other factors remaining constant,* [*human*] *culture evolves as the amount of energy harnessed per capita per year is increased, or as the efficiency of the instrumental means of putting energy to work is increased* (4).

Because of the intimate involvement of energy use with both the structure of modern post-industrial societies, and the lifestyles lived within them, it is difficult to meaningfully identify social factors which can be definitely attributed to being consequences of the use of extrasomatic energy; in a sense, the whole of society is a consequence of energy use. However some attempts have been made.

The most comprehensive attempt so far to identify social consequences of energy use was made by Budnitz and Holdren (5). They concentrated on describing the immediately observable environmental impacts of energy systems and on identifying possible social costs of these impacts. Environmental impacts were defined as what is added to or done to the environment and social costs as the nature of the damage (i.e. change) produced by the impacts. Table 1 summarises the environmental impacts and consequent social costs identified by Budnitz and Holdren.

Table 1

Environmental Impacts and Social Costs Originating from Energy Systems (6).

A. *ENVIRONMENTAL IMPACTS*
1. Accidents
2. Pollution
   - (a) Gaseous effluents
   - (b) Liquid effluents
   - (c) Solid effluents
   - (d) Heat
   - (e) Noise
3. Resource consumption
4. Environmental transformation
5. Altered opportunities

B. *SOCIAL COSTS*
1. Death and disease
2. Genetic effects
3. Loss of economic goods and services
4. Loss of environmental ("free") goods and services
5. Aesthetic loss
6. Social and political change

*Extrasomatic energy is that energy which flows through or is utilised by the human community and which is not utilised through metabolic processes within a living organism (3).

Diesendorf, M. (ed.) (1979). *Energy and People*. Canberra, Society for Social Responsibility in Science (A.C.T.)

TABLE 2. SOME EXAMPLES OF SOCIETAL CONSEQUENCES OF SOCIAL COSTS RESULTING FROM ENVIRONMENTAL IMPACTS BY ENERGY SYSTEMS.

ENVIRONMENTAL IMPACTS / SOCIAL COSTS	ACCIDENTS	POLLUTION	RESOURCE CONSUMPTION	ENVIRONMENTAL TRANSFORMATION	ALTERED OPPORTUNITIES
DEATH AND DISEASE	Death/injury from explosion of natural gas tanker. Electrocution. Cancer induced by accidental exposure to radiation, eg. during operation of nuclear power plants. Death/injury from coal mining accidents.	Pneumoconiosis (black lung disease) in coal miners. Lung cancer caused by exposure to radon gas in uranium mining. Respiratory diseases caused by gaseous and particulate effluents from burning of fossil fuels. Heat stress from release of waste heat. Noise stress. Disease caused by release of electromagnetic radiation.	—	—	—
GENETIC EFFECTS	Abnormalities in subsequent generations caused by genetic changes induced in present generation by exposure to radiation, eg. during uranium mining and processing and operation of nuclear power plants.		—	—	—
LOSS OF ECONOMIC GOODS AND SERVICES	Damage to property caused by, eg. natural gas tanker explosions, fires started by electrical faults, failure of walls of dams used to generate hydroelectricity etc.	Damage to property caused by, eg. rain containing acidic pollutants from burning of fossil fuels which can effect stonework and impair the growth of plants. Loss of services caused by, eg, release of warm water from power stations which can stimulate algal growth in rivers, requiring the diversion of resources to prevent the blocking of the river.	In addition to loss of fuels through their use in energy production, other resources are also lost, eg: agricultural land disturbed by mining or by building a power station, or by flooding for hydro-electricity generation; non-fuel mineral resources such as copper used in power lines; and energy used to produce energy such as the energy lost in pumped storage hydroelectricity in which more energy is used than produced.	Damage to property caused by, eg. ground subsidence for coal mining; surface disruption from strip mining; increased salinity of water courses as a result of diversion of water for hydroelectricity generation, leading to reduced agricultural yields.	—
LOSS OF ENVIRONMENTAL ("FREE") GOODS AND SERVICES	Release of carbon dioxide and particulate matter from the burning of fossil fuels could affect the temperature of the earth by allowing less long wave radiation from the sun to reach the earth's surface. Release of waste heat associated with energy production and use creates local heat islands with unpredictable effects on atmospheric conditions. Release of other chemical pollutants from the burning of fossil fuels and of radiation from the operation of, or accidents in, nuclear power plants could have long-term effects on the cycling of nutrients through natural ecosystems by affecting the metabolism of plants and animals. Elimination of species through altering their habitat by mining or building dams for hydroelectricity generation reduces the world pool of genetic information with unknown effects. These changes in the earth's biogeochemical cycles, ecosystem processes and pool of genetic material could have profound consequences for the production of food for human consumption and for human resistance to epidemic diseases.				—
AESTHETIC LOSS	Damage to aesthetic objects caused by, eg. explosions, fires started by electrical faults etc.	Damage to aesthetic objects caused by, eg. atmospheric pollutants from burning of fossil fuels.	—	Effects of power lines, pipelines, etc. on landscapes and waterscapes.	—
SOCIAL AND POLITICAL CHANGE	1. Change at the international level. Possible disputes between nations well-endowed with energy resources in relation to their populations, and those less well-endowed. Possible disputes between "developed" and "developing" countries ("development" is intimately related to the ownership of and ability to use energy technology). 2. Change at the national level. Political/social form of organisation of society may be dictated by type of energy system adopted - eg. use of nuclear power requires some restrictions on civil liberties to prevent access to nuclear waste products by unauthorised people. Employment opportunities may be affected by type of energy system adopted, eg. centralised systems usually create less jobs than decentralised systems. Socio-economic differences may be accentuated by differential access to energy based on ability to pay. 3. Change at the level of the individual. Possible concern and frustration with environmental impacts perceived as undesirable. Ability to relate to environment using own senses may be much reduced through transformation of personal environment by energy-using appliances. Automation of the work situation through energy-using machines could lead to feeling of anonymity, and possible decline in status with redundancy of learned manipulative skills, leading to loss of security.				

Table 2 is an attempt to take Budnitz and Holdren's conceptualisation a little further by listing some examples of possible social costs resulting from specific environmental impacts by energy systems. The examples listed are by no means meant to be exhaustive. They also demonstrate the impossibility of describing, by means of a simple table, the complexity of the interaction between energy systems and human society. This interaction cannot be represented by a simple cause-effect chain leading from observed environmental impact directly to a corresponding social cost. There are two problems with this approach: first, some environmental impacts cannot be easily observed, or attributed to originating from energy systems; and, second, the social consequences of energy use arise out of a complex network, or web, of interrelationships between energy use and society.

However, Table 2 does demonstrate some of the areas in which these social consequences occur and suggests that changes in the energy forms we currently use and the ways in which we use them are likely to cause corresponding cultural changes in present-day society, as was the case in previous human societies (7, 8, 9). It seems likely that the major changes in society over the next hundred years will occur at least partly as a result of energy policy decisions, or as a result of failure to make such decisions. Therefore it is essential that some attempt be made to understand the possible social implications of energy policy initiatives, and to take into account these implications during the policy formulation process.

Policy Responses to the Energy Crisis and their Social Implications

The 1973-74 energy crisis was the stimulus which caused governments of western post-industrial societies to turn their attention to the formulation of energy policy. In fact, almost all the energy policy measures which have been formulated in the last five years can be seen, in one sense or another, as responses to the energy crisis. There have been three main types of these responses.

Emergency measures: the first type of response

The first responses were energy policy measures designed to have immediate effect in alleviating shortages (e.g. petrol rationing, three day working weeks, etc.). The social implications of such measures were potentially enormous, and little attention was paid to these implications because of the urgency of the situation. However, the measures proved to be purely temporary and had little long term social effect.

The technically-oriented approach: the second type of response

A little while later came energy strategies* designed to guard against future failures of supply. Mostly, these took the form of extremely detailed quantitative analyses of energy resources and production, and past, present and projected future energy demand, on which were based policy measures of a technical nature designed to ensure that projected future energy demand would always be met with an adequate supply. Project Independence in the United States was an example of such a strategy; this was a series of measures designed to ensure that the United States eventually would become self-sufficient in energy supplies. Some of the social implications of the measures suggested were identified, especially impacts on the national economy and on different segments of the population, but few suggestions were made as to how these impacts would be alleviated or integrated into other aspects of the government's social policy (10). The emphasis was still very much on the technical problems involved in balancing future energy demand with supply.

*An energy strategy is a group of policy measures directed towards achieving a specific objective through influencing energy production, distribution and use. Government energy policy is normally composed of a number of energy strategies, each directed towards achieving a specific objective, together with other actions which may not be part of any particular strategy.

Another form of response to the energy crisis, also falling within this second type, is the energy conservation popularisation campaign. Such campaigns were mounted by the governments of many of the countries which experienced energy shortages; they used mass media advertisements, specially prepared booklets and audio-visual instructional materials, displays and demonstrations, referral services, and so on. The campaigns were aimed at the general public, particularly the householder and the car driver. Frequently, they followed the format of first setting out the background to the energy situation, and then using the need to extend the life of non-renewable energy resources as a further justification for energy conservation (Fig. 2). The two commonest messages throughout all the campaigns were that energy conservation means financial savings; and the communication of practical information on energy conservation techniques, stressing the ease with which these can be carried out (Fig. 1) (11). The main point about the social implications of these campaigns is that in their messages they stressed that people's lifestyles would not be greatly affected if they adopted the suggested energy saving techniques; however, the campaigns failed to mention the possible profound effects on the national economy, and therefore on society as a whole, of a long-term reduction in energy use. The campaigns were another example of a technically-oriented response to a situation which was seen primarily in terms of overcoming problems involved in balancing energy demand with supply.

Fig. 1: *Advertisement emphasising both financial saving and practical information on energy conservation techniques; from the UK campaign.*

Assumptions implicit in the technically-oriented approach

Most energy policy which has been formulated in western post-industrial societies in the last five years adopts a technically-oriented approach to what is perceived as being solely a supply-demand problem. There are four basic assumptions implicit in this approach (12):

1. *Energy demand is taken as "objective" and fixed, and largely independent of price. Since the social need for energy is taken to be indicated by the demand on the market, it too is objective and fixed.* However, demand is manifestly not fixed, as is demonstrated by the activities of energy utilities in modifying customer demand through advertising and promoting new energy-using products. Also, there is no particular reason why the demand for energy, which is directly related to ability to pay, should also be related to social need, which is also partly a product of biophysical requirements, but mainly of social conditioning.
2. *Technologies are taken as given.* In fact, the development of new technologies is a societal concern which necessarily reflects societal priorities. These priorities are expressed in the preferential allocation of resources to the development of certain technologies and the economic advantages allowed to particular technologies, through fiscal measures. At present, in Australia, societal

priorities, as expressed through government energy policy, favour the development of certain energy technologies, particularly large-scale systems, but there is no reason why these priorities, and therefore the technologies, should not change in the future.

3. *Reality is uniform or, in other words, change occurs only in the direction of established trends.* It is necessary only to examine a sufficiently long period of human history (perhaps as short as fifty years, or two generations, at present) to demonstrate the fallacy of this assumption.
4. *Economics and technology are assumed to be value-free.* In fact, the adoption of the entire market system of exchange of goods and services represents a fundamental moral choice; the basis of differential pricing of goods and services is not "economic rationality" but a series of moral and value judgments. Similarly, the institution of technology is not a morally neutral endeavour. There is a value judgment implicit in deciding to turn on a mains electric light instead of going to sleep at dusk; thereby accepting both the social consequences of the technology necessary to supply the light, such as pollution, resource usage, centralisation and growth of bureaucracies; and the social revolution which resulted from the human acquisition of the ability to replace sunlight with artificial light.

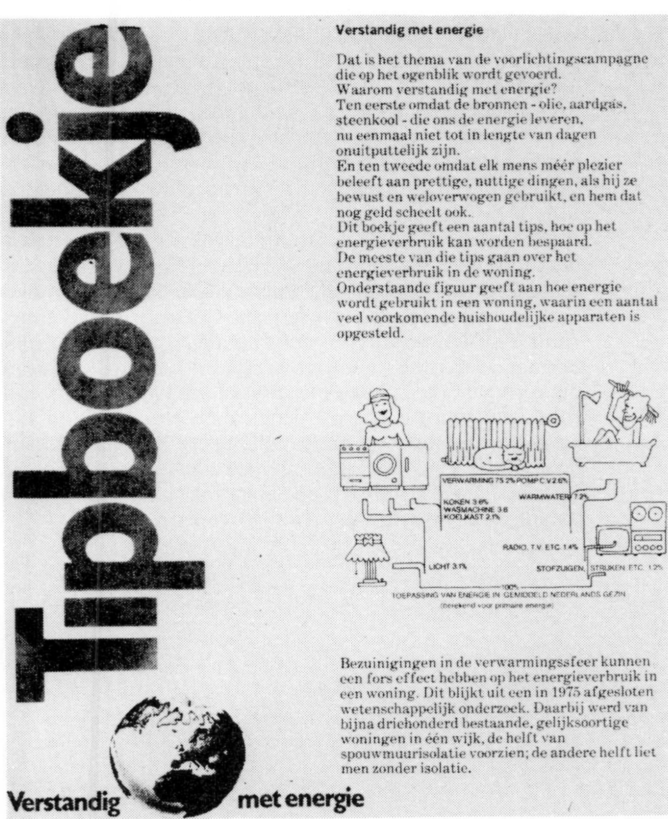

Fig. 2: *Advertisement setting out the background to the energy situation, and using the need to extend the life of non-renewable energy resources as a further justification for energy conservation; from the Dutch campaign.*

Broad-ranging strategies: the third type of response

Since late 1976, a third type of response to the energy crisis has begun to emerge which consists of more than just simply attempts at modifying existing energy systems. This type of response is manifest in the shape of more long-ranging, less technical, and more open-ended and wide-ranging energy strategies, which attempt to take into account dimensions of the limiting role of energy other than the supply-demand problem.

One example of such an approach is a Swedish study *Energy in Transition* (13). This study examines the current situation in Sweden in relation to energy production, distribution and use, and the relationship of Sweden to the world energy economy. Instead of making predictions on possible future demand for energy in Sweden, the report looks at the history of the development of the energy systems currently in use, and examines other systems which may be in use in the future. It discusses some of the social and political consequences of the use of both types of systems, and the relationship between energy and economics, energy and environment, energy and geographic structure and energy and foreign policy. Through a discussion of the energy carriers which may possibly be used in the future, the study introduces the notion of freedom of action in energy policy; that large centralised systems of energy distribution limit future options in energy policy, whereas small decentralised systems allow much greater freedom of choice in future energy policy. Finally, the report examines the relationship between energy and social organisation and suggests some suitable objectives for energy strategies developed within an overall policy goal of maintaining future freedom of action.

Another example of an alternative approach to the development of energy strategies is a study prepared for the Royal Commission on Electric Power Planning in Ontario (14). This study is concerned with the role of institutions in, and the conceptual framework of, energy policy making. The report suggests that future increases in energy demand are likely to be much lower than those predicted by the institutions which have a vested interest in increased use of energy. These predictions would only come true if present energy policy was continued without change. The report also questions the economic and technological criteria which are used as the basis of energy planning, and suggests that such planning should be also based on the consideration of much wider social aspects. The concluding recommendations of the study were based on three requirements in relation to energy policy: that the separation of the roles of energy production, distribution and use be broken down; that the relationships between institutions concerned with energy policy be reorganised to coincide more closely to the reality of energy production, distribution and use; and that a participatory, multilevel, iterative process of decision making in relation to energy policy be instituted.

Energy policy and moral values

Questions of morality arise in relation to energy policy because of the profound social implications of energy policy initiatives. Energy policy decisions have great potential for modifying the structure of society and for altering people's lives; the nature of these changes depends on the energy policy option chosen. This choice cannot be taken without making value judgments and it is therefore a choice involving questions of morality.

The first two types of responses to the energy crises, emergency measures and the technically-oriented approach, attempt to ignore value judgments by implicitly or explicitly denying the importance of the social implications of energy policy initiatives, or by dismissing them as too hard to come to grips with. The perception of energy policy as being solely concerned with problems of supply and demand, which appears to be the main standpoint currently adopted by governments in modern post-industrial societies, implicitly avoids the moral questions involved by ignoring them. However, in doing this, governments are not avoiding value judgments — they are, in fact, affirming the value system of the existing *status quo*, and this can be seen quite clearly when the four assumptions underlying the technically-oriented approach to energy policy formulation are examined in detail.

The third type of response to the energy crisis differs from the two preceding ones by carefully examining the social implications of suggested energy policy initiatives, and the value judgments which must be made to implement each energy policy option. Examples of this third type of response are still at the exploratory stage; none have yet progressed to fully developed strategies ready for implementation. However, even at the exploratory stage, this type of approach to energy policy formulation is already questioning the moral basis of some existing energy policy, particularly: current inequities in access to energy by different socio-economic groups, based on the ability to pay; and potential implications for the life of the individual and for the future structure and control of society, resulting from the adoption of energy systems which require restrictions on civil liberties. The major advantage of this third type of response is that it exposes to scrutiny the social implications of, and therefore the value judgments implicit in, energy policy initiatives, whereas there is no opportunity for these to be examined in approaches to energy policy formulation which identify supply-demand as the major, if not only, problem.

Procedures for Energy Policy Formulation in Australia

Australia was cushioned from the effects of the OPEC actions in 1973-74 by the ready availability of indigenous crude oil. Because of this, the pressures for government action on energy matters have been less intense than in other countries which experienced oil shortages as an external threat to their internal stability. However, energy policy presents just as much a moral and political problem in Australia as in any other country.

Until the mid-1970's energy was not a matter of great concern to governments in Australia and energy policy was generally *ad hoc* and unco-ordinated. However, Australia has not been totally isolated from the nascent world-wide increased interest in energy matters and government energy policy has become the subject of some discussion following the publication of a number of reports containing policy recommendations (15, 16, 17, 18, 19, 20).

Examination of these reports, and of the energy policy measures currently adopted by both Federal and State governments, show that the approach taken to energy policy formulation in Australia is technically-oriented, and energy policy is perceived as being solely concerned with problems of supply and demand. There appears to be little recognition of the fact that energy policy initiatives have social implications, and the value system resulting from the energy policy formulation process is therefore implicitly established as that of the *status quo*. The lack of any broad-ranging approaches to energy policy formulation can be traced to the methodology currently used in Australia for developing energy strategies, and to the decision making process involved in formulating government energy policy.

Methodology currently used in Australia for developing energy strategies

Examination of the various reports on energy policy provides insight into the methodology currently used in Australia for developing energy strategies. Almost invariably, this consists of making predictions about future energy demand, and then developing energy strategies to ensure that energy supply will always be able to meet this predicted demand. In this process, the methods used for predicting future energy demand are crucial. There are two main classes of these methods:

1. Simple extrapolations from historical data;
2. Models of energy demands.

Simple Extrapolations from Historical Data. Predictions of future energy demand in Australia have traditionally been based on simple extrapolations of historical rates of growth (21, 22). Until the mid-1970's, the logic of predicting future energy demand by extrapolating from historical trends was not open to debate, even when this method led to the production of demand curves rising at exponential rates. So strong is the tradition of historical extrapolation that substitution of one fuel for another is predicted by attributing different annual percentage growth rates to the two fuels (23, 24). A Report by the Institution of Engineers, Australia used the energy demand figures predicted by the Department of Minerals and Energy to predict the increase in Australian Gross National Product (GNP) and then used that predicted increase in GNP to further predict the growth in energy demand (25). The circularity of this process is obvious.

The unreliability of the historical extrapolation approach has been demonstrated by energy statistics which show that total energy use in the O.E.C.D. countries in 1975 was about five percent *lower* than the 1973 figure (26). The response to such statistics by those using the extrapolation approach in Australia was simply to alter the exponent in the extrapolation (27).

There are two major sources of the unreliability of the historical extrapolation approach.

1. *Simple historical extrapolations do not take into account factors which affect energy-using behaviour.* The historical extrapolation approach assumes that people will carry on using energy in much the same way as they have previously. However, there are many factors which could affect energy-using behaviour and lead to changes in use patterns. Predominant among those factors would be changes in the price of energy forms, but there are also a number of other important factors, such as changes in technology, social changes, and changes in the attitudes of consumers.

   An example of a change in energy-using behaviour caused by a factor other than price has been revealed by a detailed study of the trend in electricity demand in the United Kingdom, the growth rate of which has been falling for more than ten years. Examination of the statistics of appliance ownership indicates that this growth rate has fallen because of market saturation (28). Once most households have most of the electrical appliances they desire, the scope for increased domestic electricity use is limited. This development has affected total energy use in two ways. Domestic electricity use has shown little or no growth, and the reduced demand for appliances has led to reduced industrial production and therefore, there has been little or no growth in industrial use.

   Examination of the actual annual increases in electricity demand in Australia from 1951 to 1976 (Fig. 3) shows a distinct downward trend, which is particularly evident, and constant, for the last ten years. This suggests that a similar situation may be developing in Australia as has occurred in the United Kingdom. Both situations demonstrate the naivete of extrapolating historical trends to predict future energy demand, without taking into account factors causing changes in energy-using behaviour. This prediction method has led to the Central Electricity Generating Board in the United Kingdom having a fifty percent generating overcapacity (29).

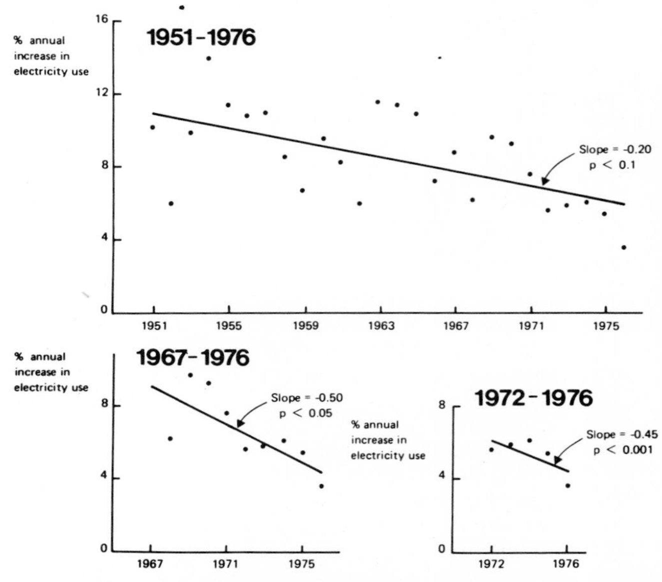

Fig. 3: *Trends in annual increases in electricity use in Australia, 1951-1976.*

Source: Percentage annual increases calculated from data for electricity use for the six Australian States (30, 31). Statistics for the Australian Capital Territory and the Northern Territory are excluded from the calculations since strictly comparable data are not available for the whole of the period under consideration.

2. *The statistics on which the extrapolations are based are inadequate.* Frequently, existing statistics on energy use are too highly aggregated, or are categorised in ways which makes their use in predictions suspect. Waverman (32) points out that in North American statistics, the category "commercial use" includes street railways, apartment buildings, the government, and small manufacturing firms as well as retail trade and services. In Australia, the non-availability of suitably categorised and disaggregated energy use statistics is particularly acute. In the United States, the Department of Energy has established an extremely detailed and extensive data base which is backed up by legislation requiring private energy companies to provide specified information. In Australia, the Department of National Development is building up a national data bank on regional end-use of energy by fuel, end-use sector and telephone area code, but there is no legislation requiring energy users to provide information to this project.

Models of Future Energy Demand. Models for predicting future energy demand have been developed in an attempt to overcome some of the problems inherent in simple extrapolation from historical energy use data. Hoffman and Wood (33) have identified four major groups of energy models:

1. Sectoral models, covering the supply of or demand for specific fuels or energy forms;
2. Industry market models, which include both supply and demand relationships for individual or related fuels;
3. Energy system models, which encompass supply and demand relationships for all energy sources;
4. Energy/economic models, which model the relationships between the energy system and the overall economy.

In Australia, the Department of National Development used sectoral models in its predictions of the future demand for primary fuels in Australia which were published in 1978 (34); previous predictions by the Department were based mainly on historical extrapolations, except for predictions of future demand for transport fuels which were always based on sectoral models. One large-scale model of the energy system type (but also including other sectors in addition to energy) has been developed to predict future total national energy demand in Australia (35). Based on the M.I.T. *World 3* model modified to apply to Australia, it predicts the national energy demand resulting from several different suggested scenarios. One of these scenarios, labelled "Standard Run", consists basically of extrapolations of historical sector growth within certain limiting interactions between the different sectors. Another scenario, labelled "Solar Energy Policy", involves similar extrapolations, but on the basis that large scale development and utilisation of solar energy will take place.

Models of each of the four types listed above have been used in several countries to supply different information about future energy demand for energy policy formulation. Despite their different utilities, all the four types of models are subject to three sets of limitations.

1. *Limitations inherent in the scenario approach.* Energy models which use the scenario approach are subject to limitations arising from the process of choosing the scenario to be studied. These models can only predict the outcome of the scenarios chosen. They cannot make any statement about the probability of occurrence of a specific scenario, such as the large-scale development of solar energy.
2. *Limitations arising from assumptions about human behaviour.* Assumptions made in energy models about human behaviour are based almost entirely on data linking historical energy use with economic factors such as prices. This causes three problems.

   First, the relationship between energy use and economics is complex. Since it is not possible to use energy without employing an energy-using device, energy demand is derived from, and ultimately connected to, the stock of energy-using goods, rather than being directly related to energy prices. Another complicating factor is that one type of fuel cannot be regarded as being a perfect substitute for another type. Economists have developed a large and complicated array of techniques for dealing with the complex relationship between energy use and economics, but the relationship cannot yet be said to have been adequately modelled (36).

   The second problem is more fundamental. In reality, there is no such thing as a perfect market because of various factors other than prices which influence consumer's behaviour. Moreover, a perfect market mechanism requires that consumers have perfect knowledge of all the consumption options open to them, and the costs of each. This is especially unlikely in the case of energy where the consumer is faced with a bewildering array of different energy forms and tariffs. Also, the ability of market prices to predict future behaviour is highly restricted, since such prices reflect only short-term expectations and are greatly influenced by past and present policy decisions (37). In summary, economic factors alone can never completely explain current energy-using behaviour, and are an even more unsound basis from which to predict future behaviour.

   Third, all social trends, of which energy-using behaviour can be considered one, generate responses and active intervention by the participants. These can feed back to affect the trends themselves (38) in ways which cannot be predicted by current modelling techniques, which attempt to explain behaviour through the use of economic factors only.
2. *Limitations inherent in the baselines adopted.* The baselines adopted in energy models are all derived from statistics of historical energy use. The inadequacy of these statistics has been mentioned previously. All predictions of future energy demand using baselines derived from existing statistics must be suspect until these statistics are improved.

Summary. A crucial first stage in the methods currently used in Australia for developing energy strategies is the preparation of predictions of future energy demand. However, both the methods used for preparing these predictions (historical extrapolation and models of future demand) are unreliable because they do not take into account all the factors which affect people's energy-using behaviour, and they are based on statistics of historical energy use which are inadequate.

Decision making in energy policy formulation

In Australia, there are four main groups involved in energy policy formulation:

1. Publicly-owned energy utilities;
2. The public service;
3. Advisory bodies and pressure groups;
4. Ministers.

The Role of the Publicly-Owned Energy Utilities. The primary responsibility of the publicly-owned energy utilities is to the success of their commercial operations. In pursuit of this success, they carry out actions which have major effects on energy production, distribution and use within their geographical area of operation. Energy utilities are major customers of the private companies involved in energy production; often a contract with a utility for the supply of fuel is the key factor which decides whether a major energy production project will go ahead. The actions of energy utilities in modifying their distribution networks can significantly change the nature of energy distribution patterns, and the actual energy carriers distributed, over wide geographical areas. Energy utility marketing activities, such as tariff structures and promotional campaigns, can significantly affect patterns of energy use. In effect, energy utilities actually contribute to the formulation of government energy policy through their actions; the bases of these actions are the requirements of the utilities' commercial undertakings.

Since energy utilities are statutory authorities, they are nominally responsible to a Minister but, in effect, Ministers have little to do with those statutory authorities which function as commercial undertakings (39). Such authorities are only infrequently subjected to Ministerial policy directives. Another way in which governments can control statutory authorities is through the control of budget appropriations. However publicly-owned energy utilities have their own source of income, derived from the sale of electricity or gas, and sometimes also from private and public loans. These financial dealings usually require approval by the relevant Minister or Ministers, but it seems that rarely is this approval withheld, or changes required in the financial proposals put forward by the utilities.

Publicly-owned energy utilities in Australia therefore act as relatively free agents in the formulation of government energy policy through their actions. This lack of governmental control has been subject to some modification since the mid-1970's following an institutional reorganisation in the States of responsibility for energy planning, and the beginnings of attempts at integration and co-ordination, but direct lines of control over the actions of the publicly-owned energy utilities do not yet seem to have been developed in any State. This reorganisation led to an interesting situation in Western Australia, where the State Energy Commission of Western Australia, which was given the responsibility for energy planning, also operates as an electricity and gas utility. In the words of the Commissioner:

> *When facing the prospect of the North West Shelf gas development the State Energy Commission was conscious of its dual responsibilities to provide policy advice to the Government of Western Australia and to carry out its traditional function of supplying electricity and gas. On the one hand the Commission has a broad overall responsibility to ensure the efficient and effective utilization of the State's energy resources while on the other it must pay due regard to the usual commercial considerations of a large trading enterprise.* (40)

The Role of the Public Service. One of the most prominent ways in which governments influence energy production, distribution and use is by the regulation of energy-related activities. The implementation of the regulations imposed by legislation is primarily the responsibility of government departments, particularly the Commonwealth Department of National Development and the State Departments of Mines. The actual implementation is by no means a straight forward matter devoid of political or policy content; in fact, the public service is effectively involved in energy policy formulation through the means it chooses to implement the regulation of energy-related activities.

The nature of some of the legislation regulating energy-related activities in Australia allows a particularly large amount of discretion in its implementation. The legislation regulating the exploration for and exploitation of fossil fuels is an example; the mechanisms and requirements for the granting of prospecting and production licences are spelled out in detail, but the decision as to which applicant shall receive a licence, and on what grounds, is essentially a political decision (in the widest sense). To be involved in the process of making this decision is to be involved in energy policy formulation. Such decisions are sometimes made by Ministers, especially where large projects are concerned, but frequently the Minister's decision is based on advice from public servants in the responsible department. In other cases, the decisions are actually made by public servants.

Government departments provide advice on energy policy by carrying out energy planning — the definition of objectives to be achieved through energy policy. This is an essential first stage in energy policy formulation; it is on the basis of objectives identified at this stage that decisions are made on the energy policy measures to be implemented. Departments may recommend specific policy measures to achieve the objectives, however the decision to implement these measures is not the responsibility of public servants, but is carried out by Ministers.

The role of the public service in energy policy formulation is therefore twofold: actual energy policy formulation through decision-making in regulating some energy-related activities; and the definition of objectives to be achieved by energy policy measures, the decision on the implementation of these measures being taken by Ministers. The role of the public service can actually be quite influential, given legislation which allows relative freedom in decision making, and a Minister who is responsive to policy advice from the department.

The Role of Advisory Bodies and Pressure Groups. One result of the greatly increased interest in energy matters in Australia from the mid-1970's is a large increase in the number and variety of bodies created to provide advice to governments on various aspects of energy policy. A recent survey identified thirty-eight such bodies, ninety percent of which had been created since 1976 (41).

Tables 3 and 4 present an analysis of the membership of these advisory bodies. The overall picture which emerges from this analysis is that the members of these bodies are predominantly an elite group composed of highly educated and technically-oriented people who have a professional interest in the energy matters on which they are advising governments. Though this is not reported in the tables, most of the members are also at least 45 years old (indeed, many have actually retired from working life) and only three, just over one percent of the total, are women. Of the major groups identified in Table 3, only the parliamentarians, and perhaps some of the public servants, could be said to represent the interests of the general community; these groups, in fact, are professionally involved in doing just that, and therefore subject to various political pressures. The other major group which could provide a detached viewpoint are the academics, but the analysis of their background which was carried out to produce Table 4 showed that the academic members are mainly technical experts in various aspects of energy production, distribution or use.

Table 3

Affiliations of Members of Government Energy Policy Advisory Bodies in Australia, December, 1978 (42).

Parliament — Federal and State	41	16%
Local government council	1	
Public service — Federal and State	34	13%
Statutory authority — Federal and State	62	24%
Private company — energy industry	39	15%
Private company — other	23	9%
Private consultancy	10	
Tertiary educational institution	38	15%
Trade union	3	
Judiciary or legal practice	2	
Consumers' association	1	
	254	92%

This domination of the advisory bodies by technically-oriented, professionally-involved people seems to result in the role of the advisory bodies being restricted to providing technically-oriented professional advice on energy policy. Such advice is appropriate in the case of bodies which allocate funds for energy research and development, but those bodies concerned with general policy advice also seem to conform to this mode of operation. Another possible role for advisory bodies on energy policy would be to act as channels through which the opinions and attitudes of the general community could be incorporated into the policy formulation process, but there is no evidence of this occurring anywhere in Australia.

Another source of advice to governments on energy policy is pressure groups who seek to promote their particular view of how energy policy should be conducted. It is difficult to gather information on the activities of such groups, but it should be noted that companies in the energy industry can now seek to

Table 4

Backgrounds of Members of Government Energy Policy Advisory Bodies in Australia, December, 1978 (43).

Major Discipline Studied for Tertiary Educational Qualifications		
Engineering	96	38%
Physics/Chemistry	30	12%
Geology/Surveying	14	
Biology/Medicine/Veterinary Science	6	
/Agriculture/Forestry	6	
Architecture	2	
Economics/Commerce/Law	38	15%
Political Science	3	
Education/Geography/Philosophy	5	
Major Field of Experience (No Tertiary Educational Qualifications)		
Public service	5	
Corporate business	12	
Small business	10	
Farming	13	
Armed services	1	
Media	1	
Politics	1	
Trade union	5	
Community work	1	
Not Known		
No reply	10	
Refusal	1	
	254	65%

influence government energy policy formulation through their employees who are members of the recently-created advisory bodies. In fact, there may be some reason to believe that, at least in the case of the energy industry, the creation of these bodies represented the formalisation of a consultative relationship which already existed.

The Role of Ministers. Government Ministers in Australia, under the Westminster system of government, are charged with the responsibility for making decisions about policies. Individual Ministers interact with their departments in the development of policy proposals which they then take to Cabinet. Once they are accepted by Cabinet, and if the proposals require legislative action, legislation to implement the proposals must be approved by Parliament; otherwise the proposals can be implemented by executive action following Cabinet approval. Ministers make decisions about policy on essentially political grounds: individual Ministers on the basis of whether a particular proposal is likely to be accepted by Cabinet; and the Cabinet on the basis of whether the proposal is in accordance with general party policy, and whether it is likely to be acceptable to Parliament and ultimately to the electorate.

Once it reaches the level of the Minister, however, policy formulation ceases to be a visible process. It is not possible to predict with certainty whether a particular energy policy proposal will end up as legislation or executive action, because this depends heavily on the personal idiosyncracies of the responsible Minister and of his colleagues in Cabinet, particularly the personal and collective value systems under which they operate and the political realities of the situation in which they find themselves.

Because of the multifarious institutions which can influence energy policy, it is difficult for a Minister to obtain a comprehensive overview of the situation. Under the Westminster system of government, Ministers have no special qualifications or expertise in the areas over which they have policy control, and they usually move from one area to another fairly frequently, though this has not been followed in all cases in Australia. Ministers, are, in fact, acting as laymen, accountable for their actions both to their colleagues and to the electoral process, and this is seen as a strength rather than a weakness of the system. In the area of energy policy, however, with its complexity, profound social implications and therefore the tendency to affect many other policy areas, and long lead times for policy initiatives to take effect, the relative inexperience and temporary responsibility of Ministers could be seen as being disadvantageous to effective policy formulation. It will certainly mean that Ministers responsible for energy policy will need to rely heavily on advice from more knowledgeable sources.

Summary of decision making in energy policy formulation

The basic decision making process in Australia which results in governments taking specific actions to influence energy production, distribution and use is carried out by government Ministers. However, the publicly-owned energy utilities and, to a certain extent, the public service, can carry out energy policy formulation directly through their actions. Ministers receive advice on energy policy from the public service and advisory bodies; the departments define objectives to be achieved through policy measures, and the advisory bodies provide technically-oriented professional advice. There seems to be no channel for the opinions and attitudes of the general community to be incorporated into the energy policy formulation process. Because of the many different bodies and activities which influence energy policy formulation, Ministers have difficulty in maintaining an overview of the totality of the energy situation. The energy policy decisions they eventually make depend on their personal idiosyncracies and those of their colleagues.

Consequences of Energy Policy Formulation Procedures in Australia

The overall effect of the detailed procedures adopted in Australia for energy policy formulation is that virtually no consideration is given to the social implications of the policy decisions made. Both the methodology used for developing energy strategies, and the domination of the channels of information and advice flowing to government Ministers by technically-oriented and professionally-involved people virtually ensure that social factors are either not considered or are assigned a very low importance in decision making. In other words, the political processes involved in energy policy formulation in Australia are preventing the value judgments implicit, in the energy policy initiatives being suggested, from being exposed to public scrutiny.

The problem with such a system of formulating energy policy is that it is basically inflexible. The domination of the process by people who have had many years of technical training in existing energy technologies tends to restrict the scope of suggested energy policy initiatives to modifications to existing technologies, and the development of new technologies, to solve what is seen as an essentially technical problem of meeting future energy demand with an adequate supply. Such a conceptualisation of the problem ignores both the fact that suggested technical solutions may have profound social implications, and also the possibility that the problem itself may be more amenable to a different approach incorporating a social dimension — such as an examination of what our needs for extrasomatic energy really are, rather than assuming that past, present and projected future statistics on energy use are an adequate definition of those needs. Such an approach, through questioning the basis of the conceptualisation of the problem, is inherently more flexible than an approach which is locked into one particular conceptualisation.

The inflexibility, and hence the inadequacy, of current procedures for energy policy formulation in Australia would be overcome if the domination of the technically-oriented approach to policy formulation could be broken. This would involve the development of broad-ranging energy strategies by teams of people trained in areas other than engineering and the "technical" social sciences such as economics. The development of such strategies would require investigations to be made of energy use in Australia — particularly detailed studies of the way people actually use energy as well as improved and disaggregated statistics on energy use (44).

The membership of the government advisory bodies would have to be radically altered to include more people aware of the social implications of energy policy decisions, more women and younger people, and also representatives from the general public. In fact, since energy policy decisions potentially could have profound effects on the lives of every member of society, there is a good case to be made for processes for public participation to be built into the energy policy formulation process, as has been done in several other countries (45, 46). If this were done, the primary role of the energy policy advisory bodies could be to prepare a number of alternative energy strategies to initiate, and to be reviewed in, public debate. These strategies would have to be presented in terms understandable by the general public, with the underlying value judgments and possible implications for future society of each strategy clearly spelled out. It should then be possible for an informed public debate on energy policy to take place, in which a consensus could be reached on preferred energy futures. This should not only be a once-only process; as conditions changed it would be necessary to review and modify existing strategies or develop new ones, and these modifications and new developments could be submitted for public debate. In this way an ongoing, continuous, iterative process of energy policy formulation could be established.

Conclusion

In most western post-industrial societies energy policy and people are at present not very closely related. Policy initiatives are suggested and implemented with little regard for their effect on people, and the general public is not much concerned or involved with energy policy. However, the crucial point is that any action taken on energy policy must recognize that the problem of energy is not just simply one of balancing future energy demand with supply. Energy is intimately connected with all social processes, and energy policy decisions will ultimately affect the basic structure of human society and hence people's lives. It is essential that social consequences of energy policy decisions be considered and accounted for in the development of energy strategies. This is unlikely to happen if energy policy formulation remains, as it is at present in Australia, the exclusive preserve of a highly educated elite, professionally involved in the technology of energy production, distribution and use. In the words of Lynton Caldwell:

> *No lasting solution to the energy problem can be found within the confines of policies [i.e. strategies] concerned only with energy per se. The proper context for . . . national policy for energy is the broader field of social and environmental policy with reference to the quality of life. . . . [E]nergy policy that fails to consider the purposes to which energy is put, and their subsequent chain of consequences, can never be adequate, even for energy narrowly defined* (47).

References

(1) Meadows, D. H., D. L. Meadows, J. Randers, W. W. Behrens III (1972). *The Limits to Growth.* London, Pan Books. 205 pp.

(2) Caldwell, L. K. (1976). Energy and the structure of social institutions. *Human Ecology* **4**, 31-45. (This quotation is from p. 32.)

(3) Newcombe, K. (1975). Energy use in Hong Kong: Part 1, An overview. *Urban Ecology* **1**, 87-113.

(4) White, L. A. (1969). *The Science of Culture.* Second Edition. New York, Farrar, Straus and Giroux. 444 pp. (This quotation is from pages 368-369.)

(5) Budnitz, R. J. and J. P. Holdren (1976). Social and environmental costs of energy systems. *Annual Review of Energy* **1**, 553-580.

(6) Budnitz and Holdren (1976) *op. cit.*, p. 554 (modified).

(7) Carruthers, J., J. Clark, M. de Smet, F. Freckleton, R. McClelland, G. Pike and M. Roe (1976). Historical aspects of energy use by mankind, pp. 27-67 in: *Energy and How We Live.* Canberra, Australian-UNESCO Committee for Man and the Biosphere.

(8) Cottrell, F. (1955). *Energy and Society.* Westport, Connecticut, Greenwood. 330 pp.

(9) White (1969) *op. cit.*

(10) Federal Energy Administration (1974). *Project Independence Report.* Washington, D.C., U.S. Government Printing Office. 338 pp.

(11) Crossley, D. J. (1977). The role of popularization campaigns in energy conservation. *Energy Policy* **7**, 57-68.

(12) Hooker, C. A. and R. van Hulst (1977). *Institutions, Counter Institutions and the Conceptual Framework of Energy Policy Making in Ontario.* (A study prepared for the Royal Commission on Electric Power Planning.) London, Ontario, Dept of Philosophy. University of Western Ontario. 214 pp.

(13) Lönnroth, M., P. Steen and T. B. Johansson (1977). *Energy in Transition.* Stockholm, Secretariat for Future Studies. 134 pp.

(14) Hooker and Van Hulst (1977) *op. cit.*

(15) Australia. National Energy Advisory Committee (1977a). *Proposals for an Australian Conservation of Energy Program.* Canberra, the Committee, 34 pp.

(16) Australia. National Energy Advisory Committee (1977b). *Proposals for a Research and Development Program for Energy.* Canberra, the Committee. 40 pp.

(17) Australia. Senate Standing Committee on National Resources (1977). *Solar Energy.* Canberra, Australian Government Publishing Service. 92 pp.

(18) South Australia. State Energy Committee (1976). *Report of the South Australian State Energy Committee.* Adelaide, Government Printer. 178 pp.

(19) The Institution of Engineers, Australia (1977). *Conference on Energy 1977. Towards an Energy Policy for Australia.* Canberra, the Institution. 242 pp.

(20) Victoria. Ministry of Fuel and Power (1977). *Victorian Government Green Paper. Energy.* Melbourne, Government Printer. 96 pp.

(21) Australia. Department of Minerals and Energy (1973a). *Forecast of Consumption of Primary Fuels, 1972-1973 to 1984-85.* Melbourne, Fuels Branch.

(22) Australia. Department of Minerals and Energy (1973b). *Australia's Fuel Requirements to the Year 2000.* Melbourne, Fuels Branch. 47 pp.

(23) *Ibid.*

(24) Australia. Department of Minerals and Energy (1974). *End Use Analysis of Primary Fuels Forecast.* Melbourne, Petroleum Branch. 47 pp.

(25) Urie, R., I. W. Meldrum, D. W. Hocking, J. B. Kirkwood, R. N. Morse, R. J. Foster, D. W. George, S. F. Harris and J. L. Symonds (1977). Long-term primary energy prospects and usage patterns, pp. 1-22 in: *Conference on Energy 1977. Towards an Energy Policy for Australia.* Canberra, The Institute of Engineers, Australia.

(26) Morrell, F. and F. Cripps (1976). The case for a planned energy economy. *Science and Public Policy*, **3**, 505-528.

(27) Australia. Department of Minerals and Energy (1975). *Forecast Demand for Primary Fuels 1974-1975 to 1984-1985.* Melbourne, Petroleum Branch. 32 pp.

(28) Casper, D. A. (1976). A less-electric future? *Energy Policy* **4**, 191-211.

(29) *Ibid.*

(30) Electricity Supply Association of Australia (1953, 1957, 1961, 1965). *Statistics of the Electricity Supply Industry in Australia.* Melbourne, the Association.

(31) Electricity Supply Association of Australia (1969, 1973, 1977). *The Electricity Supply Industry in Australia.* Melbourne, the Association.

(32) Waverman, L. (1977). Estimating the demand for energy: heat without light. *Energy Policy* **5**, 2-11.

(33) Hoffman, K. C. and D. O. Wood (1976). Energy system modelling and forecasting. *Annual Review of Energy* **1**, 423-453.

(34) Australia. Department of National Development (1978). *Demand for Primary Fuels Australia 1976-1977 to 1986-1987.* Canberra, Australian Government Publishing Service. 38 pp.

(35) Mula, J. M., R. A. Ward, B. S. Thornton and C. Malanos (1977). *Solar Australia: Australia at the Crossroads.* Sydney, Foundation for Australian Resources. 122 pp.

(36) Waverman (1977) *op. cit.*

(37) Hooker and Van Hulst (1977) *op. cit.*

(38) Encel, S. (1977). Forecasting and the social sciences. *Search* **8**, 185-189.

(39) Australia. Royal Commission on Australian Government Administration (1976). Appendix 1. K. Statutory authorities. pp. 311-374 in: *Appendices to Reports Volume One,* Canberra, Australian Government Publishing Service.

(40) Kirkwood, J. B. (1977). North West Shelf gas effect on Western Australia's fuel economy, paper presented at the *West Coast LNG Symposium,* Perth, 15 December. 25 pp.

(41) Crossley, D. J. (1978). *A.E.S. Working Paper 6/78. Energy Policy in Australia: The Social/Institutional Context and Procedures for Policy Formulation.* Brisbane, School of Australian Environmental Studies, Griffith University, 74 pp.

(42) Crossley (1978) *op. cit.* (condensed from table on p. 27).

(43) *Ibid.*

(44) Crossley (1978) *op. cit.*

(45) United States. Executive Office of the President. (1977). *The National Energy Plan: Summary of Public Participation.* Washington, D.C., Energy Policy and Planning, Executive Office of the President. 40 pp.

(46) Saddler, H. (1978). *Public Participation in Technology Assessment with Particular Reference to Public Enquiries.* Canberra, Centre for Resource and Environmental Studies, Australian National University. 32 pp.

(47) Caldwell (1976) *op. cit.*, p. 43.

Institutional Aspects of Energy Futures: The Role of the Australian Atomic Energy Commission

Ann Moyal

Science Policy Research Centre
School of Science, Griffith University
Nathan Qld, 4111.

The Commonwealth Government has declared its intention to forge a "national energy policy". Responsibility has been assigned at the highest level to the National Energy Advisory Committee (NEAC) responsible to the Minister for National Resources established in February 1977 and to NERDDC, the National Energy Research, Development and Demonstration Council appointed in June 1978 to advise the Minister for National Development on the co-ordination and development of a national program of energy research, development, and demonstration, and to allocate funds between energy needs (1). In addition, ASTEC, the Australian Science and Technology Council, established by Act of Parliament as a permanent statutory body reporting to the Prime Minister, has contributed some overall recommendations on the structuring and institutionalisation of a national energy policy in its Report on *Energy Research and Development in Australia* (2) and sees its own role as working in close liaison with NEAC. In commending ASTEC's Report, the Prime Minister, stressed the present need to utilize Australia's science and technology resources for energy development "efficiently and effectively" and encourage their use "rationally and wisely" in Australia's long term interests (3).

Such rationalisation will require some complex co-ordination and decision-making. There are at present some 26 Commonwealth and State Ministries touching aspects of energy responsibilities, over 37 Statutory Authorities linked with energy research, and at least 13 Advisory Committees (national parliamentary, and interdepartmental) dealing with energy questions in Australia (4). Rationalisation, as ASTEC and NEAC have emphasised, must rest on the establishment of national priorities and policies, the shaping of existing programmes to conform to these priorities, and the allocation of resources to achieve agreed objectives (5).

In this paper, I will focus on one major energy agency, the Australian Atomic Energy Commission which, as the second largest civilian scientific institution in this country, must figure prominently in any appraisal of the future institutionalisation of energy policy. Its history and its capacities relate centrally to perceptions and planning of national energy developments.

The AAEC reached its quarter century this year. After CSIRO ($144.3 million) and Defence Science ($84.4 million), its budget accounts for the third largest share of annual national funding for R & D, a sum set at approximately $26 million in the Estimates for 1978-9. The evolution of the Commission has been a particular one. From its inception in 1953 until 1972, its development has been linked with a consistent view of nuclear development in Australia and the involvement of Australian industry in the Commission's active role in nuclear work. The AAEC was established by the Menzies Government with strong Opposition support in 1953. Its purposes were twofold: (i) to make use of Australia's important, but then scarce, uranium resources to gain access to the emerging body of nuclear knowledge abroad, and (ii) to launch Australia as an independent and creative participant in nuclear science and technology. The Atomic Energy Act, 1953, entrusted the AAEC with promoting the discovery, mining, treatment and processing of uranium ore, the operation and construction of plant for the liberation and conversion of atomic energy, and the training of nuclear researchers. It was also apparent from the stringent secrecy provisions and penalties fixed for disclosure of information under the Act, that early planning assumed the possible development of atomic energy for defence and military purposes.

Since its foundation in 1955, the Research Establishment at Lucas Heights has carried out research into nuclear science and engineering, organised graduate scholarship schemes and collaborative research through AINSE (the Australian Institute for Nuclear Science and Engineering), offered its research reactors as a facility for some university research, and through the high flux Australian reactor, HIFAR, provided for the expanding production of radiosotopes.

Historically, there have been two highpoints in the AAEC's evolution: first, the research undertaking that occupied research teams from 1956-1964 in an attempt to develop a high-temperature, gas-cooled power reactor concept using beryllium oxide as a moderator which was seen as a potentially adaptable model for small power units appropriate to Australian needs; and secondly, the planning of a nuclear generating power reactor at Jervis Bay based on imported technology which occupied research teams at Lucas Heights from 1965-1972. Both developments proved abortive, the first because the beryllium oxide research, already explored and dropped by such advanced nuclear countries as France and the USA, proved too challenging; and the second because, though it carried the backing of two Liberal Prime Ministers, Holt and Gorton, eager to bring Australia into the nuclear age, it was terminated by Prime Minister McMahon on the economic grounds that the operation of a nuclear power plant in Australia must be based not on government subsidy, but on costs comparable with conventional power (6).

The Federal Government's decision to foreclose the planning for a nuclear reactor raised the question of the relevance of a nuclear mission-oriented laboratory in Australia. There were some pluses on the balance-sheet. There had been some return to general revenue from uranium exploration and sales for which the AAEC held responsibility; research had been carried out in the nuclear disciplines, Australia had "got its feet wet" in certain nuclear areas and technologies, and the Radioisotope Division provided an expanding and important service to the medical and industrial community.

But on the negative side, the AAEC had, in effect, lost its mission. *"The team is there"* Sir Philip Baxter former AAEC Chairman wrote in 1973, *"but it will pay no major dividends on the investment unless Australia, last among the nuclear nations, does proceed to make use of this major new technology"* (7).

The point was crucial for Australia. The scientific community and the public might well have expected some serious reappraisal of the Commission's role. In the United States, the successful Atomic Energy Laboratories, which had been directly involved in the development of nuclear reactor prototypes, were changing missions under new legislation and diversifying their activities to embrace research in

Diesendorf, M. (ed.) (1979). *Energy and People*. Canberra, Society for Social Responsibility in Science (A.C.T.)

environmental, pollution, computer and other specifically applied fields. Similar diversification and change was also occurring in the British Atomic Energy Authority where the former reactor research laboratory, Harwell, was being adapted to serve wider national scientific needs (8). As Harvey Brooks, eminent American scholar of science and government summed up, *"The time had come when some of the great research institutions we have built primarily as the captive of particular missions and particular agencies should be regarded more as a national resource and should be used for the solution of national problems with less consideration of the mission of a particular agency."* (9).

The point was not taken in Australia. Critics of the generous government funding of AAEC which compared most favourably with that granted to other scientific claimants (10), advised that the Commission's historic role in providing expertise on overseas nuclear development and on safety and licensing arrangements in the event of Australia wishing to take up the nuclear option, could be more cheaply and effectively realised by the secondment of a corps of staff to nuclear laboratories and regulatory bureaus abroad. Yet significantly, the new Labor Government, critical in Opposition, failed to act or to open up discussion of the role and direction of the country's nuclear energy establishment.

Since 1972, certain concentrations have developed at the AAEC. Research on uranium enrichment, begun in 1965, now represents the largest single developmental undertaking at Lucas Heights (11). The Commission became a member of the Association for Centrifuge Enrichment in 1973, and present research includes work on machines and components related to the gas centrifuge technology and an experimental program on laser technology. Since 1976, the Commission has been engaged in consultations with Japan on a feasibility study for the establishment of a uranium enrichment plant in Australia, the final report on which has been submitted to Government. In addition, a pilot plant feasibility study is under way on the installation of an Australian plant to convert uranium ore to uranium hexafluoride. As the Government's Safeguards Policy announced last year indicated that Australian uranium would only be sold in a form that attracted International Atomic Energy Safeguards (12) — a form that requires conversion to hexafluoride — this development is particularly pertinent. It was further intimated last year that the Department of National Resources (now Trade and Resources) might combine with the AAEC to build a plant capable of producing 5000 tonnes of hexafluoride a year at the cost of $50 million (excluding operating costs) for operation in the early 1980's (13).

Two other areas of research which lay outside the original Atomic Energy Act have also been initiated at the Commission: the first in solar energy research where three programmes (on sodium sulphur cells, the direct conversion of solar energy into electricity and hydrogen by exposing various metal catalysts in sunlight under water, and research on total energy requirements of a residence from solar energy) make up a modest ½% of total AAEC annual funding (14); and secondly, fusion research. Professor Stuart Butler, recently appointed Head of the Nuclear Science and Technology Branch, now Director of the Research Establishment, pressed in December last year for the establishment of a $10 million fusion facility at Lucas Heights and publicly advocated that the AAEC Research Establishment should become "the centre of research into all sources of energy" in Australia (15). Fusion work is, as yet, confined to a collaborative undertaking between the AAEC and three universities, Flinders, Sydney and the Australian National University. Under this arrangement a small number of AAEC researchers are farmed out to the universities for research; but it is not without significance that the Atomic Energy Act 1953-73 was amended in June 1978 to enable the AAEC to extend the definition of its atomic responsibilities from nuclear fission to "thermonuclear fusion or other nuclear transmutation" (16).

The AAEC also derives powers under the amended Act for the mining of uranium for civilian nuclear power development overseas, prescribed as "acquisition, production, transportation, treatment, possession, storage, use and disposal of a prescribed substance" (Clause 10).

Given these developments, where does the AAEC stand in respect of a national approach on energy priorities in Australia? Within its institutional framework, the AAEC has remained conspicuously free from Ministerial directives and from demands for accountability on its research front. Alone among such major scientific instrumentalities as the CSIRO, the Bureau of Meteorology, all of whom have been subjected to government commissioned scrutinies in the past two years, there has, as yet, been no Commission of Inquiry into the AAEC.* Programme and policy decisions have, historically, been carried out by the Commission in an environment of secrecy and, since the greater part of research conducted at Lucas Heights is published in the form of in-house papers and reports, there is, and has been, little opportunity for the normal operation of peer review.

An assessment of capabilities and performance of the AAEC must hence become a central question in determining a rational and effective balance between Australian energy needs. There is current expectation that the Commission can no longer be permitted to determine its internal objectives behind closed Ministerial doors. The organisation is essentially seen by some as a "memorial to old problems" in Australia. There are, resultingly, questions about the technological capabilities of an aging research staff who, since 1972, have experienced some demoralisation and inflexibility from lack of identified goals. There is also apparent conflict of outlook between the "old guard" hierarchy at AAEC, conceptually committed to nuclear goals, and the viewpoints of the AAEC Chairman Professor George (a member of NEAC), Professor Butler, Director of Lucas Heights, and the newly appointed Deputy Chairman, Mr A. J. Woods, Secretary of the Department of National Development (both members of NERDDC) who acknowledge the importance of bringing the AAEC into broader energy fields (17).

What modes of change, then, are open to this major energy agency? Key policy questions must be answered. What proportion of national funding should, for example, continue to be committed to the maintenance of a "watching brief" for the technology of nuclear fission and breeder reactors in Australia? Are conventional nuclear power stations envisaged as a viable option for this country (18)? If a watching brief is important for Australia in nuclear technology, what similar commitments should be made in respect of the highly complex, competing, expensive, and as yet unproven methodologies for controlled thermonuclear fusion research (19)? And, will some form of priority assessment between competing claims on the national energy purse be applied in Government decision-making about an Australian commitment to uranium enrichment development, an undertaking which has been estimated would commit Australia to an investment of research and development of $2,000 million for the establishment and operation of the plant for five years?

Change within a mission-oriented agency poses large organisational problems. ASTEC has broadly suggested that a national body such as NERDDC should concern itself with the "programme components of AAEC that fall within the energy area" (20) and that this be affected through consultation and the contracting out of some funding for research. A more dynamic, though not unrelated model of change, may be found in the U.S. organisation ERDA (Energy Research and Development Administration) which consolidated U.S. Federal R & D energy activities in 1974 (21) and supplanted the function of the U.S. Atomic Energy Commission.

In the USA this has meant that the former National Atomic Laboratories including the major reactor Laboratory at Argonne and the weapons laboratories at Los Alamos and Livermore have come within the direct aegis of ERDA and

---

*Since this paper was given, the Minister for National Development, Mr Newman, announced the appointment of a Committee to review the Atomic Energy Commission. The Committee, set up within NERDDC, is made up of the Chairman of NERDDC, Mr J. D. Kirkwood, Dr R. G. Ward, General Manager Research, BHP, and Dr L. J. Farrands, Secretary, Department of Science, and will report to the government in mid-1979.

operate across a diversified spectrum of non-nuclear and alternative energy and environment fields.

Historically the National Atomic Laboratories in America have also been closely linked with regional universities. While ERDA disperses funds for in-house Laboratory R & D, individual or consortiums of universities serve as the operating contractors of the Laboratories. The University of California is, for example, the operator of Los Alamos, Livermore and Lawrence Laboratories; while the University of Chicago and a consortium of nineteen mid-western universities (Argonne Universities Association) operate a tripartite contract of management with Argonne. The concept has some relevance for Australia. Under the Argonne tripartite contract, advisory powers have been vested in the AUA to review Laboratory policies and programmes, to establish co-operative research and educational programmes between the Laboratory and the universities and to develop long range plans for the best use of laboratory facilities (22). Within this ambit, Argonne has developed its particular concentrations of staffing and research but its facilities and expertise have been made widely available to the regional universities.

A similar approach might be broadly contemplated in Australia. AINSE already provides a vehicle whereby a number of university researchers are assisted to make use of facilities at AAEC; and some collaborative research including work in plasma physics and studies of pollution already takes place in the ANU, Flinders, Sydney and Macquarie Universities. The AAEC, however, with its valuable equipment and present plans for large-scale funding to update the research reactor HIFAR could well be restructured to provide a national laboratory run in direct collaboration with a consortium of either regional universities or a broader Australia-wide university group. A regional consortium might reasonably be composed of Sydney, Macquarie, Wollongong, and Newcastle Universities (at the last of which the part-time Chairman of AAEC is Vice-Chancellor) and the University of New South Wales.

The arrangement, which envisages a basic reappraisal of a major scientific institution, has the merit of selecting a number of important and converging proposals advanced in the past year from the Independent Inquiry into the CSIRO, the ASTEC Report Vol. 1A, and the Report of the Science Task Force, for a greater collaboration and interchange between universities, statutory authorities and industry, and for the building up of shared centres of research excellence (23). A contractual arrangement between universities and AAEC, confined where appropriate to designed energy fields, could also enhance mobility between governmental and university centres, encourage short term exchanges for research and teaching and lead to fruitful interpenetration of staff and ideas. It is not necessarily realistic to expect a corps of senior nuclear scientists and engineers recruited since the 50's to deflect their researches into new and radically different fields (24). As U.S. Laboratories have found, however, the infusion of ideas, collaborative programme planning and the process of peer consultation and review can constructively alter an internalised and sometimes stulified atmosphere of long established mission oriented groups (25).

In his opening address to the Conference, Dr Coombs has emphasised that we need to look to our institutions and the men and women who run them to be prepared to adapt to and accommodate change. The future institutionalisation of the Australian Atomic Energy Commission, with its complex policy, resources, administrative, manpower, and programme planning, may prove to be a striking and innovative case in point.

References

(1) *Search* **9** (8/9), p. 292-3, Aug-Sept., 1978.

(2) ASTEC (1977). A Report to the Prime Minister by the Australian Science and Technology Council. October. Canberra, Australian Government Publishing Service.

(3) House of Representatives Debates, 4 April, 1978, p. 908.

(4) I am indebted to David Crossley, Australian Environment Studies, Griffith University for supplying this information from his thesis in preparation on "Energy Policy in Australia".

(5) ASTEC (1977), p. 20. Similar recommendations have come from the OECD *Examiners' Report on Science and Technology in Australia* (1974), p. 34, the Ranger Uranium Environmental Inquiry, *First Report* (1976), p. 186 and the Senate Standing Committee on National Resources, *Report on Solar Energy* (1977), pp. 8-9, 12.

(6) Cf. Moyal, A. M. (1975). The Australian Atomic Energy Commission: A Case Study in Science and Government, *Search,* **6** (9), p. 365-384.

(7) Baxter, P. (1973). What is research worth? *Current Affairs Bulletin,* **49** (11), p. 344.

(8) U.K. Atomic Energy Authority, *Annual Report*, 1976-7.

(9) Brooks, H. (1968). *The Government of Science,* Cambridge, M.I.T., p. 53.

(10) The Bureau of Mineral Resources' annual budget for example dropped from $19 million in 1973-4 to $11 million in 1976-7 and is $12 million in 1978-9. ARGC (the Australian Research Grant Committee) responsible for funding the "most outstanding research being carried out in universities and research institutions" fell from $8.6 million in 1975 to $7.2 million, 1976. The AAEC, by contrast, enjoyed a rising appropriation from $16 million in 1974-5 to $21 million in 1976-7 and $26 million in 1978-9.

(11) AAEC (1976-7) *Annual Reports,* p. 42.

(12) House of Representatives Debates (1977), 24 May.

(13) *National Times,* 13 June, 1977.

(14) AAEC (1976-7) *Annual Report,* p. 65 and House of Representatives Debates, 25 May, 1978, p. 2575.

(15) Raymond, R. (1978). How we almost built the bomb but found the sun. *Bulletin*, 7 February, 1978, pp. 42-3.

(16) Atomic Energy Amendment Act. No. 31 of 1978. Clause 3.

(17) *National Times,* 7 October, 1978, p. 10.

(18) Cf. Institute of Engineers Australia (1971). *Summary Report and Recommendations,* pp. 3 and 13.

(19) Weinberg, Alvin (1974). Institutions and Strategies in the Planning of Research. *Minerva,* **XII** (1), pp. 9-17.

(20) ASTEC (1977), *Energy Research and Development in Australia,* p. ix.

(21) *United States Government Manual,* 1975-1976. Energy Research and Development Administration. Office of the Federal Register. National Archives and Records Service, pp. 448-450.

(22) Mozley, Ann (1971). Change in Argonne National Laboratory: A Case Study, *Science,* **174**, 1 October, pp. 30-8.

(23) *Independent Inquiry into the CSIRO Report* (1977). Canberra. Australian Government Publishing Service; ASTEC (1978) *Science and Technology in Australia,* 1977-78, Vol. 1A, p. 6, Australian Government Publishing Service, 1978; *Towards Diversity and Responsibility,* Report of the Science Task Force, Canberra, 1975.

(24) ASTEC (1978) in its Report Vol. 1A, for example notes that "in the course of time a lack of flexibility and of new ideas may become a problem in the AAEC where recruiting consists only of replacement for resignations", p. 46.

(25) ERDA Laboratories (1977). *Science,* **196**, 13 May, pp. 743-5.

The Newport Power Station and the Planning Process

D. G. Hill

Australian Conservation Foundation
672B Glenferrie Road, Hawthorn, Vic, 3122

This paper presents some personal impressions of various aspects of the planning process in the latter stages of the Newport controversy.

The history of the controversy, very briefly, is as follows. In 1971, the State Electricity Commission (SEC) put to the Victorian Parliament a proposal for a 1000 MW gas-fired power station at Newport. This was approved by both parties with no environmental review. Subsequently, local resident groups questioned the proposals; union bans were invoked; and, in 1973, the Environment Protection Authority (EPA) and

Diesendorf, M. (ed.) (1979). *Energy and People.* Canberra, Society for Social Responsibility in Science (A.C.T.)

the Environment Protection Appeals Board conducted hearings. The various hearings approved the project, subject to conditions. Opposition continued and the union bans were confirmed by the Trades Hall Council (THC).

In 1976, the Government brought in the Vital State Projects Act, intended to break the union opposition. Discussions between Government and THC then resulted in the establishment of the independent Newport Review Panel (NRP).

This paper deals with the NRP hearings, and subsequent developments.

The Role of the Professional Engineer

The Institution of Engineers (Victoria Division) made a submission to the NRP which said in effect that the professional engineering work done on the SEC proposal was competent, and therefore the proposal should proceed.

Accepting for the moment that the work was competent — I will come back to that — this submission indicates a misunderstanding of the decision-making process. It is one which I find widely held among engineers. That is, that the competent professional is well equipped to make a decision about a complex matter like a major power station.

In such a decision, facts are only half the problem. It was one of the virtues of the first part of the NRP that it did sort out the facts. However, once we have the facts established there are a number of value judgments to be made. In this example, they included:

(a) How do we balance the need for extra generating capacity against some risk of energy shortfall;
(b) How do we balance a risk of air pollution against a need for extra energy;
(c) What visual intrusion are we prepared to tolerate in a residential area?

I think the essence of the Newport controversy was the question of who made these and other value judgments. Notionally, Parliament did — and I agree that it should. However, in effect, the position was that SEC engineers used their professionalism to inhibit public discussion.

The attitude of the Institution of Engineers, that the SEC must be right because they are professionals, is disquieting.

Very clearly, the community must decide what is to be done by engineers; and then should expect that the job will be done competently. However the first stage is the important one.

Engineers have an important role to play in providing information as a basis for value judgments by the community. However, they must learn that it is not their role to pre-empt community decisions. There is an interesting article in June 1978 issue of AQ by Professor Peter Mason, of Macquarie University, discussing the roles of experts and facts in complex decision-making (1).

In my view, in a parallel situation, the Ranger Uranium Environmental Inquiry ultimately established the distinction between facts and values very well. They sorted out the facts and defined the value judgments which must be made as part of the political process.

SEC Competence

A central theme of the Institution of Engineers submission was that the SEC is competent. I have no wish to deny the professional competence of individual engineers in the SEC. However, I ask you to ponder on the following conclusion by the Review Panel (p. 17, 18 April Report) (2):

> *9. We have been greatly concerned, throughout this very serious and difficult review, by the attitude of the State Electricity Commission which to say the least has not been in keeping with the way in which the Panel has endeavoured to conduct its proceedings.*
>
> *The SEC has appeared to be most reluctant to concede that any of its decisions, whether technical, commercial, financial or environmental were open to challenge.*
>
> *Such an inflexible attitude does not bode well for the future. We have only reached the decision recorded above after agonising over the correct attitude to adopt in this most difficult matter. Our recommendations have to be read as a whole and strict adherence, not only to the letter but to the spirit, of the proposed conditions of operation of Newport is central to what is recommended.*
>
> . . . .
>
> *We therefore urge the Government to ensure that these vitally important matters are built in to the thinking of all its agencies especially, in this context, the SEC. If new legislation is required it should be enacted.*
>
> *What is really needed is a change of attitude so that environmentalists and the Commission itself will accept the same basic values as they approach the continuing problems of Victoria's development.*

I mentioned a query about the competence of the SEC. At the Inquiry, the SEC claimed that station emissions would not add to pollution concentrations in excess of standards. They postulated essentially two cases:

1. With a low inversion leading to high concentrations of emissions. For this case, the SEC claimed the plume would penetrate the inversion, and disperse.
2. With an inversion high enough to trap the plume. For this case the SEC claimed that emission concentrations would not exceed standards.

During the various hearings from 1973 to 1976, the SEC made a number of strong statements to the effect that under no circumstances would the station contribute to ambient pollution concentrations when those concentrations exceeded standards (3).

In the event, the Australian Conservation Foundation (ACF) presented data on the plume behaviour for a number of days in January-February 1976. Of a total of 46 days examined in the period, we concluded that the plume was trapped on 24, and that on 15 of these the pollution concentrations exceeded standards. The Panel had our evidence reviewed by two outside experts and concluded "*the position of the ACF . . . was supported . . .*" (4).

This experience raises interesting questions about the objectivity of professional staff in major organisations. I find it difficult to believe that no-one in the SEC could not have shared some of our doubts about the plume behaviour. If we, and the NRP's experts could do the analysis, why couldn't the SEC, who are the "experts"?

The Process

I move now to discuss the decision making process.

The first stage of the Newport Review went well. The Panel members were competent and used technical advisers when they needed to. Its hearings provided a forum for cases to be heard, tested and developed, and it seemed to give the SEC plenty of opportunity to develop responses. All in all, a rational process. It concluded by finding that if a better site than Newport was available at a reasonable cost then that site should be chosen. That report was dated 29 March 1977. The Panel met first on 5 January 1977, 12 weeks earlier.

The second stage of the Review lasted just four weeks, with the Panel reporting that the best alternative site would involve an excess cost of $100 million, which "*could not be regarded as reasonable*". None of the evidence to the second stage was presented in public or subject to review by the public in any way. Questions of what was "reasonable" were not explored — and should have been.

In its first report, the Panel had noted that the Newport controversy arose because "*a significant segment of the public did not have confidence in the procedures used to approve the Newport project*" (4).

They suggested "that consideration be given to developing new procedures in which the public would have greater confidence", including full public review. Then they went ahead and ran a secret inquiry to produce the answer the Government wanted.

In the matter of process, worse was to follow.

The Newport Panel concluded that only 500 MW of gas turbines were required. Within weeks of that report Parliament considered a proposal for an extra 200 MW of gas turbines to be located in the Latrobe Valley. That proposal was approved on 3 May 1977 without any public review.

Having found it so easy, the SEC then put forward on 21 December 1977 a proposal for a further 200 MW of so-called "Emergency Gas Turbines", bringing the gas-fired power capacity back to 900 MW, compared with the original Newport proposal of 1000 MW.

The Government balked at this and referred the proposal to the Parliamentary Public Works Committee (PPWC) for review.

The SEC based the proposal on alternative annual growth rates of 6.8% and 5.8%. The former was precisely that which was rejected by the NRP eight months earlier.

The PPWC had no apparent technical expertise and no technical support staff. Its meetings gave a distinct impression that they considered the SEC a fine body of competent professionals and that their role was to work out whether the station went at Jeeralang or somewhere else.

They found that there was a need for the station and that the project should go ahead.

On this occasion, the ACF put a case as follows:

1. The SEC proposal did not meet what the SEC claimed was the need. The SEC in fact did not define a single "need". They put forward four alternative cases showing various shortfalls of energy over the next three years. The minimum "need" was:

1978-79	100 GWh
1979-80	1000
1980-81	300

It was to meet this (or greater) need that the Emergency Gas Turbines are proposed. There is essentially no requirement for later years, when other new stations will be available.

The capacity of the turbines to meet the need was not defined in the SEC proposal. However, it could be assessed as:

1978-79	—
1979-80	330 GWh
1980-81	800

Comparison of the "need" with the proposal shows the dismal state of planning and review in Victoria.

GWh	*Need*	*Proposed*
1978-79	100	—
1979-80	1000	300
1980-81	300	800

There is no evidence that the PPWC compared the proposal with the need.

2. The second part of our case was that an energy conservation program could provide at considerably less cost a benefit equal to that of the proposal. Interestingly enough, the SEC did not deign to tell Parliament what the operating cost of the station was. They claimed the gas contract was confidential. So much for Parliamentary control. In the event, we were able to deduce the annual operating cost when the SEC claimed that EPA requirements to control emissions would reduce efficiency. We suggested that the annual operating cost would fund an attractive energy conservation program, which would yield benefits equalling those from the proposal. The proposal output is equivalent to conservation at the rate of 1% a year.

We suggested that an energy conservation program yielding 1% reduction in demand a year could equal the benefit from the proposal. This compares with the assessment by Dr Doctor, consultant to the Newport Review Panel, that a 5% reduction in system load through early initiation of targeted energy conservation programs is well within the capacity of the SEC. It compares also with Canadian experience of conservation savings at a rate of 1½% of a year.

We were questioned by the committee on such centrally-relevant matters as ACF funding, and how I was authorised by ACF members to speak on their behalf.

In subsequent correspondence, the Premier has indicated that the Government is "committed to energy conservation" — whatever that might mean.

In the meantime, the SEC has placed its orders and is building as fast as it can.

Organisational Implications

It seems to be coming fashionable for conservationists to suggest that what is required for rational energy planning is the elimination of competing agencies. It is claimed — to take the example of Victoria — that separate agencies such as the Gas & Fuel Corporation(G & FC) and the SEC compete to expand the energy market. This competition can be eliminated, it is claimed, by creating an Energy Commission, as in W.A. Now, we find the G & FC to be a rational organisation. They make information available and discuss both policies and particular issues — at all levels. If the G & FC took over the SEC, things would be fine. But what happens in the reverse case? We would have the SEC even bigger than it is now. There would be only one source of information on energy matters. I shudder at the possibilities.

It seems to me that the age-old principle of checks and balances has a lot going for it. Very clearly, in my view, the SEC is out of Parliament's control. The limited number of staff at the Ministry are essentially ex-SEC people. Who is to control them?

One hope is that the State Co-ordination Council, in the Premier's Department, might facilitate better review of projects.

A second step, which I commend, is that the SEC might be fragmented into a series of organisations — perhaps one to construct stations and one to operate them. Retail distribution could be hived off altogether. The routine response to such a proposal is that it would involve duplication of some functions.

If those duplicated functions were at one another's throats pressing their own interests, I would have no doubt that overall efficiency would be improved. A similar situation arose with the Manapouri controversy in New Zealand in 1969-70. The proposal was essentially similar to that at Lake Pedder — to destroy a beautiful lake for marginal energy output. In the Manapouri case, there were two organisations. The Ministry of Works was the construction authority and the Electricity Department the operator. Electricity wanted the lake flooded; the Ministry had its doubts. At the Royal Commission of Inquiry, I acted as technical adviser to the conservationists. Solely because the Ministry was partly opposed to the project, we were provided with access to the computer simulation model of the scheme. With this, we were able to present alternative cases and argue for a staged development, with Stage 1 not involving raising the water level. This argument was accepted and, subsequently, the development was stopped at Stage 1. Quite clearly, in my view, if there had been only one organisation involved, we would not have been given access to the information which let us develop the alternative scheme.

Clearly, from my experience of various energy controversies, the prime requirement for public control of energy authorities is information. It is all very well to claim economies of scale for "rationalised" organisations. The advantages to the public from having several sources of information clearly outweigh any conceivable economics of organisation.

References

(1) Mason, P. (1978). Nuclear decisions. *Australian Quarterly* **50** (2) 7-21 (June).

(2) Victoria. Newport Review Panel (1977). *Final report to the Government and the Trades Hall Council.* [Melbourne] April. pp. 17, 18.

(3) Victoria. State Electricity Commission (1974). *Newport Power Station: evidence for the Environment Protection Appeal Board Hearing by Mr R. P. Llewellyn* [Melbourne] S.E.C., p. 1.

(4) Victoria. Newport Review Panel (1977). *Report to the Government and the Trades Hall Council.* [Melbourne, The Panel], March, p. 52.

How can we cure the Machines? Harmful Technology, its Reasons and Remedies

Alan Roberts

Physics Department, Monash University
Clayton, Victoria, 3168

1. Technology as a Hope and Threat

Physical exertion need not be unpleasant, as any footballer or squash player will testify. Even pick-and-shovel work, or the manipulation of heavy loads, can produce a rewarding glow in a volunteer whose muscles are adequate to the task, particularly in good weather and on a holiday basis.

It is quite different when the brute necessity of gaining a living demands such activity without respite — rain or shine, whether you feel like it or not, and regardless of the current state of your arthritis. This is close to the condition of a draught animal, and leaves only a small margin of life in which something like human freedom can be enjoyed. Reverent eulogies to the mystic beauty of labor indeed exist, but are rarely observed to flow from people chained for life to any working tool more massive than the pen.

To liberate humanity from compulsory labor is a necessary part of any real liberation; to eliminate extreme physical effort is a particularly important aspect of this process. This is why people look, and quite rightly, to science and technology as holding the promise of liberation, and why they particularly value the sources of power now available which replace the human muscle.

It is all the more remarkable, then, that proposals to expand society's power supply now evoke opposition throughout the industrialised countries, wherever the state of civil liberties allows its expression. It is not a question simply of the environmentalist lobby which, in its own right, has neither the numbers nor the social strength to impede a natural-gas station at Newport, a hydro-electric extension in Tasmania, a nuclear reactor in the USA or Germany. The effectiveness of this lobby derives crucially from a groundswell of support in the general community.

This sentiment does not aim to restore the human muscle to its historical role; it is directed against quite specific forms of power technology. Nor is the broader sentiment of which it forms part a revolt against *all* technology; it is evoked by quite definite examples of a particular technological path, which are seen as betraying that promise of liberation which technology still holds out. (The straw men who are alleged to abominate all technology would need to have a peculiar program indeed. What are they supposed to advocate — ban the wheel, burn all the digging sticks?)

It is true that this sentiment is often given a misleadingly general expression. In Australia, for example — as in most countries where the wedding of science to the way of life was celebrated long ago — the honeymoon is over. Now the air is thick with recriminations against the former idol, looked on with distrust and fear by a wide cross-section of people — ranging from residents threatened by a freeway's bulldozers, to bank clerks due now in their turn for sacrifice to the Moloch of the computer.

This is no mere lovers' quarrel. How deep the disillusion goes can be read from such data as the spread of the alternative-technology, simple-living movement, or the decline of enrolments in science and engineering courses. Advertisers appeal to it; governments must now reckon with it; giant corporations like Shell, employing social scientists to peer into the years ahead, prepare contingency plans for production and marketing in a future where — perhaps — the revulsion against "expansion" and "progress" shapes national policy.

I am not going on here to justify this revolt against technology; there are enough well-documented cases extant — e.g. exposing the illusions of the freeway solution to transport problems, or cataloguing the unacceptable aspects of nuclear power — to save me the trouble. I will simply state my belief that it is usually real evils which are under attack, and that the widespread opposition to unlimited technological progress is itself a heartening sign of a shift from the mass society and the atomised consumers of a couple of decades ago towards something much better.

But, although one can readily expose the defects of modern technology by an appeal to the brute facts, it is not so easy to explain and understand how those defects arose and why they continue to breed so profusely. It is this question, of far more than academic interest, which will be discussed here.

Of course, we do not lack for general observations about the attitudes of mind which encourage and spawn harmful technologies. These attitudes can be described in a great variety of ways, usually in terms that are explicitly evaluative and condemnatory — as indeed they should be. (The divorce of "facts" from "values" always depends upon an uneasy, fragile and artificial delimiting of the context of discussion, which could not here be justified.)

We might say, for example, that the trend of modern technology is indifferent or is inimical to the welfare of the people affected; that in worshipping abstractions like progress, size and ingenuity, it neglects the sole justification for its existence: human welfare. Or we might point to the culturally inbuilt urge to "dominate" nature, which now runs counter to the ecological imperative of nurturing the only biosphere we have. We might even analyse this attitude in terms of the Judaeo-Christian tradition, contrasting it unfavorably with the more enlightened Buddhist way. Or the "objectivist" mode of thought could be indicted in its entirety, and blamed for the creation of a social world deprived of feeling and basically inhuman, mimicking the grey and poverty-stricken universe of the laboratory.

To describe such states of mind does not have to be a false or useless pursuit. Undoubtedly a way of life needs its cultural supports, its conventional wisdom, its dominant ideologies, and so is vulnerable to critical attack on this plane. But such critiques, however true they may be, can hardly satisfy us, if our aim is not just to refute a view in theory, but to change a state of affairs in practice.

In a way, these criticisms of social attitudes fail by their very success: it is possible to formulate them in so many different ways, within such a variety of conceptual frameworks. Thus they can give a whole host of illuminating insights, but no one can really recommend itself as that which, compared to the rest, is the crucial point to grasp. They rarely address themselves to the questions that need answering if a viable change is to be achieved — questions like: Why are these attitudes so prevalent, that nurture harmful technologies? What are their economic, social and political supports? What must be changed for these attitudes to change?

There is another approach which certainly supplies these deficiencies: the traditional Marxist critique, common to a whole spectrum of political thought that is otherwise very divergent, ranging over orthodox communist parties, maoists and trotskyists.

Diesendorf, M. (ed.) (1979). *Energy and People*. Canberra, Society for Social Responsibility in Science (A.C.T.)

Marxist critiques of harmful technologies

In this approach, the root of harmful technological practices is located in the economic sphere, and stems from the private ownership of the productive machinery. Capitalism, producing for profit rather than use, will seize on even the most noxious technology if it promises fat returns. The solution is then apparent and unambiguous: get rid of capitalism.

This view may appear simple, but it cannot be dismissed as simplistic. Capitalism has a clear criterion for deciding whether or not production will be engaged upon, that controls the bulk of its economic activity. This criterion — the need to enrich a small minority — is dominant, and cannot in general be expected to yield to some other basic imperative, like human welfare or the preservation of the environment. It is not a question of hard-hearted Scrooges insisting on their rights as exploiters, but of a system which can reject the demands of its driving motor — the search for profit — only at the risk of choking it off altogether, and bringing the system itself to a halt.

I would agree, then, with the view that a capitalist society must bring forth an endless series of harmful technologies; to accept the inevitability of capitalism is to envisage a likewise endless series of rearguard actions to combat and limit the harm such technologies can do. Indeed, it is optimistic here to use the word "endless", since a possible end to the series is only too grimly apparent.

So far, so good; but it is hardly far enough. Countries which have shut off the private-profit motor seem, in general, hardly less enthusiastic in speeding towards the ecological abyss, and equipping themselves with the technological apparatus most essential to this race. Their attitude towards nuclear power, for example, tells us much. The USSR is the only country engaged in mass production of nuclear plants (Atomash is to turn out a giant gigawatt reactor each *month*), and actively planning the switch to breeders. China, an oil exporter, is negotiating with France for pressurised-water reactors of the Westinghouse model; Poland, with its enormous coal reserves, has nevertheless joined the nuclear-power club.

Of course, the orthodox-marxist schools will extend their analysis to cover such phenomena. It is hardly necessary to comment on the "loyalist" trends, faithful unto death to a particular State, who may deny — in the teeth of the evidence — that it suffers environmental and social damage through harmful technology, and/or explain how a reactor in the USSR is quite a different thing from one in the USA. After the People's Bomb, we are offered the People's Reactor. Maoists will account for Soviet technology, and its "convergence" to the capitalist pattern, by the turn of the Soviet leadership to "revisionist" ideas (some maoist trends will now say the same about China, following the triumph of "revisionism" they see in the ascent of Chairman Hua and Teng Hsiao Ping). Other schools will solve the problem by dubbing the USSR a State-capitalist country; naturally its technology will be capitalist also.

Most trends within the Trotskyist movement will relate the development of technology in such countries as the USSR and China to the usurpation of political power by a bureaucracy. This analysis, which rests on a theory of bureaucracy far less arbitrary and much worthier of respect than the accounts mentioned above, sees it as hardly surprising that an exploiting caste — the bureaucracy — will have a technological policy similar to that of an exploiting class — the capitalists.

I cannot accept any of these "supplementary" explanations as satisfactory. It is easy to understand why the drive for private profit will bring about technologies harmful to human welfare and the environment alike; but these "addenda", designed to cover countries where economic activity is not determined by entrepreneurial greed, are nowhere near as transparent or as convincing as the original analysis. Indeed, it is when we compare them with the case against capitalist technology that their deficiencies emerge most clearly. None of them can provide a clear logical thread, indisputably tying the harmful technologies to the social and material interests of a powerful group of people who simply cannot do otherwise without abandoning their privileged positions.

This is true even of the least unsatisfactory of these analyses: those which start by recognising that these countries of "statified" production are ruled by a bureaucracy. For, even if one says that this bureaucracy "exploits" the workers, the mechanisms of this exploitation differ significantly from those of capitalism, and this difference cannot be rubbed out even by dubbing their economies as "State-capitalist". Is it just a coincidence, then, that the same technological means suit their purposes also? But if we call it a coincidence, this is just the same as saying: "I don't understand it" — although, of course, we have dignified our ignorance with a four-syllable word.

Yet understanding is vital here. Unless we see clearly why harmful technologies arise, *everywhere* that they arise — whether in the USA, the USSR or ancient Mesopotamia — we cannot be sure that the society of our hopes will really see their elimination. But the scale of potential damage, both social and ecological, is now so intimidating that we must regard the achievement of benign technology as an acid test for any projected new society.

These introductory remarks should explain what this paper is about; if the size and importance of the problem is appreciated, I may be forgiven for offering only some relevant first considerations, and certainly nothing like the last word.

If we want to explain something, it is a good idea to start by describing what we are trying to explain. So first, let us see if any generalisations can be made about the kinds of technology that most disturb us.

I do not believe that this task can be achieved by simply fixing a date to serve as a watershed, so that everything up to 1930 (or 1890, or 1780) was good, and everything after that has been bad. I think one can point to harmful technologies that existed as long ago as Adam Smith, and to liberatory ones that were developed last year. The problem is to disentangle the unwelcome strand, and trace it back to its social origins.

2. Economic Efficiency or Unshakeable Dogma?

Critics of modern technology broadly agree in levelling three general charges: it is too big, too centralised and too complex.

It is not hard to expose the social evils to which these features lead. The workers expend their energies on fragmented tasks which have no meaning for them, in a production process generally so large-scale and intricate that it escapes the comprehension of all save a privileged few. The consumers are "persuaded" and programmed to serve the ends of that productive machinery, whose enormous capital requirements make it unthinkable that the disposal of the product be left to the whims of a free, unmanipulated market. Increasingly the citizens lose any degree of autonomy, and become helplessly dependent on the centralised institutions called for by that centralised, gigantic economic machine.

Such criticisms may sometimes exaggerate the degree of atomisation and de-humanisation actually achieved at present, but they are certainly not devoid of truth. Yet they tend to sin by default; one could easily conclude that these evils arise from the single-minded, deplorable but *successful* pursuit of a narrowly-conceived economic nationality: the production of more goods with less labour cost.

This goal of greater economic efficiency is, of course, the proclaimed goal of all the systems we are discussing. Capitalism prides itself on delivering the goods as no other system has done; the Soviet leadership accepts the criterion of higher productivity, as eventually determining the outcome of the struggle between their "socialism" and the world system of imperialism.

However, the fact that a country's leaders have proclaimed something is not an infallible guarantee of its truth. Most of us will have noticed such discrepancies in our lifetime; they tend, moreover, to crop up with especial lavishness when those leaders are proclaiming what their own motivations are. It need not even be assumed that they are telling conscious lies; people have a great ability to kid themselves, when the truth is unpalatable.

In any case, we do not need to delve into psychological phenomena here; it is not really their motivations we are concerned with, but their actual productive goals as revealed in practice. Are they really dedicated to economic efficiency above all?

No, they are not. To support this assertion fully would require much more time and space than are available here, but some significant indicators are worth considering.

Suppose that the aim of production really was the attainment of maximum efficiency, in strictly economic terms. Here are some activities we could expect to see:

Evidence now exists from many sources that the productivity of workers increases dramatically, when a scheme of self-organising work groups replaces the usual pattern of hierarchical control (a typical productivity rise seems to be about 20%). Seizing on these results (and they have come from Sweden, Norway, Great Britain, the USA itself), our single-minded economic leaders would be implementing a vast experimental program throughout industry, designed to pave the way for the generalised introduction of such autonomous work groups . . .

Noting the hard evidence that nuclear power stations show a catastrophic drop in efficiency as they are built larger, all the relevant representatives of "Economic Man" would have cut back on plant ratings years ago, when the data first emerged (1). Instead of the sizes now standard (1,000 or 1,100 Megawatts), more efficient reactors around 500-600 Megawatts would be in the pipeline. And of course, our devotees of economic efficiency would never have ordered larger and larger plants in the absence of operating data on even the smallest ones (as they did over ten years ago); they would have demanded hard proof that the new technology was a practical proposition. In fact, they would never have gone nuclear at all (2).

Similar conclusions would have guided them in the building of coal-fired stations, where a tough-minded analysis of the operating data indicates an optimum size (around 300 Megawatts) much smaller than the ones most favored in practice. (3)

Such samples can be multiplied. Telecommunication networks in the USA, at least for trunk line use, seem to have got too big (4) — the cost per call is increasing with size of network; Bougainville Copper built an expensive ball-mill of impressively novel and gigantic size, which failed to deliver the goods; innovative structures (like some bridges of unhappy memory) embody wonderful technology, but have a distressing habit of falling down. In all such cases, our careful and superbly rational decision-makers would be drawing the necessary conclusions.

In fact they show little sign of doing so; despite such cases where expanded scale or more complex technology is manifestly less efficient, their investment decisions seem to be guided by unshakeable dogmas: bigger is better, more centralised is better, more complex is better.

Of course, much of the time they may still be right — as they certainly were in the past in most industries, if we retain the narrow economic criterion of what is "better". It is even difficult to say how often they are wrong now; the examination of alleged "economies of scale" is studded with notorious ambiguities, and new technologies are not easy to evaluate (5). The point to be appreciated here is that the decision-makers themselves operate in a similar grey area of uncertainty, as sociologists who specialise in this field are quick to recognise. Is the kind of example cited above a rare and negligible phenomenon, or does it simply represent the tip of the iceberg? We do not know. Nor do the decision-makers (6).

In the case of capitalist countries, an obvious objection to the views above needs to be confronted. Isn't the rule of the profit motive a sufficient guarantee that — within certain limits — investment will be channelled into new capital goods that really raise productivity? If firms too often act contrary to this imperative of increased efficiency, will not their profits fall, will not more "national" capitalists be able to seize the market from them?

This is indeed true for that sector of the capitalist economy which remains competitive. But it is not this sector which is dominant today; the system of competitive "free enterprise" lives on only in the self-congratulatory and blatantly false speeches of company chairmen engaged in public relations exercises. Above all, today, we are dealing with a system of monopoly, or more usually oligopoly, in which a few giant firms dominate the market and are careful not to compete in price. So long as each of these firms takes the same path of technological development, they need not fear the consequences of a bad decision; the extra cost of a "mistake" will simply be passed on to the consumer in a price rise. And it is precisely this sector, of course, which is chiefly responsible — if not for the invention, then for the wide dissemination of a new or expanded technology.

There is another and perhaps even more important factor to be considered: the State (7). No capitalist system can maintain itself today, without constant and massive intervention by the State in the daily workings of its economy. We live in a society "stabilised" by defence contracts, investment allowances, import tariffs, direct and indirect subsidies of every kind. The nuclear power industry in the USA bears eloquent testimony to the effects of this intervention; dubious (and with good reason) about the profitability of the field, General Electric and Westinghouse were persuaded to enter it only after the offer of guaranteed State contracts — the power companies, for their part, agreed to buy reactors only when the Federal Government brandished the stick of "public power", and dangled the carrot of lavish subsidies (8).

To expound precisely how a capitalist system operates today is not possible here; I would come up against the limitations on both my space and my knowledge, not necessarily in that order. Fortunately, the discussion here does not require an analysis of such daunting scope. The needed point can be summed up very simply:

It is just not true, that the evil content of modern technology can be seen as an unfortunate by-product of the search for greater economic efficiency. On the contrary, when the trend to over-centralisation, over-complexity, gigantism, comes into conflict with a narrowly-conceived economic "rationality", it is the latter which we can find giving way.

The examples given, and others which could be cited, by no means "prove" a *universal* irrationality of the kind described. Their significance lies rather in the evidence they offer as to the relative strength of these two impulses, which is difficult to gauge in general but appears more transparently in instances of a sharp conflict between them. It is when two forces oppose each other that their relative pulls can be gauged; it is in a time of crisis that the priorities assigned to different goals must be made explicit.

In some cases, the deviation from economic rationality can be easily understood. Whether in capitalist or "socialist" countries, we could not really expect the political or economic leadership to grasp eagerly at the opportunities for increased productivity, that arise from eliminating hierarchical control of the work process. As a writer in the *Spectator* observed over a century ago, such arrangements "do not leave a clear place for the master". (Quoting this, Marx added: *"Quelle horreur!"*)

The way in which the work process is organised must of course be considered as a part, and not an insignificant one, of the prevailing technology. But inefficiency due to the other features looked at above — size, centralisation, complexity *for their own sake* — are not as readily understood as is the organisational aspect. Evidently we are dealing with something very like an ideology; at least, with a set of prejudices and dogmas that the facts find difficult to shake. These predilections on the part of the decision-makers are significant phenomena to which a number of different disciplines are relevant: psychology, even psychoanalysis, sociology, economics, politics, perhaps even philosophical theology . . . The reader will be gratified to learn that any such comprehensive analysis will be cravenly sidestepped here.

Why do decisions about new technology so often take the unwelcome shape sketched above? This question could be answered in a hundred different ways at a hundred different levels, but only one relevant point will be made in what follows next. It is a simple and indeed an obvious point; but sometimes there is value in stating the obvious.

3. *Who Benefits from Harmful Technologies?*

Big, centralised, complex. On whose ears do these words fall sweetly? What social groupings could be expected to favour technology having these characteristics?

Let us first consider — and then push to one side — the most obvious candidates: scientists, engineers and technologists generally. They must be considered because their special interest in such developments hardly needs to be argued for. Any new advance, even if the novelty is only in size, must call on their abilities, must depend on their achievements; their social importance will increase, the resources under their control must expand, they may even make a little more money. Looking at it a bit less crudely, we might note that such motivations as the spirit of scientific curiosity, or the engineer's compulsion to make a dream into an objective reality — driving forces by no means to be despised — will predispose them towards climbing new Everests, simply because they are there — or can be built.

But if they must be put aside in our considerations, it is because that is just how they are generally treated when conflict arises over their projects and proposals. Their social power, when it comes to decision-making of the kind we are considering, is minimal. To the boards of giant corporations they are simply employees who must know their station, and they rank somewhere below the marketing branch. To the cabinet minister or the Politbureau member, they are advisors on a leash, in a relationship usually tinged with some contempt for their political naivety.

The technologists do not rank high among the decision-makers, even when the question is one of introducing new technology. If we want to locate groups responsible for major technical innovations, we do wrong if we assign the technologists much more responsibility than the construction workers who build its housing.

But there are other groups, similarly benefitting from these trends in technology, who cannot be so easily dismissed. They include the executive officers of large corporations (private or State), the wielders of political power (whether in parliamentary or single-party system), the administrators highly enough placed.

If we recall the social evils of modern technology denounced by its critics, it will be apparent that every single one of them is only an evil if viewed from "below". If one ascends sufficiently high in the social, economic or political structure, they are each transformed into nearly-unmixed blessings.

The withdrawal of initiative and understanding from the work force? Only another way of saying that more responsibility accrues to the management level. Manipulation of the consumer? But this lightens the task of the economic hierarchy; politically, it results in a population easier to administer and less liable to irritate with "unreasonable" demands. The loss of citizen autonomy and community, through centralisation? But there is nothing wrong with this increased dependence on distant authorities — nothing wrong, that is, if I happen to be part of such an authority, so that my power increases with your dependence.

The point hardly needs laboring further. We are dealing with societies that incorporate a ramifying network of hierarchy — hierarchy in the productive sphere, for example, with power and authority increasing steadily from near zero, as we move up from shop floor through plant manager to managing director; a structure of inequality that characterises all the major social and political institutions. The distortions we see in the pattern of technology are precisely those which suit the purposes of hierarchy in general, which expand particular empires and funnel control increasingly to the top.

But it is within this herarchy that the decisions on new technology are effectively made: the all-important decisions which allocate resources and lay down the main lines of research and development, thus determining what will be technologically feasible in a few years' time; further decisions on the investment of capital, on the relative rates of growth of different branches of a firm, on the State encouragement awarded to different forms of industrial undertakings. It is not hard to understand, then, why technological deformations take the particular form that they do, or why this form is so similar in the hierarchical societies of both "East" and "West".

Appreciating this, we might well wonder, not why technology is deformed, but why the deformations have not swollen to far more monstrous size. (One short answer might be: wait and see!) But of course, the hierarchy's decision-makers operate within severe constraints, the most important of which is that proclaimed desideratum of "economic efficiency". We have already noted, however, the significant range of cases in which these constraints may be evaded, and a freer rein given to the centralising, expanding forces which suit a hierarchy's book.

Most important here is probably the field of research and development, where criteria of immediate efficiency hardly arise. Will the power supply develop in the direction of nuclear, or of solar? The question is easy to resolve: simply allocate billions to the development of nuclear reactors, while measuring funds for solar research in petty thousands. Result: there will be functioning nuclear reactors before solar power is anything more than a pipe-dream. Now the decision can be made on grounds of economic practicability alone: solar is simply not a present alternative. The rules have not been breached . . .

It should be emphasised that nothing above implies the existence of a conscious conspiracy, with ruthless hierarchs gathering in a smoke-filled room to plot fresh moves towards centralisation, size, complexity — rubbing their hands with glee, perhaps, as they chuckle: "Now they will be even more helpless and dependent!" No, such a conclusion would greatly underestimate the complexities of the human mind, and the mechanisms by which what suits us becomes what is right. Was Charles Wilson consciously cynical when he declared that "what is good for General Motors is good for the country"? It would be overly naive and simplistic to assume this; it may well have been a profoundly sincere profession of faith — even a patriotic one . . .

Nor should the general concept of "hierarchy" delude us into imagining that all hierarchies are equal in power. In a capitalist country, each hierarchy — political, educational, social — will in practice subordinate its goals to those of the economic system and its profit motive; by a not-so-curious coincidence, the values and attitudes it has historically formed will attend to this requirement almost automatically. In the Soviet Union, it is the political hierarchy which has both the first word and the last word; again, no coincidence. In this respect, as in many others, there are profound differences between the two types of hierarchical rule which cannot be glossed over, despite the similarities that — as we have seen — allow their technological convergence.

To sum up, then: in both "East" and "West", decisions on the nature of new technology are made by groups which — whatever their differences, both within each camp and between the camps — share a common interest in maintaining and strengthening hierarchical structures of social relationship. The constraints within which they operate — that they must aid the cause of private profit, or raise the efficiency of "socialist" industry — still leave a significant amount of leeway for them to exercise choice of the path that technology will follow.

It may be emotionally satisfying, but it shows little realism, to abuse them for exercising this choice in the repugnant ways that they do. Would we really expect a group with special interests to exercise what power they have *against* those special interests0 They may, of course, contain deviant individuals who, from humanist principles or ecological enlightenment, identify themselves with, and act in the interests of, a broader group than that to which they belong — even, perhaps, a group as broad as humanity itself, present and future. Such deviance might be described — and I think a seemingly-outdated phrase can be justified — as showing the influence of higher moral principles. But one would go sharply against the evidence, with more optimism than realism, if one fancied that such deviant behaviour was becoming the norm.

Ineffective methods for countering hierarchies

If you believe there is something in the argument so far, then there would appear to be only two broad lines of solution to the problem. And let us recall again: in a world where the integrity of the ozone layer is threatened, where the very oceans can have their vital ecological role destroyed by oil spills and undersea mining, where the carbon dioxide content of the atmosphere rises inexorably — in this situation, there can be few problems so urgent and all-encompassing.

Two choices, then, are all that broadly appear: either eliminate all hierarchies in society, or eliminate their control over new technology.

To do away with hierarchy is, of course, the traditional anarchist prescription. I am very much in sympathy with this — not just as a desirable long-term goal in a socialist (without quotes) world, but as a short-term perspective providing a touchstone for political action. Where I would part company with the anarchists is in their estimate of the speed and thoroughness with which this change can occur.

It is true that experiences of collective action and achievement, which abound in times of "mass upsurge", can produce extraordinary changes in people, and reveal in one illuminating flash what vast potential has lain for generations under the cloak of repression and conformity, what urges towards a life based on new values have smouldered unseen beneath the vacuous features of the Consumer. If we think that consumerist society has abolished these possibilities of upsurge, or doubt that people retain the capacity for significant change, we have only to think back to the staggering events of May 1968 in France.

But we should not overestimate the possibility of wiping out the past at a single stroke — that past which, as Marx said, weighs like a nightmare on the minds of the living. The values of hierarchy are too deeply embedded for this; they will persist (we would hope in severely diminished degree) and will retain the ability to flourish again if given the chance. Can anyone seriously envisage, for example, the swift disappearance of the oldest and most pervasive hierarchical structure of all: that of patriarchy, the domination of male over female? Or that based on inequality of schooling? It is only the critical enfeebling of hierarchy, and not its elimination once and for all, that we can plausibly hope for in any not-too-distant future.

(For a convincing and enlightening account of the forces making for the continual re-establishment of hierarchy, the best source is probably not a political treatise or a work in social psychology, but a novel: Ursula Le Guin's "The Dispossessed".)

We are left, then, with the second line of attack: to see that those hierarchies which exist do not control the development and spread of new technology.

In general, prescriptions for achieving this constraint seem to me wildly and unjustifiably optimistic; they seriously underrate the magnitude of the problem. Much "ecological" writing, for instance, seems to imply that the necessary controls can be achieved while a hierarchy as dominant and well-implanted as the capitalist class continues to exercise its legal rights. It is only a matter, it seems, of enlightened people making their case to the nation's legislators — perhaps assisted by a mass movement — and securing the passage of the needed laws. They do not seem to appreciate the strength of the people and interests which give their name to "capitalism"; they never ask what the legislative bodies are *for*. That such activities can secure worthwhile results, both ecologically and in raising public consciousness, seems undeniable; that they represent a satisfactory solution — and not a commitment to a continuous series of rearguard defensive actions — is a quite untenable view.

In a different corner of the political spectrum, we are told — explicitly or implicitly — that benign technological policy can be achieved in a single-party state. (The latter feature is often not spelt out, but can be plausibly inferred from references to "the" working-class party.) The social power of the political hierarchy which can erect itself in such a system is well enough known; but this hierarchy is apparently expected *not* to push its own interests, but rather those of humanity, because its members hold to certain conceptions about history and politics. That a privileged social grouping will act, as a whole and in the long term, not to further their special interests but rather as prescribed by certain beliefs they profess, is a bold revision of the marxist approach at its most basic and valuable. For all its bold revisionism, however, this view cannot survive the most elementary criticism — or even a careful reading of the *real* history of the Bolshevik party (for all its admirable qualities) in the period immediately following 1917.

It seems to me that, if we wish to learn how a hierarchy can be deprived of power over technology, we must ask ourselves why it holds that power in the first place; and that is the question we will now examine.

4. The Development of Hierarchical Control over Technology

The link between hierarchy and technological control is not a logical, unbreakable one, for there have been societies in which this link was tenuous or even essentially missing. These societies provide a test case for any explanation of the powerful control that hierarchies exert in this field today — it should also explain those periods when the control was absent, and preferably without any additions or modifications to the theory.

If we look back to feudal England, for example, we find massive changes over the centuries in the predominant productive sector of the day: agriculture. The technology of the countryside altered so considerably in the later Middle Ages that one writer in this field (Kerridge, in "The Agricultural Revolution") denies the conventional wisdom that a great "agricultural revolution" occurred in the 18th century; his case is simply that this revolution had been essentially accomplished two centuries earlier. Kerridge also produces evidence that the commons system, although much less amenable to change than its private-farm successor, was not by any means static in its technology. Important advances in farming methods occurred in the commons period, even if the speed of progress was slow compared to that of succeeding centuries.

These technological developments do not seem to have been influenced very much, if at all, by the existing hierarchies, clerical or secular; they certainly did not generally occur on their initiative. Indeed, it is difficult to see why they should greatly interest themselves in what the commoners did with their own labour and resources in their own time — so long as rents and tithes were duly paid, either in money, produce or days of labour (according to the district and the period).

The difference between this feudal arrangement, and that in modern societies, is easily seen to stem from a fundamental economic factor: the right to dispose of the product. Except for certain customary deductions (like tithing), the feudal producers had the right to enjoy and dispose of the outcome of their labours in their own time — which was the bulk of their working time. Roughly speaking: the producers substantially retained the ownership of their product.

In this respect, their situation was close to that of the producers throughout the greater part of human history — that is, throughout the hunting, food-gathering and early agricultural and pastoral periods, in which the producers were naturally assumed to "own" (individually or, more often, collectively) the products of their labour. The feudal assumption differed radically from that generally underlying the preceding eras of slavery (though a slave could exceptionally become wealthy), and differed again from that of the succeeding centuries, when capitalist forms of property became the norm.

In these latter periods, the value of the labour embodied in an object, whether value in use or in exchange, is seen as rightfully belonging to someone other than the people who actually exert that labour power. Custom or law assign that value as the property of a definite "owner", who may be a person, a group of people, or a legal entity. As would be expected, this assignment away from the actual producers needs a finally available enforcement through institutionalised means of violence, guaranteeing the enjoyment and disposition of the product to its selected "owner" — in slavery, the master of the slave; in capitalism, the recognised holder of the productive resources used.

Entitled to the product, the assigned non-producers are also entitled to any surplus which remains when the costs of production (including the cost of maintaining the producers) have been met. Once exchange value becomes an important element in productive relations, and a money economy has developed, they are also entitled to the monetary expression of this surplus. If production is to be expanded, it is with this surplus that the necessary resources will be obtained — that is, funds for investment, and the decision over their manner of use, will also lie within their control. Thus, when the introduction of new technology becomes a significant social process, it is a process that they will direct.

At this level of abstractness, it does not matter much how the lucky non-producers are selected; they can be slave-owners, gentlemen farmers, a joint stock company, General Motors, the Ministry of Heavy Industry, or people of a particular blood group. However, if technology is to bear the stamp of those qualities discussed above, a further property is needed: the recognised owners, or the people acting on behalf of the recognised corporations (private or public), must be a small minority of the population. They will be foolish indeed if, with this head-start over the rest, they cannot accrue material and social privileges, constitute themselves as an elite, place themselves near the top in a hierarchy. The consequences of this, in biasing the trend of the new technology they will favor, have already been noted.

In this perspective, the feudal case is readily understood. Despite the existence of a formidable hierarchy, the producers themselves were able to introduce significant changes in technology — marked by a maddening conservatism, but also by an admirable solicitude for what we would now call the environment. Their prejudices and values were those of producers, careful for the interests of themselves and their descendants; but these constituted the bulk of the population. An elite will generally have a self-image somewhat different, and value above all the well-being of people like themselves; they are leaving something to their descendants only if they maintain and strengthen the hierarchical structure in which alone they can exist. But the feudal lords and bishops were not remiss in their duty, when they permitted technological advances that neglected to fortify their own privileges, and those of their descendants by blood or faith. Their writ just did not run so far, for a large part of the product lay outside their control — which remained with the people who actually produced it. In more modern times, this defect has been rectified.

Today, this control over the development of technology is infinitely more important to a hierarchy than in previous history. It seems reasonable to conclude that, to retain this trump card, they must retain control over the disposition of the product — either by legal ownership, as in countries like the U.S.A., or by the use of political power to dub themselves the true representatives of the owner, as in countries like the U.S.S.R.

But what of the converse to this proposition? I mean: if the control of the product does not belong to the producer, but is exercised in practice by a small minority, can the latter be relied upon to bias the direction of technological advance so as to serve their existing or developing interest?

Two questions are involved here, which it is advisable to disentangle: first, would they want to do it, and second, would they be able to do it.

On the first question, that of their will, I would confidently answer: yes — for reasons similar to those sketched in section 3 above. To the second I would reply: almost certainly. I do not share here that faith in the workings of representative institutions, which allows one to believe in constitutional arrangements that force a minority exercising power — in this case, over the products of other people's labour — to do so in the general good. I think that one can have too much reliance on paper provisions, and that it is salutary here to compare reality and the actual amount of popular power, with the verbal assurances of a parliamentary democratic system, or with the even more comforting phrases of the Soviet Constitution.

The conclusion to which these remarks lead should now be clear: that there is no remedy for damaging technology, short of complete self-management in the sphere of production — including, above all, the full disposition of their product by the producers themselves.

"Self-management" has become an "in" word in recent years — indeed, the ranks of its verbal supporters were swelled in 1978 by the addition of no less a figure than the Prime Minister of Australia. This may be considered a classic example of hypocrisy as defined by Oscar Wilde: the tribute paid to virtue by vice. But many adherents to the concept who are far from hypocritical would still recoil from the "extreme" character of the definition above.

For the contrary idea — that some other people or some other body is rightfully entitled to the product — did not arrive on the scene five minutes ago; it was gradually developed over centuries and embedded in our culture in a thousand ways. How deeply rooted it has become, and the thoroughness of our acculturation to it, is shown by one extraordinary fact alone: that it has never been questioned by the great bulk of opponents to the capitalist system.

For most socialists of the last century have accepted, not just the general idea, but the actual form it has taken in capitalist society. They have assented, not just to the rightfulness of assigning the product away from the producers, but to the grounds which capitalist legality lays down for choosing who the recipients shall be. That is to say, they have usually agreed fully with their opponents, that the rightful recipient should be the owner of the productive capital — machines, raw materials, land. Their disagreements have usually been over the identity of this owner.

For supporters of capitalism, private ownership was desirable; for most socialists, the only permissible owner was the State. That this owner then took the product — rather than the producers themselves — then went without saying; it was common ground.

Given this deep penetration of the contrary concept, it may be understood why self-management as defined above can seem a proposition of Martian weirdness. Each of its real difficulties and problems will be seen, not as posing a historical task which humanity must tackle, but as refuting the whole idea in one move. In the next and final section, I will try to indicate — with unavoidable brevity — how some of the main objections may be met.

5. Producer Control?

The particular evils of technology that we have been surveying — gigantism, over-centralisation, over-complexity — do not seem likely to continue in a system of producer control. There are no ideological or material reasons why producers should favor such trends, when their economic efficiency is either doubtful or squarely negative. Thus, even if producers' self-management breeds its own distortions and insufficiencies, there is no reason to expect these errors to have the destructive and even fatal character of the policies peculiar to a group with vested interests in hierarchy.

Nevertheless, we cannot rest comfortably on such a general ground for approval, no matter how fundamental and decisive we may consider it. Obvious questions about the workings of such a system will occur in most people's minds, and they should not be brushed aside.

With many workers, it is a matter of adding value to material, rather than producing a completed commodity; here, as with service workers, we can ask what is the content of a slogan, "the product to the producers". And what of the revenue needed for undertakings of national scope, for social welfare payments, for the establishment of new industries? What of the welfare of the community, if held to ransom by the particularist greed of a small body of workers? What of the industrial stagnation that could follow, if workers decided that a Christmas bonus was preferable to funds for new investment?

The first response to such queries should really be a series of counter-questions: If the producers are not to claim the product, what people or what body is to do so? What measures can realistically be expected to prevent the dire consequences of such an expropriation, as outlined above? The difficulties and conflicts that would flow from producer self-management would at least take place in a world physically able to support life. Can any proposed alternative guarantee this minimum need?

Unless these questions can be satisfactorily answered, we have no alternative but to admit the inevitability of a self-management solution. The queries above must then be seen as statements of real problems that must be grappled with, rather than as knock-down arguments demanding acceptance or refutation. Taking them in order:

It is true that the value added by many workers does not result in the appearance of a physically distinct commodity. But this has never been an objection to the working of a contract system, where the body of workers concerned is paid for the actual value they add, rather than maintained under a wage system. Nobody finds it mystical or unrealisable, that they should thus sell a "commodity" having no definite physical form. The crucial point is whether this initial sale is properly *their* right, or somebody else's.

As for the financing of undertakings on a national scale: there seem no objections in principle — certainly none that can be derived from the arguments above — to a deduction for these purposes from the realised value. We have seen how even the substantial cut taken by the lord did not deprive the peasant of the right to *most* of his product, which still provided enough surplus for some kind of investment under commoner control. The imposition of company tax still leaves a capitalist corporation in substantial control of its product, and able to finance investment and new technology. The institutional arrangements for collecting such funds, in a self-managed economy, would of course depend on the shape of the broader society outside.

More generally: nothing said here should be taken as opposing the need for economic and social planning. Unless, of course, one assumes that any plan must necessarily be imposed from above, and then have "the force of law", to quote the ominous terminology of the U.S.S.R. The themes dealt with here relate, not to the need for planning in any modern economy, but to the agreed basis on which the planning should proceed: with the product already handed over to a planning elite, or remaining in the hands of the body of producers responsible for its creation.

It would be no change, that a group of workers managing production could hold the community to ransom; capitalist firms already have exactly this power. But of course, the community can also hold those workers to ransom — presumably, for instance, they would like to buy at a reasonable price the bread that other workers bake. In a society, not of atomised consumers but of autonomous work collectives, such defensive actions would be so easily organised that they would not need to be called upon; common sense would prevail.

If workers had the say, would they usually prefer immediate consumption to saving for investment? A study by Marglin suggests that indeed a hierarchical form of control over the surplus results in a higher rate of investment, and a speedier growth in the G.N.P. I think it quite likely (though not certain) that lower growth rates would characterise a self-managed economy; it is now up to the critics of this feature, bearing in mind ecological necessities and the real content of the increased national products we have witnessed, to explain why this is a bad thing. Rather: why this is such a bad thing that it warrants the retention of hierarchical control — with all that implies — over the surplus invested (9).

* * *

These remarks, inadequate as they are, must suffice here. They are on a par with the insufficiency of the whole paper, which suffers from an unavoidable abstractness: a single feature of social reality has been abstracted from its context and considered in relative isolation. But, in thus separating off the nature of technological change and its tight connection with the way production is organised — self-managed or hierarchical — and the social layering it favors, I have hoped to retain sufficient of the strong connections in the social fabric that actually matter, to give some useful insights.

Really, the significance of self-management in production cannot be adequately discussed without looking at the wider social fabric, and the need for replacing hierarchical structures in every social sphere. The reader must look elsewhere for more satisfactory discussions along these lines.

One important point should, however, be mentioned in conclusion. The concept of self-management is not some bright idea which just occurred to me, or to a few other individuals having privileged access to theoretical truth. It is the most convenient term with which to characterise the thrust of a great number of social movements, involving not hundreds but millions of fairly ordinary people. In the past twenty years or so, it has emerged from the ante-room of theory into the arena of practice; for this we must thank, not the stimulating efforts of a handful of propagandists, but the blunders and manifest irrationality of the various hierarchical systems.

Their misuse of technology is one of the major irrationalities stimulating opposition; it becomes more apparent as the means at their disposal grow in power and destructive potential. A theoretical study of the type above can reveal the connection between hierarchical control and technologically deformed ways of life. But this connection has already been felt to some degree, and continues to be even more strongly appreciated, by millions of people who lack the taste or the schooling for theory, but can nevertheless change the world in practice. Ideas with political implications — ideas, for example, about energy use or technology in general — should be judged, I believe, by the role they play in this massive historical process. Truthful analysis and theory can be of great value in generalising, illuminating and stimulating this movement, in assisting the millions to build a world of self-managed socialism.

References

(1) The extraordinary decline in efficiency of nuclear-power stations as their rating increases was shown in a least-squares fit to the operating data up to the end of 1975, for Britain and the USA, in a submission by the author to the Ranger Enquiry, 1976. Improved but still similar performance is indicated in Charles Komanoff, *Nuclear Plant Performance Update, Data Through Dec 1977* Komanoff Energy Associates (475 Park Ave. South NYC 10016).

(2) For the ordering of large reactors in the near-absence of data on the functioning even of smaller ones, see the author's contribution to *Atoms for the Poor,* a study of nuclear power for the Third World, funded by Community Aid Abroad in 1978.

(3) For the optimum fossil-fuel station size, see Abdulkarim and Lucas, *Energy Research,* Jan. 1977.

(4) On telecommunications networks: see Mantell, page 23, in *Digest of the Conference on the Economies of Scale in Today's Telecommunications Systems,* Mayflower Hotel, Washington, DC, September 13, 1973. IEEE Cat No. 73 CHO 830-0-SCALE.

(5) For some idea of the difficulties in relating "economies of scale" to "factor indivisibility", see chapter 2 of S. Ling, *Economies of Scale in the Steam-Electric Power Generating Industry,* North-Holland Publishing Company, Amsterdam, 1964.

(6) For some views on the kinds of "prejudice" that guide managers innovating subject to uncertainty, see Bela Gold, *Explorations in Managerial Economics,* MacMillan, London, 1971. We read on page 216, for example: ". . . some managers feel that because the future will undoubtedly be different, survival requires participation in the stream of change even if they are not sure of where it is leading and cannot contrive persuasive estimates of attractive returns from individual projects; and a rather similar view holds that innovations which are technically sound will eventually pay . . . Keynes argues that many major business decisions are taken 'as a result of animal spirits — of a spontaneous urge to action rather than inaction' ".

(7) For a detailed account of the functioning of these three sectors — competitive, monopolistic, State — see James O'Connor, *The Fiscal Crisis of the State.*

(8) For documentation on the early history of the nuclear power industry, see the author's contribution to *Atoms for the Poor* (note 2 above).

(9) The argument that hierarchical control of the surplus means an increased investment rate appears in Stephen A. Marglin, What do bosses do?, Part 2, *Review of Radical Political Economy,* Spring 1975.

The Consumption Society — Energy, Environment and Our Future

Louis Arnoux

Programme Manager
New Zealand Energy Research & Development Committee
University of Auckland, New Zealand.

Criticism has pulled out the imaginary flowers adorning our chains not for man to carry the chain prosaically, without consolation but in order that he throw it away and pick up the lively flowers . . .

Marx, K., *Criticism of the Heglian Philosophy of the Law* (1843) Spartacus no. 33 p. 49.

1. Introduction

Our modern life is facilitated or plagued, or both, by an increasing number of objects.

Most of them are industrially produced. They are bought to be used personally or to be given as gifts. They are consumed through a usually complex process and finally, sooner or later, thrown away or destroyed.

And indeed, once a household is equipped with a car, a TV, a refrigerator, a washing machine, an electric cooker, its energy consumption becomes fairly rigid. In this respect we have to consider not only the direct energy consumption, that is to say, the energy used for cooking, cooling, heating etc. but also the indirect energy used to build the appliances and the house.

For example, half of the total energy cost of transport by car in New Zealand is indirect. It is the energy used to produce the steel, to make the car and transport it, to build the roads, to maintain them etc.

Added together this represents an enormous amount of energy, which is often used with a very low efficiency. For example, in the food sector the energy ratio is below 0.2 in many industrial countries. That is to say the energy used as food at the consumer level as related to the non solar inputs all along the food system (4, 33).

It therefore seems worthwhile to consider a bit more closely the consumption phenomenon and try to understand the basic mechanisms at play here.

For the last four years I have been doing research in this field, mainly on food and food related consumption as one of the basic functions of life in all societies.

Over the last two years I have started confronting the results I had obtained in the food area with those obtained from different approaches in various countries and other sectors, for example, housing, clothing and fashion, health and medicine, transport, leisure activities etc. In common with an increasing number of other research workers I found that all approaches are pointing towards the same basic mechanisms and trends (17, 18, 19, 20, 6, 39, 41).

Indeed, a new understanding of modern society seems to be emerging. Of course, one may argue that countries are different and what is going on in Europe may differ in many respects to what is happening in America or in New Zealand. However, it is also clear that all modern countries are industrial societies. The same type of products are produced and marketed everywhere, the same books are read, the same types of buildings and shopping areas are seen in all modern towns; people tend to have the same pattern of life and are faced with basically the same economic, social, psychological and ecological problems.

Since I arrived in New Zealand I have been doing short test studies through interviews, semiological analysis of medias, and analysis of statistical data on consumption and production and I have found no marked differences with what I studied in France. On the contrary, some aspects are clearer here than in France. This is the case, for example, of the house-with-a-garden-around, TV, a-car-or-two, way of life in the Auckland area. This is still a dream or a trend in many parts of Europe.

Such basic characteristics of industrial societies can be explicited through a few examples.

2. What Are We Consuming?

A. Modern food — how much and why?

Let us consider first the food area. Food in a modern way of life implies many activities and objects: meals, kitchen, food products, supermarkets, prices, diets, wastes, pollution, energy, etc.

In fact, with such a diversity and in order to cope with it, to manage, organise and regulate the food systems it is tempting to reduce people's behaviour into figures. It is tempting to observe the phenomenon through the mirror of statistics.

There are masses of analyses of this sort. The most common ones deal with the evaluation of the households' food budgets. What meaning, however, is it possible to extract from those aggregates? Is summing up individual consumptions not hiding the internal logic for the perpetual growth and change of consumption? Common statistics are showing a static or diminishing trend for the share of food consumption in the overall household budgets. In New Zealand it is now around 17 to 18% and in France around 25% of the household budget. In underdeveloped countries it is typically more, often up to 50% or more. But what is the meaning of such figures if to eat in a modern town we have to have a refrigerator, an electric cooker, all sorts of kitchen appliances, a car for shopping around etc. Are the products we are eating the same in the different social classes and are they the same as those eaten 50 years ago? In practice, quite often, they are different in many respects such as processes and packagings, quality, prices, etc.

Furthermore, if we sum up all expenditure necessary to feed ourselves in modern society we will find some apparently surprising results.

In France, for example, the ratio is then between 34 and 42% of household expenses as opposed to the 25% usually advocated.

In New Zealand a first estimate using the 1977 Year Book data shows a ratio between 30 and 35% of total expenses. This is without taking into consideration medical expenses related to generalised hypercaloric and hyperlipidic diets. In practice this means households are spending, for eating, about the same proportion of their resources as in the past or as in most underdeveloped countries. Only the *nature* and the *meaning* of this consumption differs.

The proportion of non-food (all that is necessary for modern eating but is not eaten) is increasing and is costing more in terms of money and energy. This non-food section comprises generally about 30% to 40% of the expenses linked with food consumption (3).

Diesendorf, M. (ed.) (1979). *Energy and People*. Canberra, Society for Social Responsibility in Science (A.C.T.)

Is this just a question of progress? Are we buying this mass of objects and matter because it is useful or necessary?

A review of the work accomplished on this topic will easily show the importance and the weight of symbols and images carried along by the products (13). This is particularly true for food products and drinks.

For example, the only necessary drink from a physiological point of view is water. However, New Zealanders are drinking about 290 litres of industrial drink per capita per year (beer, wine, juices, milks, etc.), plus about 4kg of tea and coffee. That gives us an estimate of more than 500 litres per head per year, 25% of which is beer. That is to say 1.6 litres per day. Thus more than 60% of our daily physiological requirements are merely the result of cultural values, images, social status, publicity etc.

This has important energy consequences. There are, for example, about 17 megajoules of direct energy consumption per dollar worth of beer in the USA.* World beer consumption involves processing some 13 million tonnes of raw material, of which most is destroyed in the process. Three to 4 million tonnes for example are disappearing in the air through the respiratory processes, only to maintain a few nice images in our brains.

B. Transport — the myth of Narcissus

But let's consider another example, in the transport sector. It will show us that shapes of manufactured objects also have social and psychological meanings. This has been particularly well analysed by Baudrillard. The wings of cars and the changes of their design over the years is a good example. As wings they are the sign of man's victory over space. A pure sign because it has no connection with this victory. Car wings do not add to the speed and they increase often the weight of the car. The real victory is within the engine, wheels and suspension. The wing is not the sign of the real speed but *means* a sublime speed.

In that sense the car is miming a superior organism flying gracefully or a powerful flying object (plane, rocket etc.). The engine is the real technically efficient component, the wing the imaginary one. Its design expresses the phantasm of speed (a phallic one). But it is a formal speed, congealed, nearly visually consumable. Its view is not the term of an active process but the enjoyment of a static image of speed. In some ways it is an end and passive state of degradation of the energy into a pure sign; a sign where the unconscious desire repeats the same static discourse over and over again.

From that point of view the car is preeminently an object in the sense that it constitutes an abstraction of any practical use into speed, glamour, passion, possession, phantasmatic projections etc.

As such it can be viewed as adding a new dimension of transcendancy to the domestic interior (household), a dimension that was lacking. But it adds this new dimension without questioning the domestic system.

People become free to go where they want to, they can enjoy the speed and do so *without any effort.* With the car the private daily life takes the dimension of the world without ceasing to be daily life. Furthermore, if transport is a necessity and speed a pleasure, the possession of a car is more. It is a kind of patent of citizenship.

For all of these reasons cars can be considered as sublime objects and not just useful necessary appliances.

One of the most important components in this respect is the mobility without effort provided by modern cars. It procures a kind of unreal happiness, a suspension of the existence, a feeling or irresponsibility that most countries are painfully trying to counteract by road regulations and public information. Speed integrates space and time and reduces subjectively the outside world to two dimensions, to an image seen through the window and contemplated passively as we contemplate T.V. Therefore, a car is more than just complementary and opposite to the house. It is also a home, but a special one, a sphere of close intimacy but without the usual restraint of intimacy. It is a place with a very intense formal freedom and an extreme functionality. It is a place of an extraordinary compromise — to be at home and further and further from home. In that sense the car is the centre of a new highly sublimated subjectivity (12, 18).

If domesticity seems to fold up within society, the car, because its pure functionality is bound to the mastery of space and time, seems to unfold its glamour beyond society.

This bipolarity (the car excentric but complementary to the home) can also be connected to the social sex roles and their images: the domestic universe is one of food and household appliances, a feminine one. Man's kingdom, in modern ideology, is outside, a world where the car is a powerful sign.

In that world the exercise of power is somewhat a narcissistic projection through the possession of objects such as cars. That is to say, their singularity is to be possessed by somebody. They are not cars, they are always somebody's car and have therefore number plates. They are all *unique cars* from their owner's point of view and this allows these very persons to recognise themselves as absolutely singular beings, powerful beings. They master speed, a phallic symbol. The eroticism of cars is not therefore a kind of active sexual approach as it is often suggested but a passive one, a narcissistic seduction and communion in the same object.

In that sense a car is not a feminine object. If publicity connects it with women, this is just part of a general trend where the object — woman — is the basic pattern for persuasion. All the objects feminise themselves to be bought. But this is just a peripheral phenomenon.

The deep fantasy linked with the car is of a different order. Its basic image as shown by many analyses is a phallus, an object of manipulation, of fascination related to speed and power. Its erotic value plays the same role as the images (real or psychic) for masturbation.

This brief analysis which was first developed by Baudrillard shows how a common transport device can have very strong and subjective meanings (6, 12). Transport by car is not only a matter of rational behaviour, it implies the existence of a whole system of images and values which are conditioning the way people are viewing themselves and playing their social role. Such a system also determines their energy consumption. The design of new energy conserving transport systems, for example, community as opposed to individual transport systems, should take this into account.

Another example in the same field shows how publicity is playing with subconscious imagery.

A car advertisement was broadcast on New Zealand TV over the last Christmas period in December 1977. In the spot a set of new cars was displayed on top of a hill in the middle of tussock grass in the wilderness of hinterland hill country and viewed from a small airplane flying close to the ground.

I am purposely going to present its analysis as if it was a kind of simultaneous translation into intelligible language of what the advertisement is saying in a hidden way to the viewer's subconscious.

The view of the area from a small plane is a familiar one for New Zealanders (of planes top dressing, films, tourist tours etc.). It evokes the image of wild free Nature with a capital N. It is also risky to fly so close to the ground. You seem to be in the plane, and playing with your life a bit in the middle of nature (Nature).

Flying, speed, strange feeling of power without effort, subconsciously connected with sex; the feeling is enhanced by the rhythmic movement of the plane passing over the tussock grass. You're making love with nature (Mother Nature?) and risking your life slightly in the process.

And then the cars appear, shiny, glamorous, long and polished. Their shape is like congealed speed, penis like.

The masculine voice is explaining it all. They are soft and comfortable inside, luxury, as safe as your home but mobile. You don't risk very much in fact, the suspension is good, but actually driving them on the tussock would give you a view quite similar to the one from the plane; you would think you were flying over the land, over everywhere you want to go, everywhere over Nature with your car, your flying penis-like car.

*This means about 600g of coal per dollar worth of beer.

These displayed cars are beautiful objects. A car may be used for a purpose; for example, to go to work, or to go on holiday, but an object is just there to be possessed by you and in doing so it says to you that you are yourself and nobody else. You are the bright owner of such a beautiful object. An object is like a mirror in a narcissistic relationship with yourself, it has no other use.

Beautiful objects indeed on the top of that hill, and there is not just one, there is a whole collection of them. Because the possession of one object is not sufficient. It has to be a unique object, yours, different from the others, therefore you have to be able to make a choice, to pick out your one.

This also suggests the idea of a collection, perhaps not yours, but you will be part of the lucky ones who are owning cars from this collection, who are sharing the same narcissistic feelings for these objects.

And these cars are coming from all over the world, says the masculine voice, like powerful slaves in the Roman Empire, they are bought to you by X, your dealer.

They look a bit feminine in this way although they are basically penis-like, but there is no contradiction in that. In modern societies the stereotype of *the* beautiful object to be possessed is the object-woman and indeed modern fashion, modern attitudes, tend to transform women into penis-like objects*too. This is what this advertisement is saying to your subconscious while you are calmly watching your TV programme. And it does not matter if you are a woman either as the ruling values and images are male ones. If you are a woman you are for the advertiser an object-woman, you behave yourself, you have no particular values or feelings, you may only have male ones, or you are a housewife and you are not concerned by this advertisement.

Anyway this is just about the final slavery part, most of the advertisement is simply narcissistic, this is the great advantage of it, it works for everybody, male, female, teenagers, etc.

It is very representative of modern publicity and shows most of the features outlined more broadly within Baudrillard's analysis and connected with the consumption of objects.

It is also worth comparing this myth of the car with its practical efficiency as a transport means. A study conducted in France by J. P. Dupuy has shown surprising results (14). By adding the time spent to earn the money to buy the car, plus the time necessary to buy it and maintain it and then drive it and relating this to the mileage driven for a year one obtains a generalised speed. Table 1 gives the main results; the car is not faster than the bicycle but the speed is tied up with the social status. In New Zealand for an average citizen I have found very roughly an average speed of 17 km/hr. The myth is clearly very different from reality but we behave according to the myth.

Table 1

Generalised Speed (km/h)

Social Groups	*bicycle*	*small car (Citroen 2 CV)*	*medium (Simca 1301)*	*large (Citroen DS21)*
Manager (Paris)	14	14	14	12
Office Worker (medium town)	13	12	10	8
Factory Worker (medium town)	13	10	8	6
Farm Worker	12	8	6	4

km/h = yearly mileage/time necessary to earn the money to buy the car, to maintain it and to drive it.
Source = CEREBE — (J.P. Dupuy) — 1974

C. Eros

Another example of the importance of unconscious images in publicity and consumption can be found in a recent TV spot on a liquid foam bath product.

The film shows the product. It is a cylindical container. It then shows the water in the tub, a hand presses on the tube and a jet of viscous liquid falls into the foamy water, the camera insists on the final drop flowing slowly out of the tubular container. Interspersed with this are pictures of the sensuous woman undressing herself as reflected by the tiles on the wall along the tub.

* (10) p. 156. The body a charnelhouse of signs.

The sexual and voyeur context of the whole scene is particularly strong. It is in fact almost pornographic but as it is subliminal almost nobody notices it. Only the subconscious mind records it.

More generally speaking, because precisely publicity is concentrating on a message about objects its analysis is very powerful in helping us understanding the meanings of our consumption (6, 35, 42, 46, 47).

Indeed it is not a phenomenom peripheral to the consumption process. Publicity is part of it as a consumable cultural object.

3. Publicity = "A Logic of Faith and Regression"

Even so people tend to react cautiously and deny the fact they are forced to buy all sorts of things by direct or indirect means.

The efficiency of publicity for specific products is very difficult to measure and is often indeed very low, but from a social point of view this is not the major aim of publicity.

What is important is not so much the injunction to buy a given brand but something more important for the whole of society. To become aware of it we have to pay attention to what is indicated by the massive flow of advertisements.

What we are in fact consuming through this flow is "the luxury of a society which displays itself as bestowing goods upon its members and which present this as the core of its culture." ((6) p. 196). It is a kind of show, the show of society caring for us with a mothering attitude.

We are all subconsciously impregnated by this enormous show. We do not believe publicity is efficient. The publicity on a specific product in fact does not convince anybody but it helps in rationalising the act of buying it. For many children at the end of the year, Santa Claus stands as a symbol of miraculous parental care for them even when they do not really believe in its existence anymore; publicity acts in the same way.

We are all subconsciously sensitive to the subliminal idea of protection and gratification. This is especially true in urban environments heavily loaded with all sorts of stresses (43).

There the individual is particularly sensitive to the care taken to solicitate him, to persuade him, to care for all his needs.

In a society where direct traditional human relationships tend to be destroyed and where the individual himself tends to feel isolated and transformed into a piece of a complex machinery, the objects through publicity, are talking to him, are loving him and because he is loved he feels he is still a person in some way.

At least a presence cares for him. The more people are isolated by modern life the more they tend to be sensitive to this relationship. It puts them into an infantile position. "It is a logic of faith and regression." ((6) p. 198).*

Actually such a phenomenom is not without some analogy with the Freudian concepts or oral and sadistic-anal phases and regressions.

In fact this is not only true of publicity it is also true of all modern mass-media, but its analysis is particularly clear in the case of publicity (35, 39).

Throughout this complex process, slowly but surely. individuals are conditioned and forced to consume, to swallow the whole society as a gigantic show.

If the multiplicity of products suppresses the idea of scarcity, publicity gives a feeling of warmth and security. Furthermore it is free, it is offered to everybody. In that sense it is also a means of reviving the archaic rituals of the feast and the gifts.

Other important aspects of publicity, all interrelated, are its play components and its erotic components. They are particularly powerful. In Germany, the word for publicity is *die Werbung,* which means the amorous quest.

But if in a way publicity replaces the festivals of pre-industrial societies it is forced upon people and it is only gratifying in imagination, in anticipation. As a dream of a feast it is frustrating. The desire involved is never really fulfilled but puts the individuals into the right mood to buy.

* An approach quite similar to this was recently very brilliantly presented in Auckland by I. Illich.

In fact its function is really comparable to a dream. As a dream publicity is entirely based on subjectivity and imagination without any negativity or relativity.

As Baudrillard puts it — "as dreams are protecting our sleep, publicity is facilitating the spontaneous absorption of social values and individual regression into a social consensus." ((6) p. 205).

Whatever the slogans are they are always saying basically the same thing: "society cares for you, adapts itself to you, therefore you *have to conform to it* and buy."

It is a regressive process which masks real social relationships at work, in transport, while shopping or in the home.

We are put in a schizoid situation where production is differentiated from consumption in the same individual. It is even more than consumption. The desire to possess an object represents "a regression to the phase where the child identifies his mother with what she gives to him," and is possessed as much as he possesses ((6) p. 207).

However as the object offered to the sight is never really given, this creates the permanent rousing of a desire which is permanently suppressed.

The desire is liberated by the imagery only to trigger up reflexes of anguish and culpability linked with its abrupt emergence. It is then cooled down into an infantile regression and the buying of something obviously perceived as necessary once the norm has been interiorized.

4. *Modern Life = the Myth of Consumption and the Consumption of a Myth*

As I said earlier such an analysis could be and has been extended in various research work to all the social relationships in our lives (7, 8, 9, 24, 36, 41).

Society then does not appear as a homogeneous system anymore but basically as a dual structure, as shown in the following diagram: a society of the show, of the spectacle, a myth of consumption masking the emptiness of human relationships; men reduced to objects and signs.

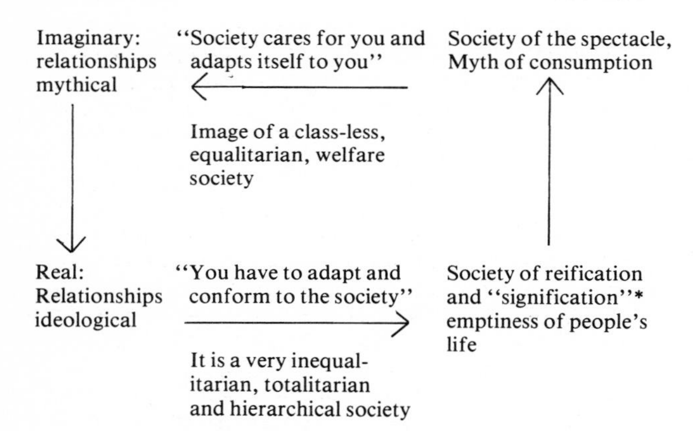

Through publicity and consumption thus the whole society is permanently voted for by its members. In this way it is performing a function quite similar to the totem in other societies.

Publicity is therefore an extremely powerful tool to impose and maintain the present structure of modern society.

From this point of view mass consumption and publicity can be considered as playing a political role. They are substitutes for preindustrial ideologies and norms. In enforcing social behaviour without too much need for direct repression they even appear to be more efficient.**

Publicity goes even further. Indeed once people are maintained in a childish position it is rather simple to give them the liberty to behave as children without feeling guilty about it. This is well explicited in Dichter's (16) work on the "strategy of desire". But this liberty is only a liberty to consume objects according to the norm. It therefore rejects any other social form as abnormal, marginal, archaic or integrates them as part of the show, as part of fashion.

The norm however, is never clearly defined. It is permanently moving through fashions but we are all subconsciously pushed to conform to it. We all know what is right or wrong. We know the scale of value.

It is the first time in history that such a norm has been imposed and has been understood universally†. It is certainly not more arbitrary than any other social code, be it birth, casts, classes, etc. It has destroyed all previous discriminatory systems (10, 23, 33, 39, 40, 47). But conversely this is obtained through a tremendous impoverishment of human relationships. The same process means also a reification of people in the sense that they are possessed by their objects as much as they possess them.

This is certainly not a really unifying process. On the contrary this unique system of reference seems to increase the desire for discrimination. New concepts, new ideas, new behaviour, new fashion are permanently created to maintain the social scale of value operating.

From that point of view the consumption phenomenon does not correspond to a welfare state, nor to the satisfaction of need but to the organisation of a growing mass of objects into a signification.

As we have seen earlier, to be included into a consumption process one object has to become a meaning be it toothpaste (white teeth, fresh, cool breath, youth, eros etc.) or a car, or a piece of furniture. Its use becomes secondary and relative to its signification. Another good example is cigarette smoking. People will smoke a particular *brand* even though the cigarettes are much the same — i.e. kids will smoke Marlboro, aspiring executives Benson & Hedges.

But to the individual also the consumption process, as an obligatory path, signifies as well to accomplish himself and to abolish, to destroy himself in his relationship with society through a series of objects be they kitchen appliances, a car, a TV set or stereo set.*

What is consumed therefore is not so much the objects themselves but the relationships. What is important is not the beer drunk but the meaning around that event.

In preindustrial societies too, objects and their uses were heavily loaded with symbols and controlled by precise rituals.

But in our society these old systems are progressively being destroyed and replaced by the single norm (10, 34). We have a good example of this process in N.Z with the progressive disintegration of the traditional Maori society into modern Pakeha life style.

It appears to be even more than a norm. It functions as a code, the code of the social status which puts people in kinds of pigeon holes.

For example, you do not just need a watch. You have to a have a watch adapted to your function. If you are an important person you have to have "a personalised wrist-size time and data system" not just a watch . . . and vice-versa. You have to have a "Hewlett Packard 01 computing time instrument." Or you will remodel your house according to the latest edition of "Sunset Ideas for Remodelling Your House." Whether the

*"Signification" in the sense of being reduced to a sign and also being notified or intimated to conform to models and/or to behave according to an order or a set of rules or code of conduct.

** However, nothing is perfect. The same period has seen the explosion of mass media, advertising, and the development of planning, organisation, technocratic control and management of all aspects of people's lives. I will deal with this aspect later.

† It is also the first time in history men are "consuming" something, everything. This word was never used in this way before.

* See (10) The symbolic exchange and death. The reverse is true, to be a human being again one has to die symbolically as a worker/consumer. See later

social status is based on power, authority, responsibility or information it does not exist without signs and ultimately as an amount of money, a monetary sign.

In short, we are living a sort of "close encounter of the third kind" with ideology.

In the first instance values of objects can apparently be determined according to their uses.

This has been retained as a first order of ideology, the ideology of the "natural of law of value", when value was primarily determined by the market and not by use values anymore.

Today the market value, the "exchange value", the concept of "supply and demand", is used as second order of ideology and is masking the "structural law of value", the value governing social status and power, the third and last order of ideology.

In that sense everything becomes a matter for consumption, a matter for supply and demand, everything becomes functional governed by economics, goods but also feelings, culture, health, love, sex and even death (10).

Such a process is particularly obvious in the food area with the transition from homemade products based on use values or as Illich says on vernacular values to the food-town where the only logic is the dollar print out at the bottom of a piece of paper ribbon.

But we can analyse it in all aspects of our lives. It can be summarised into four points (24).

1. The desire is no longer channelled by traditional ritual systems since these have been destroyed, but it is not left free to annihilate itself into pleasure and joy.
2. On the contrary, this desire is recaptured by society (desire for power, for social recognition etc.) and finds as its only outlet the possession of objects, magic signs masked by their relative usefulness.
3. The only meaning resulting from this situation is that of the social hierarchy and its code of values, the structural code of power.
4. As a result of this process the consumers through the objects they are buying appear more and more isolated, reified, signified by the code but this very process is masked by the explosion of imaginary productions linked with the objects produced. (See Table 2).

Table 2

The Structural Value of Objects

Logics	*Characteristics of the Relationships*	*Characteristics of the Objects*	*Order/Domain*
1. Functional logic of *use value*	practical operations	tools	usefulness
2. Economic logic of *exchange value*	equivalence between the terms exchanged	goods and services	market
3. Logic of *symbolic exchange*	ambivalence of the relationships (love and aggression)	symbols (for example, a gift, a wedding ring)	person to person relationship
4. Logic of *differentiation*	maintenance of a difference between the subject and the other members of the society according to a code of value.	sign (for example $)	social status

Differentiation
Symbolic Exchange
Use Value
Exchange Value

The objects in a consumption process are those, material or not, (beings, cultural objects, ideas, behaviours, language, tools etc.) which "are freed:
from their psychological determination as symbols
from their determinations as tools
from their commercial determinations as products
and therefore freed as signs and recaptured by the formal logic of fashion and power that is to say by the logic of differentiation" ((8) p. 64 to 66).

This destruction of the structure of traditional rituals and symbolism allows an endless cycle: production → representations and desire → needs → production etc.

In such a cycle the needs have no independent existence. They are the result of the conjunction of industrial production and desires: desires which (as many research results in Human Sciences have shown) all resolve finally into a desire for a status, a desire to assert oneself through a domination of ones' social environment (24, 27, 38, 40).

From a sociological point of view such a mechanism is absolutely endless. The technological and sociological aspects (imaginary and symbolic) interact strongly in a dynamic process of accumulation of industrial power and products.

It is a system of perpetual consumption and defecation, corresponding to a regression to a sort of oral-anal stage.

Beyond the myth of consumption, one can only find power relations, the development of industrial power structures and the participation in the production of more and more alienating social relationships through the possession and consumption of more and more objects.

This also means an endless cycle of energy extraction and use. In that sense the energy crisis would be probably better termed entropy crisis.

Everything seems to happen as if the goal of modern society were to consume more material to produce more waste at a high energy cost simply to extend itself further and further and maintain a code of value.

This casts a different light on wastes or pollution, be it solid, liquid, gaseous or thermal. These so called negative effects are part of the system. They cannot be corrected without deeply altering the system (4)*.

5. The Management of a Myth

What can we do then? Obviously many people are not satisfied by the present situation. But what is the meaning of words like "to manage the environment", "to manage the economy", "to match energy supply and demand"? To read the mass of printed paper emanating from organisations involved with management it seems simply, implicitly or explicitly, palliating some effects considered as negative by these same organisations or by others without changing the basic structure of the system.

Examined in their detail, negative valuations are all directed at the more efficient control, organisation, and domination of nature and society. There are always partitions, divisions, "bars" to quote Baudrillard: man/man, man/nature, society/nature, rulers/ruled. They correspond to the hierarchical code of power. Most policies are based on such dichotomies.

Unfortunately neither nature nor societies, which are part of nature anyway, behave in this way. In nature, everything is based on never-ending interactions between cycles. In this particular respect one of the most important features of man and human societies is the symbolic exchange, the exchange of symbols. When a bar is put up somewhere, and the symbolic exchange is broken down as it is in modern society, problems accumulate as fast as industrial products and we spend most of our time trying to solve them.

Without entering into long developments we have to acknowledge that so far we have not been able to manage our own affairs properly, let alone manage our environment. We know very little about it. Take 10 economists and ask them why there is inflation and underemployment and what is to be done, you will only obtain nothing but an uproar.

What is implicit then, behind this vast effort made by all countries to manage society and its environment? Research work conducted by the Epistemology Research Centre of Lille University (France) in which I took part tried to answer this question as far as pollution was concerned (4, 28, 29, 30, 31).

It was found that the assessments and the policies related to pollution control and management of the environment were loaded with sociological and psychological elements which were definitely outside the scientific field, or simply outside the rational level of thought even in the most technocratic spheres.

Many of these features were in fact carried over from previous social forms, from various epochs but most of them were implying pre-industrial concepts, some related to magic behaviours (1, 39). Baudrillard also analysed consumption in general as a process related to magic and aiming at "conjuring up the reality into the signs of reality" ((7) p. 30)**.

*Remember the energy ratio in food: 0.2. Such a ratio is directly linked to the present consumption process. It can be altered only marginally through energy conservation measures if the consumption relationships remain unaltered.

** A good example of such magic processes is the "magic of the airport". Tribesmen had noted that white men managed to trap airplanes by building airports. Therefore they cleared a piece of bush, they built a mock up plane with boards and branches, they displayed lights at night hoping for the planes to come. Of course as no plane landed they thought they did not have the magic quite right and they kept trying.

We are doing exactly the same thing with our objects. We display them around us hoping happiness will come but as it is never quite there we keep changing the objects (fashion) and adding more until we die (7).

It appeared indeed that the various discourses analysed in these assessments and policies corresponded to the setting up of a new system of taboos, that is to say establishing a new set of guilt feelings and enforcing this within the education process. This system of taboos seems to be based on the fear of death and on the consequent unconscious exclusion of death from the social scene in industrial societies. This exclusion is just another bar, perhaps the most fundamental one (2, 10, 38, 48).

Of course these taboos open new fields for industrial production and the control of society by administrations.

In fact they correspond to new markets. Such is the case for the development of so-called "clean technologies", or anti-pollution techniques or alternative energy technologies. They imply more research work, more employment, more consumption. But even this seems to me to be just another illusion, another facade. I believe it is merely a mimicry of power, a show of power.

Indeed from a practical point of view we can evaluate the impact of specific projects. We can add up the water treatment plants built over the years, the amounts spent to fight against all sorts of pollutions. We can analyse the energy supply and demand and deduce from all these data some indicators and recommend new courses of action. But none of this will significantly alter the fundamental dynamics which have been briefly analysed in this paper. It does not destroy the set of bars. *It does not alter the code.*

From a global point of view such an attitude towards energy and the environment appears as a tranquillizer. Society tranquillizes us by drawing our attention to particular aspects such as energy, and isolating these from the other effects of the industrial system. By doing so it limits our awareness of the main question which is not confined to any particular aspect.

On the contrary the real problem seems to be the relationship between all effects and the system producing them. That is to say, that through the official acknowledgement of the environmental and energy questions and by treating them as separate matters the industrial system avoids the more fundamental question of its own future as a social form.

But conversely, goods and their embodied wastes and pollution, and energy, through their mass, their toxicity or possible shortage, through their apparent lack of purpose do worry people. They are disturbing because they act as a reminder of the emptiness of the system. They evoke death. Therefore, after having directed people's attention on to wastes, pollution or energy questions, the main concern of the industrial system can only be to occult those questions, to remove them, at least symbolically, through some sort of ritual . . . And, if it is not possible to succeed entirely, the system can only give *a reassuring show of such an effort.* A good example may be found in many discussions about energy and energy strategies in all countries. In that sense, from a technocratic point of view, the future of the management of society and its environment, that is to say drawing upon their resources and correcting to some extent negative effects, seems to be very brilliant. It may solve some specific technical problems but it is not going to improve the whole social situation very much. To manage the environment implies that one has to know how to manage oneself. It also implies one has to see oneself as an integral part of ecological and social cycles, *not as a kind of independent boss or landowner.*

6. Collapse of the World or Collapse of the Code?

I do not think such a management is possible for a civilisation which is entirely involved in the production-consumption process, a society which is based on a set of bars, with the accumulation of goods and power on one side of the bar. As we have seen and as Baudrillard analyses it in more detail in one of his latest works "The Symbolic Exchange and Death" modern society is controlled by a single code of power (10). This code has destroyed the symbolic exchange common to all previous societies but this very symbolic exchange is haunting modern society like the fear of its own death. This fear is creeping everywhere under rationalised issues, energy, pollution, economic crisis, cancer, health, etc.

It seems to be justified. The brief analysis of publicity given in this paper and many more in the literature are showing that we tend to be controlled by models and contingency (10, 26, 34, 44). The reference to economics acts only as an alibi, a mask. But [as Baudrillard says] "Any system leading towards a perfect functionality is close to its death. When the system says A equals A it is close both to absolute power and absolute ridiculousness" (10).

Risks of fast changes appear indeed to be more than risks.

I am not so much referring here to social changes as to people's understanding of their life and their view of society and the world. It is something probably more basic than social changes.

Such a prospect may appear to be agreeable or not depending on each person's conscious or unconscious prejudices, but whether we like it or not we are proceeding very quickly toward such changes. A mass of data from many disciplines shows it (40, 44). It tends to show that a growing number of people seem to be becoming sick of being reduced to brainless, soulless zombies, to *functional users* of worktime, of leisure time, of public transport, of environment, of energy. Many people are progressively and often subconsciously realising that they are working hard and using a lot of energy to accumulate nothing but the vague meaning of a social status and that by doing so they are nothing else but living-corpses. When they realise that, the world collapses for them. It may be traumatic. I have seen people suddenly unemployed, becoming physically sick because they had been taught to believe they had to have a job to be a man and a citizen. Eventually they recovered but then their world view was different, they had dropped their old beliefs.

In all parts of the world the same type of evolution seems to be occurring. I have observed it in Europe and in NZ, others have studied it in America with the same basic conclusions. What tends to collapse in people's minds is the imaginary and subconsciously integrated model, of the white, occidental, adult, male superiority symbolised by the monolithic structural code of value.

This can be found all around the world through the crisis and tensions around teenagers and adults, student movements, unemployed people, women's liberation movements, ecological movements, neo-archaic evolutions, racial relations, sexual freedom, mental health, etc. It seems to go round and round, to sweep across industrial society like shock waves (39).

From this we can realize that the real problem is not so much to manage the present situation as to go along with changes. To go along with the collapse of the code.*

7. The Egg — an Old Problem

The example I have given in this paper and many more which are available in our daily life or in literature show clearly that in practice neither individuals nor society are rational. They also show that developments are almost never linear and direct cause-effect processes occur very rarely. Conversely most social phenomena involve a great deal of non-rational behaviours and catastrophies.†

Indeed a rational society would be a monster. Therefore what is considered today as irrational, abnormal should not be neglected. Rationality is just the shell of the egg. It gives its shape to the living mass inside. It protects it. But this very mass produces the chicken, not the shell!

Rationality is necessary and adequate to solve precise, practical, technical problems: a solar heater, an energy farming system for example.

But experience has shown that pure rationality is fairly inadequate to investigate possible futures. Indeed pure rationality does not exist. It is always relative to the state of

* "The present system is the Master: it can like God, bind or unbind energies; what it cannot do and what it cannot avoid either is to be reversible. The cumulative process of value is irreversible. Only this very irreversibility is deadly for it" ((10) p. 12).

† See Thom's work: Catastrophe Theory a new branch of mathematics (45).

society's development. What seems to us rational and logical would often be considered as mad by other or previous societies.*

Our present logic and rationality has to be broken, hatched in the same way the chicken gets out of the egg, if a new viable society is bound to develop in the future. That is to say a society which will cope with our present unsolved problems.

Through this process a new rationality will appear which will probably be entirely different from the present occidental one (26).

Marginal groups, new embryonic societal forms are very important in this respect in that they tend to be based on symbolic exchange. They seem to be more viable than the present system from many points of view.

Of course this is not a single straightforward process. It implies many conflicts, many tensions (the chicken may not be viable, actually many are not). Here may reappear rationality, man's possible actions to master his future and survival.

By acquiring knowledge on the development of new societal forms at the "underground" irrational, unconscious, marginal level and through the investigation into the tensions and conflicts between the various societal forms and groups, we can not only guess at possible futures but also evaluate their viability and help soften the process of change.

In a way, what is important is not so much to consider todays problems, they are yesterday's problems, but to consider tomorrow's problems (44).

It seems to me that here are the priorities, to soften the mutations, to facilitate them; but cautiously. One does not build a society with theories but through practice. This is exactly what the members of the embryonic and somewhat marginal social forms are doing.

In fact I believe this new Society is being created now under our eyes out of what we consider to be marginal, irrational, a-social, abnormal, non-economical etc. . . . It is developing in a way very similar to the first elements of industrial society which developed as a parasite of feudality during the Middle Ages (5).

The concept of Middle Ages is actually a good one and is also used by several research workers such as Edgar Morin, Umberto Eco or Furio Colombo, to describe the period we are living in, a kind of Modern Middle Ages. Morin says "I understand modern society not as a society with residual archaisms but as a society generating a new archaism, a society which is not repelling myths for rationality but generating new myths and new irrationality. A society which is not mastering our problems and crises but which is creating new problems and new crises" ((39) p. 262).

An old world is dying, a new one is in gestation but is not born yet. Its birth may prove to be difficult.

Postscript

A long time ago a man put a goose in a jar. The goose grew and became too big to get out of the jar. The man did not want to break the jar nor to hurt the animal. What would you do if you had to get it out of the jar?

(Zen Koan)

Be realistic, ask for the impossible.

(Graffiti such as this one have been seen on many European walls since 1968. There is a certain resonance with the old koan. How can we stop being over-fed, over-energetic objects trapped into a code of value and become human beings again?)

* See for example the declaration of Chief Seattle "How can you buy or sell the sky, the warmth of the land? The idea is strange to us . . ." 1854-USA.

References

(1) Achard, P., et al., *Discours biologique et ordre social,* Sevil, Paris, 1977, 284 p.
(2) Aries, Ph., *L'Homme Devant la Mort,* le Seuil ed. Paris 1977. 640 p.
(3) Arnoux, L. et Meunier, A. Relations Psycho-Socioeconomiques au Sein Des Systèmes de Consommation. *Annales de la Nutrition et de l'alimentation* Vol. 30 no.2-3 Paris, 1978. pp.379-382.
(4) Arnoux, L. et Meunier, A. *Les déchets Agro-Alimentaires.* INRA, Lille 1976. 134 p.
(5) Barel, Y. A. *La Reproduction Sociale.* Ed. Anthropos. Paris, 1973. 558 p.
(6) Baudrillard, J., *Le Système de Objets.* Ed Gallimard, Paris 1968. 245 p.
(7) Baudrillard, J., *La Société de Consommation.* Gallimard. Paris, 1970.
(8) Baudrillard, J., *Pour une critique de l'économie politique du signe.* Ed. Gallimard, Paris, 1972. 268 p.
(9) Baudrillard, J., *Le Miroir de la Production.* Casterman ed. Paris, 1973. 147 p.
(10) Baudrillard, J., *L'échange Symbolique et la Mort.* Gallimard. Paris, 1970 347 p.
(11) Caugant, G., Mirabel, A., *Les Comportements Alimentaires des Ménages* CEREBE, Paris 1974 57 p.
(12) Caugant, G., and Zaslavsky, J., *Jeu automobile et espace-temps industriel,* CEREBE, Paris, 1974.
(13) Chatelet, N., *Le Corps a corps culinaire.* Seuil ed. Paris, 1977. 185 p.
(14) Debouverie, Y., and Dupuy, J. P., *l'automobile chronaphage,* CEREBE, Paris, 1974.
(15 Deleuze, G., and Guarttari, F., *Rhizome,* ed. de Minuit. Paris, 1976. 74 p.
(16) Dichter, E., *Handbook of Consumer Motivations/The Psychology of the World of Objects,* N.Y., McGraw Hill, 1964, 486 p.
(17) D'Iribarne, Ph., et al., *Les Consommations Alimentaires.*
(18) D'Iribarne, Ph., *Le Gaspillage et le Désir.* Fayard ed. Paris 1975. 154 p.
(19) D'Iribarne, Ph., and Bidou, C., *Habitat et Bien Etre,* CEREBE, Paris, 1974.
(20) Dupuy, J. P., and Karsenty, *L'Invasion pharmaceutique,* Seuil ed., Paris, 1974.
(21) Galbraith, J. K., *The affluent society,* Penguin Books, 1970, 295 p.
(22) Galbraith, J. K., *The New Industrial State,* Penguin Books, 1970, 422 p.
(23) Garine, I. de, Le Comportement Alimentaire dans les Pays non Industrialises. *Annales de la Nutrition et de l'alimentation.* Vol 30 no. 2-3 Paris, 1976. pp 453-467.
(24) Guillaume, M., *Le Capital et son Double* F.U.F. ed. Paris, 1975. 172 p.
(25) Kenneth, E. F., et al., *The Unsteady State, Environmental Problems, Growth, and Culture,* The East West Center, Hawai, 1977, 287 p.
(26) Koestler, A., *The Ghost in the Machine,* 1967.
(27) Laboritt, M., *La Nouvelle Grille,* Laffont ed. Paris, 1974. 385 p.
(28) Latouche, S. et al., *Introduction à l'Analyse de la Pollution comme Expression Sociale.* CEREL, Lille, 1975. 20 p.
(29) Latouche, et al., *Psycho-socio-économie de Déchets Solides.* CEREL, Lille 1976. 27 p.
(30) Latouche, et al., *La Représentation des Déchets Solides dans le Discours des medias.* CEREL, Lille, 1976. 176 p.
(31) Latouche, et al., *Le Système de Représentation de Déchets Solides et de la Pollution.* CEREL, Lille, 1977. 176 p.
(32) Lattes, R., *Mille Milliards de Dollars,* Editions et Publications Premières Paris, 1969 217 p.
(33) Leach, G., *Energy and Food Production,* IPl Science and Technology Press Guildford, 1976, 137 p.
(34) Marcuse, H., *One Dimensional Man,* Beacon Press, Boston, 1964.
(35) Marshall MacLuhan, *Pour Comprendre les Media,* Seuil.
(36) Meister, A., *l'inflation Créatrice,* P.U.F., Paris, 1975, 310 p.
(37) Mendes, C., *Le Mythe du developpement,* Seuil, Paris, 1977, 278 p.
(38) Morin, E., *L'Homme et la Mort.* Seuil ed. Paris, 1970. 351 p.
(39) Morin, E., *L'esprit du temps 1. névrose,* Grasset, Paris, 1962, 281, p. 2. *nécrose* — Grasset, Paris, 1975, 268 p.
(40) Morin, E., *Le Paradigme Perdu.* Seuil ed. Paris, 1973.
(41) Mumford, L., *The Myth of the Machine,* London, Secher & Warburg, 1967, 2V.
(42) Packard, V., *The hidden persuaders,* Longmans Green & Co., London 1957.
(43) Reisman, *Abundance For What,* N.Y., Doubleday, 1969, 610 p. and *The Lonely Crowd,* N.Y., Doubleday, 1956, 359 p.
(44) Rosnay, J. de, Le Macroscope, Seuil, Paris 1975, 295 p.
(45) Thom, R., *Modèles Mathématiques de la Mosphogénese,* UGE, Paris, 1974, 319 p.
(46) Warneryd, K. E. et al, *Mass Communication and Advertising,* Stockholm School of Economics, Stockholm, 1967, 115 p.
(47) Williamson, J., *Decoding Advertisements,* Marion Boyaros, London, 1977.
(48) Ziegler, J., *Les vivants et la mort.* Seuil ed. Paris, 1975. 315 p.

The High Energy Society — Adaptation to an Illusion

John H. Price

29B Mary Street
Hawthorn, Vic. 3122.

Decrease does not under all circumstances mean something bad. Increase and decrease come in their own time. What matters here is to understand the time and not to try to cover up poverty with empty pretence. If a time of scanty resources brings out an inner truth, one should not feel ashamed of simplicity. For simplicity is then the very thing needed to provide inner strength for further undertakings. Indeed, there should be no concern if the outward beauty of the civilisation . . . should have to suffer because of simplicity. One must draw on the strength of the inner attitude to compensate for what is lacking in externals; then the power of the content makes up for the simplicity of the form . . . Even with slender means, the sentiments of the heart can be expressed.

— *I Ching*

1. Introduction

From a different perspective or framework to the usual, I shall attempt in this paper to aid the 'understanding of our time' and suggest an approach to some of the seemingly intractable problems that currently afflict us.

Groupings of all species depend for their survival on the ways that they have evolved to satisfy their needs. Most have adapted to their natural environment which provides the produce which is gathered for consumption in various ways. Such ways are inherently uncertain, for famine can result from the perfidities of storm and tempest, uncertainties of seasons, and over-exploitation of the environment.

The human species *seems* to have overcome these 'natural' ravages and uncertainties and no grouping more so than those of us who enjoy life in 'advanced' industrial countries. Yet we still live in times of great stress and uncertainty and are faced with very serious difficulties in finding and deploying substitutes for fossil fuels to drive our productive system. Our environment is being seriously degraded and our distributive system is straining under the newly coupled phenomena of monetary inflation and unemployment.

Our presumption that our societies are 'advanced' blinds us, making the inter-relatedness of these and other problems difficult to perceive. Being to the front on that inevitable progressive road, we see each as an isolated obstacle to be overcome. We treat our problems as though they were independent diseases to be cured so that our otherwise normal and healthy state may be improved. But frequently the solutions proposed for one problem serve to aggravate others. Perhaps many of them are not diseases in their own right but symptoms of a far more serious malady which we can only begin to perceive if we shed our illusions about the basic normality of our state and our pride in its 'advancement'. If we do so we may find the *prevention* of the problems to be an easier solution than the many isolated *cures* that we usually seek.

The illusions that we must shed find their basis in our experience which has led us into many of our habits, dependences and expectations. The changes in productive and distributive organisation that have occurred in the last three hundred years or so have had a profound effect on the nature of human experience. We have adapted to those changes so that we now depend upon a System of human creation to provide us with the requisites of life. Since many of the changes require the use of finite fossil fuels, no matter how secure we feel in our dependence, it can only be sustained for as long as the fuel lasts. In this respect our reality itself is but an extremely convincing *illusion.*

First I shall describe how individual experience blinds us to the inherent fragility of our productive and distributive technique at the same time as it leads us to expect it to continue to satisfy our needs. Then I shall explore two aspects of the collapse of our illusion of security. The first of these — in a sense the cause of the collapse — consists of the economic difficulties that flow from shifting dependence from cheap to more expensive energy sources. The second comprises the declining faith, that people have in the System, which arises from the impacts that these difficulties have on experience, perception, and attitudes. In this latter aspect lies the promise of quite radical social change which could result in the gradual disappearance of many of our most taxing problems. And this through prevention rather than cure.

2. The Illusion

In our affluent age we live in conditions of unprecedented comfort in our boxlike homes complete with a spectacular array of 'taps' — gas taps, hot and cold water taps, electricity taps and entertainment taps of various kinds. Through them our needs are supplied with minute personal effort. Compared to life in the hostile natural environment, our lives in our homes must seem like Heaven. In Figure 1 our consumer heaven is represented by the outer circle with a box representing home, the retail trade window, the service station box complete with its taps, and a wáste chute nearby. From these we are served with all we require; consumer goods, the most exotic range of foodstuffs in and out of season, insurance schemes, the machinery to allow travel about our country or around the world, state welfare, and so on. In our homes or in our cars we are able to live, whatever the weather, in conditions of optimum temperature and humidity. The fruits of Heaven, this consumer paradise, are rich indeed!

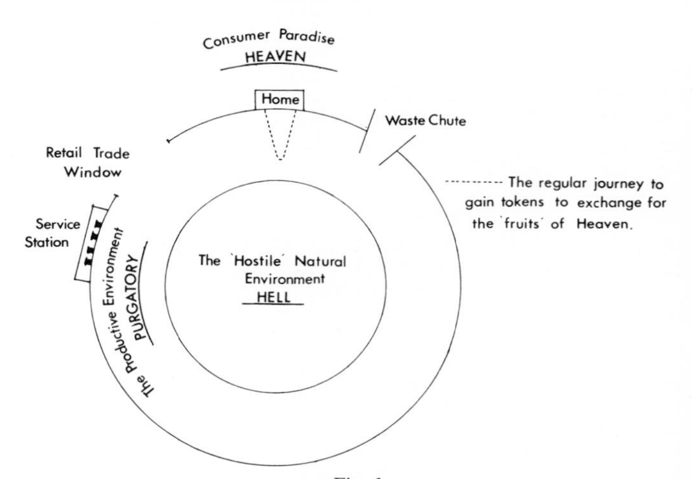

Fig. 1

This diagram represents the relationship between the consumer and productive environments (Heaven and Purgatory respectively) that we experience and the natural environment (Hell) upon which we depend. That dependence is obscured by a proxy dependence on the tokens that we acquire for our endeavours in Purgatory. This in turn leads to the distortion of political process which is predominantly concerned with the flows of those two products of the productive environment: the goods and services that we depend upon in Heaven and the monetary tokens that we so absolutely require to exchange for them. In terms of the diagram, 'development' involves the separation of the circles with the productive environment 'progressively' insulating us from experience of the natural environment.

Diesendorf, M. (ed.) (1979). *Energy and People.* Canberra, Society for Social Responsibility in Science (A.C.T.)

Through our experience we are led to expect the continuity and expansion of the range of all of these goods and services. Indeed, our expectation is so strong that we have adapted to depend upon continued supply, and hope to increase our access to these goods. This access is gained through the money we earn in return for our productive endeavours which help to create this Heavenly bonanza. For, regrettable through it may seem, the magic of our Heaven requires an extremely complex process to fashion it from the 'resources' of our 'hostile' natural environment.

The latter, with all its irregularity and uncertainly, is like Hell when compared with our Heaven. It is represented in the diagram by the inner circle. It is the Earth; the beings that we share it with; and ourselves, on the rare occasions that we think of ourselves as part of nature.

The gap between the circles, Heaven and Hell, represents the productive environment — Purgatory perhaps. Here we collectively create the magic we enjoy in Heaven. In a sense it is our insulator from the Earth's hostility for here the beast is tamed. Each working day we leave our little boxes and travel, usually in our cars, to our jobs. We do so to acquire those monetary tokens that we so absolutely require to gain access to the goods and services upon which we depend. It is here that we experience the productive process that makes it all happen; here where we learn of its inherent difficulties and techniques; here where we learn directly of the constraints imposed upon the process by the finite nature of the Earth, and by the damage we do it.

But the nature of 'work' (2, 3, 4) is such that we learn little of these things. The process is so complex and the tasks that we perform so trivial, that we never experience a full productive process. Most of us 'spend out time' performing seemingly mindless tasks that bear scant relation to the final 'product'. Typical examples are the worker on a production line or in front of a machine punching computer cards. Even the 'learned', the people who we might expect to appreciate how it all works, are 'blinkered'. They specialise in disciplines and tend to look at the minutest 'parts' with apparent disregard of the 'whole' (5, 6). The disciplines have been broken down into sub-disciplines so that physicists, for example, when they attend conferences, rarely understand what more than a few of their colleagues are talking about. It is no wonder that advice proffered by experts (7) is confusing to mere mortals. At the level of the individual we fully experience the fruits of Heaven, which are the goods that emerge from our trivial productive activities. That experience has shaped our habits, dependences and expectations. As our productive system has increased in its complexity, so we have become more insulated from the Hell which we still collectively exploit to satisfy our needs. As individuals we now need money to gain us access to goods. In a society less insulated or alienated from the natural environment, we would have required much more directly the goods or services themselves. We assume, and our past experience has justified the assumption, that provided we have 'earned' enough of those monetary tokens, the System will continue to provide.

This alienation of consumer from producer is not limited to individuals. The producing units within our societies are themselves consumers of each other's products. Firms must assume that their suppliers (overseas or at home) will continue to supply, and also that they will overcome any difficulties of production. With enough tokens all consumptive needs will be provided for.

With money the primary need that leads to the satisfaction of all other needs, our formal political systems have been similarly distorted in their activities. Politicians are elected to regulate the affairs of state and to protect the common good. But whilst we, as individuals, measure our wellbeing in terms of the access we have to the goods and services, inevitably politics is dominated by the flows of those two products of our industrial system, the monetary tokens and the goods and services to which they gain us access. Economics now dominates politics. In Figure 1 the arena of politics is at the interface between Heaven and Purgatory. Major issues — unemployment, monetary distribution and inflation — mirror our dependence, since all impinge on the distribution of goods and services. This distribution is disrupted when strikes occur. Governments react generally by opposing strikers, seeing that the common good depends on the uninterrupted flow of goods and services. Justifiably or not, their action mirrors our dependence. Social status is seen in terms of monetary earnings or in terms of the goods and services we purchase. And we can go up the social status 'ladder' via the education process (8, 9). Governments seek (adequately or not) to give equality of opportunity through compulsory schooling and funding for tertiary institutions. Here we are ordered into relatively closed bodies of knowledge, the disciplines, and graded according to our competence within them. The speciality we develop is useful to employers who use the grades attained to help with their selective procedures. And we receive a perfectly acceptable rationale for discriminatory distribution of responsibilty and its reward.

Not surprisingly we feel that somewhere someone (one of those superior to us) has control of society's destiny. I may *feel* that scientists, for example, are the most important members of my community, but scientists will *know* that they are not and *feel* that someone else is. In this feeling lies the root of an all pervasive 'psychology of powerlessness' (10). Any obvious incompetence is blamed on the government, which, being advised by unidentifiable but certainly omnipotent experts, must be ignoring their advice. All this is a perfect recipe for universal abnegation of personal responsibility.

Education which, one might have hoped, could provide us with a useful grasp of how our System works, has become the means by which we establish claims to the jobs that gain us access to the goods produced by that System. And even if we do become concerned with aspects of our social organisation, the sharp edge of that concern is blunted by our knowledge that others are more competent than we are. Furthermore, our range of possible actions is sharply curtailed by our deep dependence. We might, for example, be concerned about the degradation of the environment. But if prevention of that degradation means a reduction in our living standards, as measured by the purchasing power of our incomes, then our likelihood of expressing that concern will be reduced. So will the likelihood of getting others to share it. Much easier is it to continue to enjoy our consumptive behaviour and assume that someone else knows more about our concern than we do.

Thus we easily accept the 'rationality' of hierarchy (2, 3) in social organisation, particularly in the organisation of our workplaces. The discrete division of labour which impoverishes our work experience is compensated by the money that we earn. Furthermore, our effectiveness and loyalty to 'the firm' holds out the promise of more. We all end up being good people working for the System upon which we depend. Here we may help make the important decisions that shape our social direction for it is decisions in the productive arena that determine the range of products and services from which we, as consumers, can 'freely' choose.

These productive decisions (11) are heavily influenced by the profit motive which practically dominates consciousness within this environment. Much flows from these decisions as a bi-product. Where we live, the amount of compulsory travel, the distance between us and our friends, what is or is not good art, the entertainments we passively indulge in, our leisure activities and much more are strikingly influenced by ourselves in our working lives over which we have little control. The domination of the profit motive in decision making here has been a major feature in the establishment of our dependent relationship with the productive System.

Not surprisingly, this motive in turn has distorted our collective relationship with the natural environment. Again "*the sole criterion to determine the relative importance of . . . different goods is the rate of profits that can be obtained by providing them*" *(12)*.

As one top executive of a major U. S. oil company has said: "*It does not mean a thing to say to a private oil company that there is a great need for oil. You have to have incentive. If it turns out that phosphate rock is more profitable, we'll put our money there"* (13).

In 'The Closing Circle' Barry Commoner (14) has shown that much of the U.S. economic growth that has been a feature of the period since the Second World War has involved the displacement of old technologies by new for the production of goods that fulfil essentially similar functions (15). Soap powders have been displaced by detergents; natural fibres by synthetics; steel and lumber by aluminium, plastics and concrete; railroad freight by road freight; returnable bottles by non-returnables, and so on. These new technologies, and others established to provide them with raw materials, are in large part responsible for the seriously increased environmental degradation which was also a feature of the period (16). Yet each of the changes has been economically profitable. The labour inputs to production have been reduced with a consequent increase in the complexity of productive process and its dependence on finite fossil fuels to provide both energy and feedstocks. For example, a synthetic fibres industry has displaced the sheep and cotton plant but is critically dependent on fossil fuels. Economically 'sensible' developments such as these have radically altered the nature of our relationship with the natural environment both through increased degradation and changes in the characteristics of the resources from which our products are fashioned.

Uncertain dependence on the products of *living* things, human and non-human, has been replaced to a large extent by a more controllable dependence on *dead* organic or inorganic matter. So far, the substitution has worked, providing us with the material requisites of life and comforts, but at the cost of obscuring our relationships with each other and the natural environment upon which we still ultimately depend.

The experience of the joys of secure supply, and the 'consumption' of seductive new ways of satisfying 'old' needs, forms the basis of our belief in our 'advancement'. The costs of the weakened or distorted relationships pass relatively unnoticed; these are incurred as the inevitable price we pay for the inevitable progress of the System that makes the security possible.

Being 'advanced' and requiring money as our primary need, we easily slip into seeing monetary incomes as a measure of our standards of living. But when one makes comparisons between countries this becomes nonsense. It is said, (17) for example, that 700,000,000 people in the world live on monetary incomes of less than $50 U.S. per annum. If we allow a generous factor of ten to account for currency exchange aberrations then these people live on real incomes of less than $500 U.S. per annum. Could we do it? No way . . . unless we were able to produce and satisfy a large measure of our needs in ways that did not require the use of money as a medium of exchange. For this to occur producer and consumer would have to be geographically much closer to one another. Production would have to be on a much smaller scale and decentralised.

Now consider a similar comparison with energy use as the measure. According to U.N. statistics (18) for 1971 the average per capita rate of energy consumption in the U.S. was 10.3 kW, in the U.K. 5.0 kW, whilst in underdeveloped Nepal it was 8 W. A Nepali, on average, uses about 1/1300 the amount of energy used by an American. Most people are prepared to accept that the U.S. is unnecessarily wasteful in its energy use habits. We therefore will consider, a little more critically than usual, energy consumption in the U.K. About 1.5 per cent of Britain's energy consumption (19) (in 1972) was in the agricultural sector so the per capita rate of energy consumption for British agriculture was about 7.3 W; only marginally less than the *total* per capita energy consumption of a Nepali for all purposes. When one takes into account that the British figure includes energy consumption 'up to the farmyard gate', omitting necessary energy use for transport and packaging, the comparison is more striking. Even more so if we note that Britain imports about one half of her food requirements effectively giving it a 100% energy subsidy to agricultural production from abroad.

What do these sorts of comparisons mean? Usually we immediately conclude that the Nepali's life must be desperate. But the comparisons are essentially meaningless, because they purport to compare not what they actually compare. The reason for this lies in what data collectors choose to, or more correctly, *can* include in their statistics. Per capita energy use for *all* countries is necessarily underestimated, because estimates *can only* include sources that are *monitorable;* these are the sources that are supplied from centralised locations and distributed in *known ways* to users. The productive and distributive organisations of 'underdeveloped' countries such as Nepal depend upon the use of essentially non-monitorable, diffusely available, and, incidentally, *renewable* energy sources.

In reality these types of comparisons do not indicate relative well-being or living standards, but rather the relative degrees of centralisation of economies and the extent to which the industrial mode of production (which is centralised) dominates and displaces the alternatives. In terms of the use of critical finite resources the diffuse production and consumption patterns characteristic of 'underdeveloped' countries are vastly more efficient than our own. Having spent some time in Nepal and other such countries, my own judgement is that the people of these countries are happier than most Americans that I have met. But let's not quibble about subjectivity. The individual Nepali is certainly not 1300 times less happy than the individual American.

With the observable and monitorable increase in industrial production, there has been a decline in the *invisible* production so essential to those living in 'underdeveloped' states. This comprises the 'produce' that comes to people via relationships with each other and the natural environment.

Control of the environment through the use of our energy subsidy has fostered the Illusion that problems of production (20) have been solved. However, the Illusion is collapsing, as many in our communities find it increasingly difficult to satisfy their basic need for money to gain access to the goods.

3. The Collapse of the Illusion

We feel secure in our belief that our needs can be satisfied if only we have enough monetary tokens. But money is a unique human tool for exchange. Our non-human resources simply cannot use money to 'satisfy their needs' so the price of anything we buy is but the sum of the increments of the labour costs, profits, taxes and rentals earned and taken in its production and distribution. The money goes to people as individuals and groups; none goes to the natural environment that might be degraded in the process.

Monetary profit dominates the decisions that shape changes in productive technique, so it is inevitable that some proportion of profits will be invested to increase future profitability. The result is that, at any time, two categories of production take place. Besides the goods and services that we consume immediately, the perishables and consumer durables, production goes on to increase either the range on offer or to improve efficiency of *future* production. But it is efficiency in terms of production per unit labour input because labour is the only resource which costs money.

In itself there is nothing intrinsically unreasonable in seeking to reduce this labour input. The fashioning of tools is an example of this type of production. It is only when the tools (or machines) require an external energy subsidy as an input that the unreasonableness *can* come in. It depends on the *nature* of the subsidy. If it is derived from a *constantly available* source such as the Sun there is no problem. *Renewable* sources, though, are derived from living things and the rate of production using them must be constrained to within their capacity for regeneration if production is to continue. Trees and 'beasts of burden' are examples of these sources. And *non-renewable sources* pose even greater difficulties, for their use necessarily establishes the need to find more or some substitute source; if they are required to *run* productive systems.

These qualitative differences are not easily accommodated by a system that treats human labour input as the only relevant input to production.

Prior to the Industrial Revolution the bulk of the energy subsidy used for production was derived from renewable sources, and many civilisations fell because of their misuse of the resources that fuelled their growth. Industrial England (21) was apparently saved from collapse to 'primitive agrarianism' by the discovery that coal could replace wood in the process of smelting iron.

Given the abundance of coal in England there was no particular reason why its exploiters should have been unduly concerned about times of future scarcity. Besides, it had particular attractions that must have appealed to entrepreneurs. Like the other non-renewable energy sources, oil, natural gas and uranium, coal is found in concentrated deposits. By contrast, constantly available and renewable energy sources are diffusely available. Capitalists sought to accrue profit through increased labour efficiency brought about by supervision of production and the discipline of workers (22). There was a clear need for:

"machines whose *appetite for energy* was too large for *domestic sources of power* and whose mechanical superiority was sufficient to break down the resistance of *older forms of production,"* (23) (my emphasis).

Of necessity, those 'domestic sources of power' are constantly available or renewable because of their even distribution. They powered those 'older forms of production'. (exit the invisible non monitorable system that makes nonsense of many of our statistical comparisons between countries. See above).

The need for the machines was satisfied, as was the need for coal, which clearly must have gratified those who expropriated the coal deposits to mine and feed that 'appetite for energy'. And mines had an appetite for capital too great for any but the wealthier entrepreneurs. What a great investment! The demand for those non domestic sources of power rapidly accelerated, as demonstrated by the fact that coal production in Britain (24) increased from 10 million tonnes in 1800 to 60 million tonnes in 1850, and to 225 million tonnes in 1900. More coal was consumed in 1899 and 1900 that in the whole of the eighteenth century.

But the use of the finite coal entails a cost that could not have been appreciated or considered when investments were made. Many of the productive tasks we set ourselves as end-use consumers could be performed using either constantly available and renewable, or non-renewable energy sources. To use either type of source to perform these tasks capital, labour and resources must be invested to be capable of converting the source from its raw useless form to useful forms where required. It follows that if these investments are large and remain successful, habits, dependences and expectations will be formed. These will be directly or indirectly satisfied through the use of that resource. This would be of no concern if the choice were a constantly available resource; it need not (given adequate conservation) if the resource were renewable. But it certainly would be of concern if the choice is non-renewable. Dependence matters if its satisfaction reduces the possibility of its future satisfaction. Coal cannot be burned twice, but, for practical purposes, the Sun will shine for ever. The cost of resource depletion is borne, not by the user (nor the entrepreneur who makes choices on the users' behalf) but by future potential users, adversely affected by the depletion of stocks required to satisfy habits.

Costs, like this one, which are not included and paid for in the price of commodities, are termed 'externalities'. If such costs are borne by others *concurrently* with production then they can, in principle, be 'internalised' with the initiator of the cost being required to pay taxes, or in some other way, to compensate. A factory may be required to pay to clean up some bi-product environment damage. But what of externalities that are inevitably borne by future generations? These types of costs, which are well exemplified by the choice of fossil fuels to power productive and distributive systems, must be the most awkward of all externalities.

In practice, they have been ignored. The prevalent faith in technology and in its inevitable improvements has been so strong that it has been simply assumed that future generations will have developed sufficiently to solve any of the problems which will arise as resources become scarce. The unquestioned ethic or assumption underlying the apparent rationality of economics is that *the interests of the present are overwhelmingly greater than those of the future.*

Has not the coal-fired salvation of Britain posed similar problems for us? And just what is to be our 'coal'? We today bear the costs of the externality inherent in her method of avoiding collapse. By defining the locations of the work-places, the factory led to the urbanisation so characteristic of 'development'. With it the capacity of communities to provide invisibly, and in large measure, for themselves was reduced. Our expensive welfare systems, education systems, transport systems, communications systems, and so on, have become so absolutely required. Even national security has been weakened through this economic 'rationality'.

With out vastly increased ability to move things about cheaply, anything that can be obtained more cheaply in another country is obtained despite the distance. As one senior (U.S.) oil company executive (25) said, explaining his company's decision 'to go abroad':

"Important . . . was the apparent profitability of foreign oil operations . . . Rates of return . . . from international operations proved considerably higher than returns from U.S. operations alone."

And Britain, with her energy industries nationalised, followed suit. She allowed her coal industry to 'run down' in favour of Middle East oil. The economic difficulties faced by 'advanced' industrial countries when the O.P.E.C. countries (26) sharply increased the price of crude oil, are future costs entailed in rational decisions such as these. Interdependence can sometimes prove costly. The imposition of these costs only makes sense if we accept the assumption underlying our economic activity that present interests are all that count.

Looked at in entropy terms (27, 28, 29) this tacitly held assumption leads us to recognise, value and give a price to the order that our activities bring about, but the entropy law — the second law of thermodynamics — leads us to see that " . . . *every action, of man or any organism, nay, any process in nature, must lead to a deficit for the entire system"* (30). The externalities represent in total an amount of disorder that exceeds the order that we value. And much of our 'valued' produce ends up disappearing down our waste chute to present us with disposal problems. I do not mean to belittle the tasks that we set ourselves. Goods and services can, and do, enrich our lives and increase our comfort. Nonetheless, commonsense recognition of the entropy law would lead us to perform our tasks in ways that minimise waste; ways consistent with the tasks that we wish to perform, that reduce as much as possible the amount of disorder we cause as a bi-product.

An economic system that makes no allowance for the qualitative differences of the resources upon which it depends, and which values only the human inputs to production, must ultimately be unstable if its fuelling energy source is finite. So it must also be if one of the demands it makes is that the natural environment has an infinite capacity to regenerate the constant non-marketable conditions required for human existence; for example, clean air, fresh water and a stable climate (31).

Concern expressed about environmental degradation and impending scarcity of some critical fossil fuels indicates a growing awareness of some of the external manifestations of this instability. But once that assumption underlying economic activity is tacitly accepted, the rationality of an economic system based upon competing self interest is inherently logical and difficult to contest. If only we are prepared to pay, all of these symptoms of impending instability *seem* to be within the capacity of the economic system to absorb.

But for all its inherent logic, the economic system itself is unstable on this finite planet. To maintain its logic the energy sources which fuel it must not only be available on demand, but also be reducing or stable in price. To see this, imagine a simplified world with only one fossil fuel deposit. If this deposit were exploited for personal gain, the entrepreneur would exploit the most accessible and preferably the highest grade of this resource first. Thereafter the amount of work required to win and use the resource would tend to increase. Technological improvements would help to offset the consequently increased costs, but clearly the resource's finite nature demands that the trend must ultimately become apparent.

In Britain coal was initially taken from close to the surface but soon pumps were required to remove water from the deeper mines. Then not only were costs incurred at the time of winning the coal but also in the *prior* construction of equipment to make production possible and preferably more efficient. These latter costs had to be paid for either from past profits or from money borrowed against the promised future production (and profit). In either way these costs were paid from capital.

In our simple imaginary world the demand for capital to enable continued exploitation can be seen easily to rise if production of the resource is to continue. But we live in a world blessed initially with many fossil fuel deposits. As the capacity to exploit the first was demonstrated so was the capacity to exploit others as well as the profitability of doing so. Machines, fuelled by the resource, effectively 'reduced the distance' between deposits and the points at which they could be processed and used. The result was that the tendency of capital costs to rise was offset by the tapping of 'fresh' deposits elsewhere. But ultimately, in a finite world, the number of deposits elsewhere must reduce to zero. Then the trend of increasing capital costs must become apparent. The capacity of entrepreneurs to exploit resources elsewhere can have only *delayed* the inevitable trend. In fact, for most of the period since the Industrial Revolution capital costs have declined, but now we face the prospect of a cake without the cream. Needless to say, we are really addicted to that cream. The onset of this trend is manifest in an apparent absurdity (32):

"*. . . As the rate of production increases, the petroleum industry (in the U.S.) becomes correspondingly less able to generate enough capital to supply its own needs . . . In order to acquire more capital, they want larger profits; in order to gain larger profits, they demand higher prices and more tax benefits . . . 'investment incentives' . . . For the period 1970-73 the capital required* (33) *to support energy production amounted to about 24 per cent of the capital invested in U.S. business as a whole. It is estimated that this will increase to 33.3 per cent for the period 1975-85 . . .*

Most of the needed capital will have to come from borrowed funds . . . even though the price of energy and industry's profits have sharply increased in the last few years. According to a recent review, the U.S. energy industry will need some $900 billion over the next ten years, 'nearly half of which will have to be raised externally' . . . Thus if the energy industry is to earn enough to support its own growing needs for capital, the price of energy must increase. On the other hand, if the price of energy does not increase enough to generate the needed capital, and the industry must borrow it, interest rates will rise. Either outcome, or any combination of them, can only lead to inflation" (34).

In Fig. 2 is plotted the energy produced (in the U.S.) for every dollar invested in energy production for the years 1960-73 and, even though there are insufficient points upon which to base definite conclusions, the trend indicated is alarming. A steep decline in the energy produced per dollar invested is evident from about 1968 onwards.

The relative capital costs of generating energy using tried and contemplated substitute technologies (35) lead me to believe that the trend is accurate and likely to continue. To produce the same energy generating capacity as from Persian Gulf oil reserves:

from North sea oil reserves	between	8 and	40	times more capital is required;
from new U.S. oil reserves	"	10 "	40	" " " " "
producing oil from coal	"	20 "	95	" " " " "
using nuclear fission	"	66 "	300	" " " " "

If this was not enough to cause considerable anxiety, the lead time factor compounds the problems (36). The production technologies required for these substitutes are considerably more complex than those which they are intended to replace. Increasingly, investments are going to have to be made earlier than the anticipated beginnings of production.

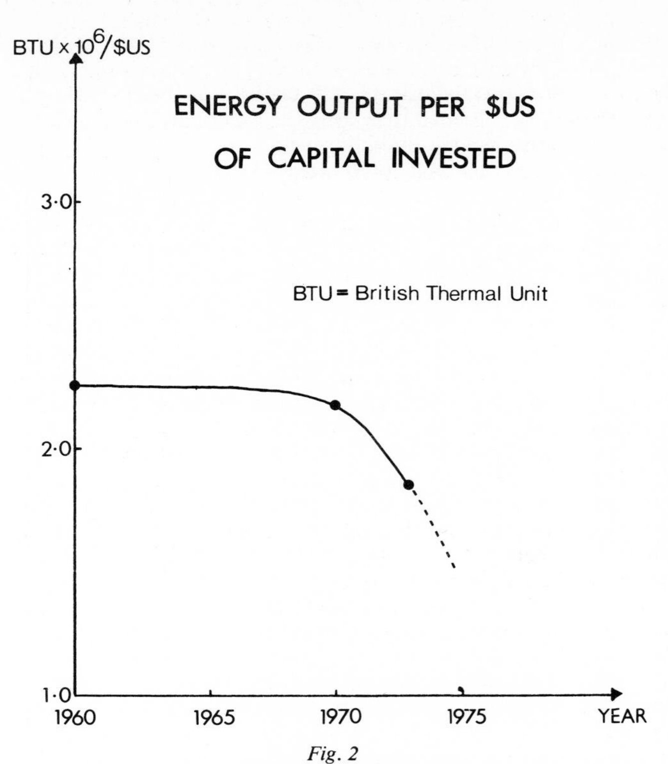

Fig. 2

U.S.A. energy production per dollar invested, by year (1960-73). The data, taken from Commoner (33), have been adjusted to eliminate the effects of inflation and are in terms of 1973 dollars.

It is quite clear that we 'need' centralised and finite energy sources to *run* our productive and distributive systems as they now are. But their use *commits* a proportion of the monetary *capital* for its future *running*. Whilst that proportion so committed was reducing it was economically sensible to increase our dependence upon this form of resource capital. The capital so generated could be used to increase the range of goods and services available and 'sop up' displaced labour. Now that the committed proportion is increasing the absolutely necessary energy conversion facility is turning into a monetary capital 'sink'. The use of energy guzzling machines must result increasingly in a decrease in the range and/or quantity of goods and services available and/or leave people unemployed.

Now we can begin to see a major factor contributing to the 'strange' economic uncertainty being experienced by 'advanced' countries, and the industrial sectors of developing countries. Our understanding of the phenomenon of development has been shaped in the period when energy costs, in real terms, were declining — when, in fact, energy was simply taken for granted and treated as a commodity like any other. With the unseen and increasing energy subsidy, firms could increase their profitability by displacing people with machines, in the knowledge that such displaced people would find other employment easily. But now the capital required to set up these new work places is in short supply. Firms still try to increase profitability by 'tried and true' methods but their machines, increasingly complex and sophisticated, are also a drain on available supplies of capital.

Governments, noticing the shortage of capital and its inflationary consequences, follow the (idealised) legal practice of looking to precedent to guide their actions in the new circumstance. Economic theory, developed in the halcyon days of declining energy costs, leads them to think that inflation is caused by excessive government spending and too high a proportion of the monetary fruits of production going in wages. The apparent reluctance of the 'private sector' to establish new jobs for the unemployed is seen as due to a lacking of business confidence, and governments seek to increase this by 'enabling' the increase of profits and the granting of incentives of one kind or another.

The real wages and salaries of workers are 'allowed' to decline through devices such as partial wage indexation, and government expenditure is cut in all areas except those which will encourage industry to resume the business of making the economy grow. But the following point seems consistently to be missed. Any society, regardless of ideological differences, that bases its productive and distributive systems on the use of finite energy sources must experience these types of problems once their continued use begins to demand increasing commitments of capital.

The phenomenon of inflation coupled with unemployment must affect the expectations and perceptions of people who are adversely affected. The first to suffer in times of recession or stagnation are the unemployed whose welfare benefits will continually be under threat as the Government tries to reduce its welfare 'burden' or allows it to reduce through inflation. For them the System is already failing in that they are suffering a reduction in the access to the goods and services to which they have become accustomed. And, for those lucky enough to be employed, there is increased threat of 'redundancy' as firms try by the old labour saving means to increase their profitability.

In times of growth people could expect promotion within their firms or bureaucracies, or could advantageously move from one to another. Thorny ethical problems like distribution of wealth or income could be treated as irrevelant or impediments to progress. Growth enabled great mobility, and people who did not advance could easily be regarded, however wrongly, as responsible for their own lack of advancement.

But now, those at the top are likely to remain at the top, regardless of competence, and those at the bottom will have to wait for the death or retirement of their not necessarily more competent 'superiors'.

Even for the employed the System, in its own terms, is not offering what it once did. Progress is no longer inevitable. Education no longer carries any guarantee of success in a status sense. Ph.D.'s literally and metaphorically drive taxis, even though they would prefer to be researching in the way promised when they began their long and arduous studies. Students are being trained to be teachers even though they may have little prospect of employment in their chosen profession. Tertiary institutions are battling among themselves (37) to think of new courses to attract students and the funds that come with them, to keep their staffs employed. And those employed, wherever they are, can but hope that they will not be the next declared redundant — a nice Systematic word! Everyone 'in' is 'holding on' in the hope that things will improve, a hope without much conviction behind it.

Yet the rhetoric they hear is one of continued 'normality'. Publically, experts continue to make reassuring noises as they act in their public relations roles for which they are paid. But privately, many are far more reserved and uncertain in their views of the future. This type of 'schizophrenia' is becoming endemic.

The Illusion is collapsing as those two things upon which it rested disappear. The first is the cheap energy sources that supplement our meagre efforts in the productive process, and the second is our confidence that the productive and distributive System will continue to provide us access to the goods and services that it magically and incomprehensibly produces.

4. Escape to a New Reality?

The availability of energy in suitable forms is critical to any productive and distributive system. Thus an energy crisis must be seen for what it is — a production crisis. It can, in principle, be overcome by the development of new energy sources that can be processed into forms that are demanded by the nature of the productive system. But, as we have seen, if these alternatives are expensive in terms of their demands for monetary capital, their use promises severe economic difficulties for the economic system that has shaped the development of that form of productive organisation.

Alternatively, we could observe that it is not consumers who require energy in the forms upon which they now depend; they require certain tasks to be performed. Observing this, we could begin to look at the productive systems of those countries less 'advanced' than ours, or at those of our ancestors who somehow survived and managed to beget us.

To quite a large extent, this small scale, localised production remains in the most 'advanced' societies. How would a society such as ours survive unless mothers reared children, or without the voluntary work done in so many areas? Neither of these activities is visible in the sense that there is monitorable monetary payment in exchange for them. But they are productive nonetheless. With our ideas of standards of living so closely linked to monetary earnings it is not surprising that:
"only recently has the value of the voluntary labour that neighbourhoods produce in quantity been recognised. Cohesive communities have an important, even critical barter economy. People watch other parent's children. Recipes are swapped. Emotional support systems are common. Bartering of goods and services goes on continuously. The strength of neighbourliness should not be underestimated . . . (and probably) . . . (communities) with a strong sense of self-identity contribute far more in neighbourliness to their local economic structure than those that lack this identity. Beyond this there is the vast range of unpaid services . . . the household economy. Several studies have found that if we gave the minimum wage for such tasks as house cleaning, child care, and the like, we find that unpaid services are worth as much as the entire national income" (38). And this is America, the country in which the atomisation of community due to industrialisation is most advanced. The inputs to these vital elements of production do not include high grade energy or resources in large quantity, but they do include human innovative capacity and imagination.

On energy grounds alone a strong case could easily be made for an increase of the scope of this non-industrial production. A potato is a potato whether it's grown in a back garden or bought from the local store, yet the economic activity involved in growing, transporting, wholesaling, packaging and retailing the latter is valued, whilst the lack of this activity in the case of the homegrown variety passes unnoticed. Nutritionally, there may be little to choose between the two potatoes but a vast amount of work has to be done to bring the bought potato to our table, whilst that grown in the garden requires relatively little. If people had the time and resources to grow their own potatoes then, to the extent that they did, the per capita need for a potato output from the System would reduce. This would be bad for the GNP, but people would still get their potatoes.

Entertainment is now provided by a number of vast industries for our passive indulgence, yet people have the capacity entertaining themselves in many varied ways. Production of all kinds of goods, and the provision of many services, are now seen as social needs demanding complex systems for their satisfaction, whereas formerly they were satisfied unobtrusively in non-industrial ways.

Our sophisticated industrial system should not be seen as an absolute benefit. With its growth much of the capacity for small units to produce requisites of life has been weakened through the break up of communities and through intimidation by new and highly advertised industrial alternatives. Industrial production does not simply supplement already existing non-industrial productive techniques and organisation; it frequently displaces them. These displaced techniques are losses that must be balanced against benefits.

Recognising this, together with the fragility and complexity of the displacement systems, we *could* consider re-establishing this invisible productive technique which has, throughout history, served the bulk of humanity well. We might *try* to follow the following maxim:
" . . . you should never assign to a larger entity what can be done by a smaller one. What the family can do the community shouldn't do. What the community can do the states shouldn't

do, and what states can do the federal government shouldn't do" (39). Applied to production as well as to political concerns, its adoption would imply that we would seek to prevent needs becoming social, rather than attempt to set up complex systems to cure or satisfy them once they have become apparent.

This would not mean that we should do away with the industrial system, but rather that the latter should be reserved for increasing the effectiveness of non-industrial production; doing what the 'smaller entity' cannot do. But, for this to happen, we would have to alter our ideas about our standard of living. For, as needs become invisibly satisfied, the necessity for industrial or paid work would reduce, and with it the total and per capita earnings of our community. Certainly the social need for centrally provided energy resources would drop as we made better use of our imagination and innovative capacities to fashion locally available resources. It follows that, as the demand for energy reduces, so will the demand for capital funds required to sustain energy supply capacity. With reduction of capital demand the inflationary pressures imposed upon the monetary economic system would be reduced, thus going a long way towards solving this all absorbing problem faced by governments in 'advanced' countries.

But arguments based on energy grounds are unlikely to influence the community as a whole to follow this course. So long as the System is able to provide and sustain our Illusion we will continue to depend upon it. In spite of the fact that I know that petroleum products are likely to increase greatly in price, and that absolute scarcity might lead or force me to alter my style of living, I do not modify my car use habits. No matter how much concern I have regarding diminution of fossil fuels, my knowledge that provided I have enough monetary tokens I can get as much fuel as I need, is sufficient to maintain my car use behaviour virtually unamended. I do not think that I am unusual in this regard. Whilst the System continues to produce the goods, all I need is money. And whilst I can trade my time for it, I will have it. Grim and justified forebodings of future crises (40) cannot be expected, alone, to lead to changed behaviour. Whilst we still can get those monetary tokens — our primary need — 'business as usual' is simply easier than change. But with unemployment increasing and stagnation dimming the future prospects of those employed, the System's holding power is weakening. No longer is it so easy for people to accept inconvenience at the hands of the System. The inevitability of progress is not now so easily verified from experience.

One inconvenience experienced by the employed is the consistent and compulsory time commitment which has to be made in service to the System. Consequently many people have opted out of the System and now attempt to live self-sufficiently on small holdings. Others attempt to increase their involvement in their work by running small businesses or by making and marketing craft products. Many would like to be employed for shorter hours and proportionately lowered earnings, so that they could spend more time doing as they wish.

This latter group is interesting. It is impossible to estimate in number because, in terms of the way we observe things, its members are invisible to all but their friends and some acquaintances. They work perfectly 'normally' in their work places, and their styles of living are not distinctive. For most, life styles are very much determined by the nature of their employment, if only by the consistency and regularity of its time demands. Invisible though they are to 'outsiders,' I know that most of my circle of friends and acquaintances belong to this group that would like to have flexibility in the time commitment they make to the System. Most would seriously consider job sharing if it became more possible. How many of the 'outsiders' might be secret members of our group?

It is impossible to know but occasionally — perhaps at a lecture by some critic of our social system such as Ivan Illich or by some advocate of simpler living — many more apparently normal people emerge to share our revulsion of the materialist system to which we conform. Frequently at events like these, some bemoan the acquisitiveness of others without having any means of knowing whether or not they also would like to express themselves in other ways. Sometimes public opinion polls (41) indicate that people would like to live more simply, but what people *do* is more apparent than what they say they would *like to do.* All of us tend to judge others by what they do without tempering our judgements with an appreciation of how constraints force us to act in non-ideal ways. We understand our own divergences from the ideal because we know, to a degree at least, what our ideal is; but it is difficult to extend our understanding to others who act in non-ideal ways — as we do.

Individuals in this 'group' who would like to gradually reduce dependence on the System, are trapped by the all-or-nothing time commitment demanded*. We are all actors and actresses in our way, with the employers as piper playing the tune to which we dance. We are all, in varying degrees, experiencing a type of 'schizophrenia' as we submerge private feelings and standards to gain the security that seems to come only through the work we do for the System.

In stagnant times such as these, we fear redundancy and redouble efforts to maintain our employers' favour. The danger is that our society will split permanently into new class divisions — the employed and the unemployed — with the employed desperately hanging on to jobs to which many feel only partially committed. Working the mandatory time with apparent enthusiasm, many privately applaud such ideas as job sharing and permanent part time work with security (43). These ideas and the possibilities that their implementation would open up are difficult to discuss in workplaces; such talk can so easily be taken as disloyalty or as indicating a lacking of desired commitment to the 'firm'. We are left with the irony that many would like to find ways of working less whilst others, many of them desperately, need work. The two groups find it impossible to accommodate each other's need.

If society recognised the perfectly reasonable right that *individuals* within it *should be free to be employed for as long as they want on a pro rata basis* then, in co-operation with employers, the two groups might find ways of satisfying their complementary needs.

Furthermore, a mechanism would be established through which we could, collectively and individually, find some balance between our need for access to the goods and services provided by the System, and the need we have for time to provide directly for ourselves. The 'smaller entities' could then choose to do the 'assigning'. The choice of the nature of the means of production would shift more to our 'private selves' and away from our 'public selves'. We may even find that *sharing* is much more satisfying than competing and owning. Then we would begin to see that co-operation with others is easier than is implied by the apparent necessity to enforce co-operation through rigid hierarchic organisation.

Earlier we saw that an ethic or assumption underlying economics is that the interests of the present are overwhelmingly greater than those of the future. Once it was tacitly accepted, economic activity led rationally to the choice of finite fossil fuels as the energy source with which to fuel the social productive system. An alternative ethic which could underlie economic activity would be that the interests of the future are paramount over those of the present. In this case fossil fuels might not have been used at all. Both ethics are extreme and absurd but one has been implicit in our mode of development. More rational would be a middle way that sought to benefit the present without prejudicing the future. This would lead to the recognition of both the usefulness and finite nature of fossil fuels which would be used reluctantly and preferably to the benefit of both present and future people. We would regard these resources as capital and use them to increase capital, not to 'pay' running costs. Consider a practical example. In terms of this ethic we could consider, in principle at least, setting up canal systems to satisfy transport needs. This

*So actors and actresses find themselves playing seductive roles for advertisers, implying that life can only be glamorous if we buy the product proffered. Their own lives need not accord in any way with the roles portrayed. It is the image they portray that seduces (42). We do not see them outside the roles.

would benefit both ourselves and future generations, for the canals would remain as a capital asset. Contrast the rationality of this approach to the use of the remaining fossil fuels with that of continuing to burn precious fuel in the motor car. The latter course keeps us and future generations dependent on a transport system that will run only as long as the finite fossil fuels last.

With industrial production and its products and services most conspicuously the consumers of fossil fuels, this ethic would require decisions concerning an ethical balance of dependence upon industrial and non-industrial productive modes. In a democratic society this balance should be set by the people. It should be a political choice, but all our political institutions have evolved together with the belief that personal wellbeing demands the continuation and growth of industrial production. Politics has been concerned with the distribution of the fruits, leaving the establishment of dependences on particular resources to unfettered competition of self-interest within the productive sector.

It interests me greatly that many people now see their self interest in terms of free time rather than money. The reduction of monetary earnings necessarily would mean that what they chose to do with their time, if able to, would depend upon the 'domestic sources' of energy — those renewable and constantly available sources unable to feed the 'great appetite of machines'. And perhaps the pastimes engaged in, in this free time, would involve no energy subsidy at all. But how can these 'small entities' exert the pressure to gain the time-money trade-off when politics responds to pressure and no one can have firm ideas of their collective numerical strength?

Failing the exertion of such pressure, society remains run by people who act as though they see society's wellbeing as being consistent exclusively with that of the Industrial System. No matter how they privately feel about their own dependence upon the System, they publically work to satisfy the dependences of 'others'. Willingly or not, we are all trapped in this bind but none more so than those who have 'made it' to positions of responsibility and associated status: the specialists and experts, the 'talking heads' on radio and T.V. It is not easy for the planners, the educators and the bureaucrats to bring their 'private' selves to bear upon their 'public' performance. Economists will try to find monetary solutions for economic problems, technologists technological ones and educationalists educational ones.

The energy experts consider the energy output of various technologies, their relative 'economics' and how energy can be conserved using existing productive technique (44). They seek alternative ways of continuing to satisfy society's needs as they are currently satisfied. The possibility that changing attitudes and values have preventive implications does not enter their calculations. Experts generally are too busy looking for cures; for technical fixes. *Is nuclear fission or fusion to be our methadone?* Changing attitudes (45) accompanied, or were the precursor of, the industrial revolution and our growing dependence on fossil fuels. In the same way that the future critical dependence of industrial societies on these resources was not explicitly appreciated at the time, the resource implications of current values and attitude changes pass unnoticed now. If they gained the attention of our experts, then some of the barriers between disciplines would break down. For they would have to be more holist in approach, putting their specialist knowledge into the context of their total experience. In effect, they would have to bring their 'private' selves to bear upon the 'public'. Then the vast amount of knowledge that has been amassed through this industrial period might begin to be used with wisdom and for future as well as present benefit.

I am not suggesting that no professionals do this now, or in the past. Frederick Soddy (46, 47), a chemist, saw much of the energy problem as long ago as 1912:

"Civilisation as it is at present, even on the purely physical side, is not a continuously supporting movement . . . It becomes possible only after an age long accumulation of energy, by supplementing income out of capital. Its appetite increases by what it feeds on. It reaps not what it has sown and exhausts so far without replenishing. Its raw material is energy and its product is knowledge. The only knowledge that will stay the day of judgement is the knowledge that will replenish rather than diminish its limited resources" (48).

With self interest identified explicitly or implicitly with the wellbeing and growth of the industrial system, these holists have been voices in the wilderness. But now many are reading with appreciation the perceptive works of people like Soddy, and blame the acquisitive nature of humans for their lack of wisdom. They should welcome the current change in attitudes and aid it as much as possible, for holist voices have some chance of now being heard.

The long vaunted dream of the technological utopia has now dimmed, been seen as hollow, and for many it is a nightmare. With its passing as a goal, attitudes and perceptions are changing as reflected by the great and increasing interest in do-it-yourself books, religious works (particularly of Eastern and other mystic traditions), and the works of philosophers. With the passing of the seductive growth period and its promise readers are seeking a new context, a framework from which to see their mileiu. Many of these books suggest ways of finding it. So do the various groups that seek to form meaningful relationships out of the distortions which have been formed as an inevitable cost of progress.

The interest in religious and philosophical works would have outraged Francis Bacon. One of the early ideologues of science, he reflected the changing attitudes of seventeenth century England and helped establish the scientific method so important in our 'development'. He warned us against such holist ways of perceiving reality as these, branding them as "Idols" to be *"renounced and put away with a fixed and solemn determination, and the understanding thoroughly freed and cleansed: the entrance to the kingdom of man, founded on the sciences being not much other than the entrance to the kingdom of heaven, wherein none can enter except as a child"* (49).

Renouncing them he outlined the "*true way*" which:

"derives axioms from the senses and particulars, rising by gradual and unbroken ascent, so that it arrives at the most general axioms last of all" (50).

And he entreated us to have faith in the expert:

"One method of delivery alone remains to us; it is simply this: we must lead men to the particulars themselves, and their series and order; while men on their side must force themselves for a while to lay notions by and begin to familiarize themselves with the facts" (51).

His advice seems to have been followed to the letter with the minutest of parts being held up for objective scrutiny — the knowledge has been generated — but the development of extreme and jargon-laden speciality has been an unwelcome consequence of his way. How can the *'men on their side'* familiarise themselves with the facts and for how long do they have *'to lay their notions by'?* Our quest for detailed knowledge of the parts has led to a failure in their integration to enrich our view of the whole.

Just as Bacon's proposed way of seeing nature and its inter-relationships was a precursor of profound social change, so may be the changing ways of viewing reality now. People, through dint of circumstance, have deemed that the time 'to lay their notions by' is over and are seeking to do the integration. Experts can help but only if they demystify their arts so that they and 'men' can begin to see the connections. For:

"We can know a successful system only as a whole, while being subsidiarily aware of its particulars; and we cannot meaningfully study these particulars except with a bearing on the whole. Moreover the higher the level of success that we are contemplating, the more far reaching must be our participation into our subject matter" (52).

And the "success that we are contemplating" is a sustainable society secure in the richness of its relationships with the whole of nature.

Experts can help but the likely solution to the energy crisis, and most of the other interconnected crises, is a political one. Central to it is the growing interest in part time work and job sharing as a solution to rapidly increasing unemployment. As it strengthens it promises that the *nature* of the means of production becomes a central political issue and, once it is, the whole cycle of production — from the Earth to the consumer to the Earth — will be also. And the smallest of 'entities' will have a chance to 'assign' to themselves that production which is within their capacities.

References, Notes and Bibliography

The breadth of my reading has been much greater than its depth, but the purpose of my 'research' has been to try and find a pattern or framework that would help me make sense of my confusing milieu.

(1) Theodore Roszac, *Where the Wasteland Ends — Politics and Transcendence in Post industrial Society,* Faber, London, 1973, Chapter 1.
This work is an extremely passionate statement of the artificiality and complexity of existence in industrial society . . . It 'goes to the heart of the issues so cleverly disguised from view by those who maintain that all is well, or will soon be well, with the world'. — Robin Clarke in *New Scientist.*

(2) Harry Braverman, *Labor and Monopoly Capital — The Degradation of Work in the Twentieth Century;* Monthly Review Press, New York and London, 1974.

(3) Stephen A. Marglin, *What do Bosses do? The Origins and Functions of Hierarchy in Capitalist Production,* Review of Radical Political Economy, Summer 1974.

(4) Frederick W. Taylor, *Scientific Management,* New York and London, 1947.

(5) Michael Polanyi, *Personal Knowledge,* Routledge and Kegan Paul, London, 1973.

(6) E. F. Schumacher, *A Guide for the Perplexed,* Harper and Row, New York, 1977, Chapters 6-9.
Polanyi and Shumacher demonstrate the folly of our striving to make knowledge impersonal which leads to the splitting of fact from value and science from humanity. '. . . no one will deny that those who have mastered the idioms in which . . . beliefs are entailed do also reason most ingeniously within these idioms, even while . . . they unhesitatingly ignore all that the idiom does not cover.' — Polanyi, page 228.

(7) Roszac, op cit, Chapter 2, particularly pp. 49-53.

(8) Paul Goodman, *Compulsory Miseducation,* Vintage Books, 1962.

(9) Ivan Illich, *Deschooling Society,* Calder and Boyars, London.

(10) Paul Goodman, *Growing Up Absurd* and *The Psychology of Powerlessness.*

(11) Paul A. Baran and Paul M. Sweezy, *Monopoly Capital,* Pelican, 1968, Chapter 2, particularly pp. 49-55.

(12) E. F. Schumacher, *Small is Beautiful,* Abacus, London, 1974, p. 41.
This is an important and highly readable critique of Economics, particularly chapters 3 and 4.

(13) Barry Commoner, *The Poverty of Power,* Jonathon Cape, London, 1976, who quotes (p. 62) John J. Dorgan, executive vice-president, Occidental Oil Corporation, *Newsweek,* August 11, 1975.

(14) Barry Commoner, *The Closing Circle,* Jonathon Cape, London, 1972, chapter 8.
Both books are important aids to an understanding of the interconnectedness of the ecological, energy and economic 'crises'.

(15) ibid, pp. 141-145.

(16) ibid, p. 140.

(17) Schumacher, in an ABC *Investigations* radio programme, in 1976 pointed out the remarkable feat of survival on less than $50 U.S. "How can they do it?" He asked. There is no suggestion here that these people necessarily live good, happy lives; the interesting thing is not that, but the fact that so many of them actually survive.

(18) *U.N. Statistical Year Book,* 1972.

(19) Gerald Foley, *The Energy Question,* Pelican, 1976, p. 96 citing *U.K. Digest of Statistics,* 1973.

(20) Schumacher, op cit, 1974, Chapter 1.

(21) Foley, op cit, pp. 52-53.

(22) Marglin, op cit, pp. 80-104.

(23) D. S. Landes, *The Unbound Prometheus,* Cambridge University Press, 1960, p. 81, cited by Marglin, ibid, p. 82.

(24) Foley, op cit, p. 54.

(25) H. W. Blauvelt, then vice-president (now chairman) of Continential Oil Company, *How to Become a Foreign Oil Company,* In *Explorations and Economics of the Petroleum Industry,* Gulf Pub. Co., Houston, Texas, p. 273. Cited by Commoner, op cit, 1976, p. 56.

(26) OPEC, the Organisation of Petroleum Exporting Countries, raised the price of crude oil in October, 1973 by about 400%.

(27) Nicolas Georgescu-Roegen, *Energy and Economic Myths, The Ecologist,* London, vol. 5, nos. 5 and 7, 1975.

(28) Commoner, op. cit, 1976, Chapter 2.

(29) Terry L. Lustig, *Planning Criteria to Cope with the Entropy Crisis,* a paper given at this conference and included in this volume.

(30) Georgescu-Roegen, op cit, p. 167.

(31) Edward Goldsmith, *The Reykjavik Conference on the Environmental Future,* (editorial), *The Ecologist,* vol. 7, no. 6, 1977, pp. 206-209.

(32) Commoner, op cit, 1976, chapter 8, the quote from p. 212.

(33) ibid. Commoner computed the data cited from capital estimates made by B. Bosworth et al, *Capital Needs in the Seventies,* (Washington, D. C.; Brookings Institution; 1975), pp. 27-29, and from domestic energy production as reported in *Energy Perspectives,* U.S. Department of the Interior, (Washington, D. C.: U.S. Government Printing Office; 1975).

(34) ibid, p. 214.

(35) Amory B. Lovins, *World Energy Strategies,* Ballinger Publishing Company and F.O.E. Cambridge, Mass. 1975, p. 28.
The costs from which these ratios were derived are for 1972-73.

(36) Increases in lead times — alone without overall increases in unit capital costs — *require* additional capital to enable the continuance of growth programmes at the same rate. This can be seen by analogy to the energy analysis in Lovins and Price, *Non-Nuclear Futures,* Ballinger Publishing Company and F.O.E., 1975. See Figure 10-12, p. 139.

(37) As evidenced by the abundance of new courses advertised in newspapers.

(38) David Morris, *Neighbourhood Power,* The Ecologist, vol. 7, no. 7, 1977, p. 328.

(39) Daniel Patrick Moynihan, quoted by Morris, op. cit, p. 325.

(40) For example those of Meadows et al, *Limits to Growth,* Earth Island, London, 1972.

(41) For example a Harris Poll published in The Washington Post of May 23, 1977 reported that "American People have begun to show deep scepticism about the nation's capacity for unlimited economic growth, and they are wary of the benefits that growth is supposed to bring". Among other equally interesting findings the poll found that:
by 76 to 17 per cent those polled would opt for "learning to get pleasure out of non-material experience" rather than "satisfying (their) needs for more and more goods and services" and
by 64 to 26 per cent those polled feel that "finding more inner and personal rewards from work people do" is more important than is "increasing the productivity of the work force".
Perhaps more interesting is the Stanford Research Institute report by Duane Elgin and Arnold Mitchel, *Voluntary Simplicity (3),* republished in The Co Evolution Quarterly, summer, 1977, pp. 4-19. This report produced by a market research organisation "(awoke) more interest among . . . subscribers than any previous report of its kind." (p. 4) "Beneath (the) popular image of simpler living, we think that there is a major social movement afoot which has the potential of touching the United States and other developed countries to their cores. This is the movement towards "voluntary simplicity" . . . a way of life marked by a new balance between inner and outer growth . . . it may be the harbinger of manifold shifts, not only in values, but in consumptive patterns, institutional operations, social movements, national policies, and so on . . ." (p. 4). No wonder business was interested. The report predicts that the "fastest growing sector of the market is people who do not want to buy much".

(42) John Berger, *Ways of Seeing,* Pelican, 1972, pp. 129-152.
This book is based on an excellent BBC TV series of the same title. The fourth of this four part series deals with the images of advertising, their unreality and their effect.

(43) *The Right to Share Work* — An Interim Report on Permanent Part-time Work, Future Lobby and the N.S.W. Association for Mental Health.

(44) For example, *Time to Choose; America's Energy Future,* the Energy Policy Project of the Ford Foundation, Ballinger Publishing Co., Cambridge, Mass., 1974.

(45) R. H. Tawney, *Religion and the Rise of Capitalism,* Pelican, 1948.

(46) Frederick Soddy, *Cartesian Economics,* Hendersons, 1922.

(47) Frederick Soddy, *Wealth, Virtue and Debt,* Allen and Unwin, 1926.

(48) Frederick Soddy, *Matter and Energy,* Williams and Norgate, 1912, quoted from Foley, op cit, p. 85.

(49) Francis Bacon, *Novum Organum,* Spedding, Ellis and Heath, quoted from excerpts in The Autobiography of Science, Forest Ray Moulton and Justus J. Schifferes (eds.), John Murray, London, 1963, p. 125.
This 'unusual anthology collects the world's greatest scientific triumphs as recorded in the original words of the men and women who achieved them'. It is a very good initial source with very interesting editorial comments as a bonus.

(50) ibid, p. 120.

(51) ibid, p. 122.

(52) Polanyi, op cit, p. 381.

Besides the works cited here many books have found their way implicitly into this work. Among them are:

Ivan Illich, *Tools for Conviviality,* Calder and Boyars, 1973;
Ivan Illich, *Limits to Medicine,* Marrion Boyars, 1976;
Ivan Illich, *Energy and Equity,* Calder and Boyars, 1975;
Walt Patterson, *Nuclear Power,* Pelican, 1976;
Peter Chapman, *Fuel's Paradise,* Pelican, 1976;
Brian Easlea, *Liberation and the Aims of Science,* Sussex Uni. Press, 1973;
Barbara Ward and Rene Dubos, *Only One Earth — The Care and Maintenance of a Small Planet,* Pelican, 1972;
Eric Fromm, *To Have or To Be;*
Ilya Ehrenburg, *The Life of the Automobile,* Urizen;
and numerous articles in The Ecologist, a fine source of information in itself, and a signpost to much more.

Some Ethical Aspects of Energy Options

R. and V. Routley

Plumwood Mountain
Box 37 Braidwood NSW. 2622.

Major ethical issues intrude conspicuously into the question of energy choice. For example, the Kantian question: what ought we to do (or try to do) in the awkward circumstances that will shortly face us on the energy front? And Aristotle's question: what is a good life like? a question transposed these days into questions as to quality of life and the extent to which quality genuinely depends on quantity of energy (Aristotle thought, by the way, that a good life required only a modicum of material goods). And traditional questions such as: what are we morally entitled to do to others, and with respect to nature? a question extended to include future others and now asked seriously as regards wild nature. And more modern questions such as: which sorts of consumer demand — especially of energy-intensive goods — *should* be met?

It is evident enough then, without going into detail (1), that ethical questions have an important bearing on main issues of energy choice and that a pure social engineering approach to the problem is bound to write in much that calls for examination or rejection. But it is often supposed — mistakenly — that the ethical issues are adequately taken care of in the structures we already have, e.g. through internal political and legal structures and by international legal arrangements.

Suppose, it is said, the question of replacement of burners in a Japanese electricity station arises, and the Japanese decide to replace the burners by nuclear ones. The matter is approved in a routine political way. Who are we to say that the matter is one of wider morality? It is a matter of internal political arrangements for the Japanese. Suppose, to begin to see that this is not so, that a nation decides to install (or replace in the course of modernisation) the burners in a concentration camp, and that the matter is approved in a routine political fashion and meets all internal legal requirements. We are almost all going to say that the matter is one of wider morality. And so also is taking choices on the life, health and well being of future people in a nuclear adventure. More generally, pollution, and especially non-local types of pollution such as nuclear pollution, raises ethical issues which transcend conventional national and political boundaries. For the effects of pollution are often not locally contained, polluted air or water may move far afield to affect even features of the world, such as climate and ocean levels (2).

The idea that the ethics of such matters is satisfactorily regulated by international legal arrangements — agreements, pacts, contracts and the like or international regulatory agencies (where they operate) — is likewise unsound. For the arrangements commonly bear little relation to what is considered right or just: they may have been arrived at by expediency or, at best, through moral compromise, and they may reflect immediate self-interest rather than morality. Just as legal principles are in general neither necessary nor sufficient as moral principles, so international legal arrangements are no substitute for morality, and usually do not even offer a poor reflection of ethical arrangements.

More difficult to dispose of, and more insidious, are engineering approaches to morality built into models of an economic cast, e.g. benefit-cost balance sheets, risk assessment models, etc.

§1. It is a commonplace nowadays that there is no method of providing for future energy needs which will not involve substantial costs to someone, and that these costs must just be accepted as part of the price we pay for our advanced technological society and our high living standard. "If you want the benefits you must be prepared to pay the costs" is part of the new conventional wisdom on energy. The model is that of a simple economic transaction, for example someone going into a shop and buying paper towels — she wants the paper so she must be prepared to pay over the money representing the equivalent of the cost of the paper in unpleasant labour, forest destruction, etc.

But here, on energy, as in so many other places, the conventional wisdom is not to be trusted. For the transaction model suggests that the costs and benefits are evenly distributed, that those who reap the benefit pay the costs and vice versa. But with energy options this is very often not so, and in fact this is one of the more important ways in which the ethical aspects of the energy issue arise. The conventional transaction model is an attempt to gloss over crucial ethical aspects of the problem. Again, the energy issue raises questions about the goals (ends) and values of society, and is not just a disagreement about the best means of achieving unquestioned or accepted goals, and it has an important bearing too on the distribution of power both economic and political between various groups. Thus are involved then crucial political as well as ethical aspects, and these too the simple transaction model attempts to sweep under the carpet. The simple transaction model, despite its appeal to the technological mind, and to those who are anxious to maintain the myth of the political and value freedomness of science and technology, is quite inadequate, for it does not reflect significant distributional features of the energy problem. It is in fact an attempt to ignore, deny or gloss over the crucial political and ethical aspects of the energy issue, and to avoid facing the social and ethical choices involved.

A more sophisticated relative of the transaction model is now called risk assessment, which purports to provide a comparison between the relative risks attached to different energy options which settles their ethical status. The following lines of argument are encountered in risk assessment as applied to energy options:

(i) if option *a* imposes costs on fewer people than option *b* then option *a* is preferable to option *b*;

(ii) option *a* involves a total net cost in terms of cost to people (e.g. deaths, injuries, etc.) which is less than of option *b*, which is already accepted; therefore option *a* is acceptable (3). For example, the number likely to be killed by nuclear power stations is less than the likely number killed by cigarette smoking, which is accepted: so nuclear power stations are acceptable. A little reflection reveals that this sort of risk assessment argument involves the same kind of fallacy as the transaction model. It is far too simple-minded, and it ignores distributional and other relevant aspects of the context. In order to obtain an ethical assessment we should need a much fuller picture and we should need to know at least these things: do the costs and benefits go to the same parties; and is the

Diesendorf, M. (ed.) (1979). *Energy and People*. Canberra, Society for Social Responsibility in Science (A.C.T.)

person who undertakes the risks also the person who receives the benefits or primarily, as in driving or cigarette smoking, or are costs imposed on other parties who do not benefit? It is only if the parties are the same in the case of the options compared, and there are no such distributional problems, that a comparison on such a basis would be valid (4). This is rarely the case, and it is not so in the case of risk assessments of energy options. Secondly, does the person incur the risk as a result of an activity which he knowingly undertakes in a situation where he has a reasonable choice, knowing it entails the risk, etc., and is the level of risk in proportion to the level of the relevant activity, e.g. as in smoking? Thirdly, for what reason is the risk imposed: is it for a serious or a relatively trivial reason? A risk that is ethically acceptable for a serious reason may not be ethically acceptable for a trivial reason. Both the arguments (i) and (ii) are often employed in trying to justify nuclear power. The second argument (ii) involves the fallacies of the first (i) and an additional set, namely that of forgetting that the health risks in the nuclear sense are cumulative, and in the eyes of many people already high if not too high.

Despite a certain superficial plausibility, so-called risk assessment as a method of comparing the ethical status of energy options is little more than a bundle of fallacies deceptively packaged in pseudo-scientific wrappings. It purports to give a simple apparently precise and scientific method of evaluating the ethical status of energy options, but in fact it depends on a number of hidden assumptions about which factors are relevant and which can be ignored in an ethical assessment, which when brought out for examination can be seen to be quite unacceptable (5).

§2. The maxim "If you want the benefits you have to accept the costs" is one thing and the maxim "If I want the benefits then you have to accept the costs (or some of them at least)" is another and very different thing. It is a widely accepted moral principle that one is not, in general, entitled to simply transfer costs of a significant kind arising from an activity which benefits oneself onto other parties who are not involved in the activity and are not beneficiaries (6). This *transfer principle* is especially clear in cases where the significant costs include an effect on life or health or a risk thereof, and where the benefit to the benefitting party is of a noncrucial or dispensible nature. (Thus one is not usually entitled to harm, or risk harming, another in the process of benefitting oneself.) Suppose, for example, we consider a village which produces, as a result of the industrial process by which it lives, a noxious waste material which is expensive and difficult to dispose of yet creates a risk to life and health if undisposed of. Instead of giving up their industrial process and turning to some other way of making a living such as farming the surrounding countryside, they persist with this way of life but ship their problem on a one-way delivery service to the next village. The inhabitants of this village are then forced to face the problem either of undertaking the expensive and difficult disposal process or of sustaining risks to their own lives and health. Most of us would see this kind of transfer of costs as morally unacceptable.

From this arises a necessary condition for energy options: that to be morally acceptable they should not involve the transfer of significant costs or risks of harm onto parties who are not involved, do not use the energy source or do not benefit correspondingly from its use. Included in the scope of this condition are future people, i.e. not merely people living at the present time but also future generations (those of the next villages) (7). The distribution of costs and damage in such a fashion, i.e. onto non-beneficiaries, is a characteristic of certain widespread and serious forms of pollution, and is one of its most objectionable moral features.

It is a corollary of the condition that we should not hand the world on to our successors in substantially worse shape than we received it — the *transmission principle*. For if we did then that would be a significant transfer of costs. (The corollary can be independently argued for on the basis of certain ethical theories, in particular contract theories such as Rawls'.)

Some philosophers (8) have attempted to undercut the transmission principle, arguing that we are not morally obligated to *make sacrifices* for the future. Making sacrifices is however significantly different from refraining from passing on costs, and it is the latter which is mainly at issue. We might be making sacrifices, for example, if we made ourselves worse off than future people might be normally expected to be in order that they might be better off than us. We are passing on our costs when we make them worse off than they would normally be because of some activity which benefits us. We may not be morally obliged to make sacrifices for future people, but we do have a moral obligation not to pass on our costs, and our obligations in this respect do not *just* apply to the next generation, but to any set of people who could be affected.

In terms of the necessary condition we can undertake some limited comparison of energy options from an ethical standpoint. It is very doubtful that the main options that are being seriously considered meet this condition for moral acceptibility; in particular it is extremely doubtful that nuclear energy options do so. Nuclear energy appears to represent a classic case of passing on costs and risks to nonbeneficiaries, especially future people, because of the way in which nuclear waste created now produces risks and problems for future people. Unless a rigorously safe method of storage is employed, as many as 40,000 generations of people have to face costs in the shape of risks to health and life arising from the energy consumption of at most perhaps 10 generations. The situation is even worse when one reflects on the fact that many of the purposes for which this energy will be required are of a dispensible and unnecessary (and even undesirable) nature, and energy use of an extravagant and needless kind would undoubtedly be involved in order for the big increases in per capita energy consumption which justify much of the nuclear expansion program in the industrialised world to be reached. Not only would costs be passed on to people in the distant future to whom no benefits seem to accrue but this is done for reasons that cannot be seen as pressing or needful (9). The waste disposal aspect of nuclear power production is not the only way in which the nuclear option may pass on problems and costs to nonbeneficiaries: unless an unrealistic perfection in the handling, mining, transport and processing and reprocessing of nuclear fuels and waste is assumed, various forms of widespread radioactive pollution could occur which would affect not only those who use and benefit from the energy source but also very many who do not, especially in the third world (10).

We have heard a good deal recently from some local quarters (the PR machinery of the Australian National University) about how the nuclear waste disposal problem has been solved, and the objections on the grounds of waste disposal eliminated. Of course a number of similar claims have been made in the past, and there never was a problem according to hard-line nuclear advocates of nuclear power. There are good reasons for treating these claims with some scepticism, and not merely because of disagreement among the parties, but because what we have in effect with the final "solution" is yet another proposal for a possible method of treating waste with significant gaps in the arguments, a considerable lack of experimental and practical support, and so on (11). We must be satisfied beyond reasonable doubt that there is a completely safe procedure (12) before claims can be responsibly made that a problem of such seriousness has been technically solved. It is irresponsible, especially on the part of university authorities, to give the impression that such a problem is solved or eliminated when so much remains to be done and when reasonable doubts may still be raised as they may in this case.

An even more important reason why the claims that the problem has been eliminated have to be rejected is that even if a method of disposal can be experimentally (or even commercially) demonstrated beyond reasonable doubt to be completely safe, it is not the technical possibility of safe disposal that is important from an ethical standpoint but the actual *practical* likelihood of such a method being *used*. Firstly, it is worth bearing in mind that the new miracle method was immediately rejected by some of the pronuclear establishment as both expensive (despite optimistic cost estimates) and unnecessary, which does not give a very hopeful prognosis for its use. There are moreover reasons for thinking that governments may not want or favour permanent irretrievable disposal methods. They may want to keep open their options for employing waste either for military purposes or for use in breeder reactors or elsewhere. There is in fact little reason to believe that nuclear pollution will be treated in a different fashion from other forms of pollution, where the mere fact that there are satisfactory methods of control is by no means sufficient to guarantee their effective employment, especially if they are expensive. It would be methodologically unsound to ignore these risk elements arising from social and political factors and to regard the problem as a purely technological one which can be classed as eliminated the moment someone puts forward a promising looking technical proposal. The practical likelihood, even with a disposal method proven safe beyond reasonable doubt, remains — that nuclear power will impose costs on future people who do not benefit, that future people will either be forced to go to great expense and trouble to dispose safely of the nuclear wastes generated by our consumption, or they will have to pay the price in their own lives and health for inadequate disposal or storage of the wastes generated by us. For these reasons, we believe that the nuclear energy option remains morally unacceptable; and unacceptable not merely from a particular ethical standpoint, but in terms of common ground from a range of ethical positions.

§3. One cannot pretend that the future energy option that is frequently contrasted with nuclear, namely coal, is particularly attractive — because of the likelihood of really serious (air) pollution and associated phenomena such as acid rain and atmosphere heating, not to mention the despoliation caused by extensive strip mining, all of which will result from its use in meeting very high projected consumption figures. Such an option would also fail, it seems, to meet the necessary condition, because it would impose widespread costs on nonbeneficiaries for some concentrated benefits to some profit takers and to some users who do not pay the full costs of production and replacement (13).

These are the conventional options and a third is often added which emphasizes soft or benign technologies, such as those of solar energy. The fundamental choice, which such options tend to neglect, is not technological but social, and involves both the restructuring of production away from energy intensive uses: at a more basic level there is a choice between consumeristic and nonconsumeristic futures. These more fundamental choices between social alternatives, conventional technologically-oriented discussion tends to obscure. It is not just a matter of deciding in which way to meet unexamined goals but also a matter of examining the goals. That is, we are not just faced with the question of comparing different technologies or substitute ways of meeting some fixed or given demand or level of consumption, and of trying to see whether we can meet this with soft rather than hard technologies; we are also faced, and primarily, with the matter of examining those alleged needs and the cost of the society that creates them. It is not just a question of devising less damaging ways to meet these alleged needs conceived of as inevitable and unchangeable. (Hence there are solar ways of producing unnecessary trivia no one really wants, as opposed to nuclear ways). Of course one does not want to deny that these softer options are superior to the clearly ethically unacceptable features of the others.

But it is doubtful that any technology however benign in principle will be likely to leave a tolerable world for future people if it is expected to meet limitless and uncontrolled energy consumption and demands. Even the more benign technologies such as solar technology could be used in a way which creates costs for future people and are likely to result in a deteriorated world being handed on to them. Consider, for example, the effect on the world's forests, which are commonly counted as a solar resource, of use for production of methanol or of electricity by woodchipping, as already planned by forest authorities in California and contemplated by many other energy organisations. Few would object to the use of genuine waste material for energy production, but the unrestricted exploitation of forests — whether it goes under the name of "solar energy" or not — to meet ever increasing energy demands could well be the last nail in the coffin of the world's already hard pressed natural forests.

The effects of such additional demands on the maintenance of the forests are often discounted, even by soft technologists, by the simple expedient of waving around the label 'renewable resource'. Most forests are in principle renewable, it is a true, given a certain (low) rate and kind of exploitation, but in fact there are now very few forestry operations anywhere in the world where the forests are treated as completely renewable in the sense of the renewal of all their values (14). In many regions too the rate of exploitation which would enable renewal has already been exceeded, so that a total decline is widely thought to be imminent if not already well advanced. It certainly has begun in some regions, and for some forest types (such as rainforest types) which are being lost for the future. The addition of a major further demand source — that for energy — and especially one which shows every sign of being not readily limitable, *on top of* the present sources is one which anyone with a realistic appreciation of the conduct of forestry operations, who is also concerned with long term conservation of the forests and remaining natural communities, must regard with alarm. The result of massive deforestation for energy purposes, resembling the deforestation of England at the beginning of the Industrial Revolution, again for energy purposes, could be extensive and devastating erosion in steeper lands and tropical areas, desertification in more arid regions, possible climatic change, and massive impoverishment of natural ecosystems. Some of us do not want to pass on — we are not entitled to pass on — a deforested world to the future, any more than we want to pass on one poisoned by nuclear products or pollution by coal products. In short, a mere switch to a more benign technology — important though this is — without any more basic structural and social change is not enough.

The deeper social options involve challenging and trying to change a social structure which promotes consumerism and an economic structure which encourages the use of highly energy-intensive modes of production. This means, for instance, trying to change a social structure in which those who are lucky enough to make it into the work force are cogs in a production machine over which they have very little real control and in which most people do unpleasant or boring work from which they derive very little real satisfaction in order to obtain the reward of consumer goods and services. A society in which social rewards are obtained primarily from products rather than processes, from consumption, rather than from satisfaction in work and in social relations and other activities, is bound to be one which generates a vast amount of unnecessary consumption. (A production system that produces goods not to meet genuine needs but for created and non-genuine needs is virtually bound to overproduce.) Consumption frequently becomes a substitute for satisfaction in other areas.

§4. *Conclusions.* The social change option is a hard option, but it seems the only way to avoid passing on serious costs to the future — and there are other sorts of reasons than such ethical ones for taking it (15). The ethical conditions thus lead us into political issues, but this is not very surprising, as there is no sharp division between the areas (and political theories always presuppose an ethics). This kind of social change option tends to be obscured in most discussions of energy options and how to meet our energy needs, in part because it questions underlying values of current social arrangements. The conventional discussion proceeds by taking alleged demand (often restated as wants or needs) as unchallengeable (16), and the issue to be one of which technology can be most profitably employed to meet them. This effectively presents a false choice, and is the result of taking needs and demand as lacking a *social context* so that the social structure which produces the needs is similarly taken as unchallengeable and unchangeable. The social changes that the option requires will be strongly resisted because they mean changes in current social organisation and power structure, and to the extent that the option represents some kind of threat to parts of present political and economic arrangements it is not surprising that official energy option discussion proceeds by misrepresenting and often obscuring it.

Notes and References

(1) A more detailed case would look at what is involved in decision methods for choosing between options on energy and would show that general decision modellings, e.g. optimisation modellings for best choice, necessarily involve evaluative factors, some of them of an ethical kind. The ethical components are particularly conspicuous in elementary decision theory where the assessment of each outcome is obtained by multiplying the probability of the outcome by its desirability, the desirability being an overtly evaluative factor often involving ethical components. Any idea that decision theory is somehow a value-free way of reaching decisions is thoroughly wrong. The pure theory may be a logico-mathematical one, but its significant applications are not.

(2) The pernicious underlying assumption — that major ethical issues are not really a matter of individual concern and can, and perhaps should, be left to elected government or appointed bureaucrats — can be despatched in a rather similar way.

(3) Even then relevant environmental factors may have been neglected.

(4) There are variations on (i) and (ii) which multiply costs against numbers such as probabilities. In this way risks, construed as probable costs, can be taken into account in the assessment. (Alternatively, risks may be assessed through such familiar methods as insurance).

A variant of principle (ii) is formulated as follows: (ii′) *a* is ethically acceptable if (for some *b*) *a* includes no more risks than *b* and *b* is socially accepted. This was the basic ethical principle in terms of which the Cluff Lake Board of Inquiry recently decided that nuclear power development in Saskatchewan is ethically acceptable: see *Cluff Lake Board of Inquiry Final Report,* Department of Environment, Government of Saskatchewan, 1978, p. 305 and p. 288. In this report, *a* is nuclear power and *b* is other activities clearly accepted by society as alternative power sources. In other applications *b* has been taken as cigarette smoking, motoring, mining and even the Vietnam war (!)

The points made in the text do not exhaust the objections to principles (i)-(ii′). The principles are certainly ethically substantial, since an *ethical* consequence cannot be deduced from nonethical premisses, but they have an inadmissible conventional character. For look at the origin of *b*: *b* may be socially accepted though it is no longer socially acceptable, or though its social acceptability is no longer so clearcut and it would not have been socially accepted if as much as is now known had been known when it was introduced. What is required in (ii′), for instance, for the argument to begin to look convincing is then 'ethically accept*able*' rather than 'socially accept*ed*'. But even with the amendments the principles are invalid, for the reasons given in the text.

It is not disconcerting that these arguments do not work. It would be sad to see yet another area lost to the experts, namely ethics to actuaries.

(5) A main part of the trouble with the models is that they are narrowly utilitarian, and like utilitarianism they neglect distributional features, involve naturalisic fallacies, etc. Really they try to treat as an unconstrained optimisation what is a deontically constrained optimisation: see R. and V. Routley 'An expensive repair kit for utilitarianism', copies available from the authors.

(6) Apparent exceptions to the principle such as taxation (and redistribution of income generally) vanish when wealth is construed (as it has to be if taxation is to be properly justified) as at least partly a social asset unfairly monopolised by a minority of the population.

Examples such as that of motoring dangerously do not constitute counterexamples to the principle; for one is not morally entitled to so motor.

(7) As we have argued in detail in R. and V. Routley (1978). Nuclear energy and obligations to the future, *Inquiry 21,* pp. 133-179.

(8) For example, Passmore in *Man's Responsibility for Nature,* Duckworth, London, 1974, chapter 4.

(9) A further problem of a less obvious kind is also created for future people; the postponement of the switch away from energy-intensive economies, which nuclear power is designed to effect, creates a situation of increasing and critical dependence upon energy-intensive uses at a time when there is every prospect that they cannot be sustained for more than a short time. Nevertheless the growth of energy-intensive societies and lifestyles is encouraged and fostered. The switch is accordingly made far more difficult than it is at present, and thus we may well be placing future people in very difficult positions, with a real energy crisis. Continuance on a high energy path in the present circumstances seems then to violate the commonly recognised principle that we have an obligation to hand on to the next generation a society that is not conspicuously worse than that which we received.

(10) A few nuclear plant accidents, for example, would significantly increase background levels of radiation, so that millions of people who are not involved might have to carry risks or costs because of the energy consumption of a few wealthy nations or wealthy elites. The increased risk of nuclear war is another way in which global risks are imposed because of the determination of industrial nations to maintain and increase lavish energy consumption levels.

(11) For an excellent discussion of the limitations of the proposed disposal method see B. Martin, *Radioactive waste disposal: is Synroc the solution?* Friends of the Earth, Canberra.

(12) For the reason that the consequences of failure are so serious: see R. and V. Routley (1978) op. cit. footnote 1.

(13) Certainly practical transitional programs may involve temporary and limited use of unacceptable long term commodities such as coal, but in presenting such practical details one should not lose sight of the more basic social and structural changes, and the problem is really one of making those.

Similarly practical transitional strategies should make use of such measures as environmental (or replacement) pricing of energy, i.e. so that the price of some energy unit includes the full cost of replacing it by an equivalent unit taking account of environmental cost of production. Other (sometimes cooptive) strategies towards more satisfactory alternatives should also, of course, be adopted, in particular the removal of institutional barriers to energy conservation and alternative technology (e.g. local government regulations blocking these), and the removal of state assistance to fuel and power industries.

(14) Symptomatic of the fact that is it not treated as renewable is that forest economics do not generally allow for full renewability — if they did the losses and deficits on forestry operations would be much more striking than they already are often enough.

It is doubtful, furthermore, that energy cropping of forests can be a fully renewable operation if net energy production is to be worthwhile; see, e.g. the argument in L. R. B. Mann *Some difficulties with energy farming for portable fuels,* (the author is at the Univ. of Auckland) and add in the costs of ecosystem maintenance.

(15) Certainly it is the only sort of option open to one who takes a deeper ecological perspective.

(16) Thus it is argued by representatives of such industries as transportation and petroleum, as for example by McGrowth of the XS Consumption Co., that people want deep freezers, air conditioners, power boats, . . . It would be authoritarian to stop them satisfying these wants. The argument conveniently ignores the social framework in which such needs and wants arise or are produced. To point to the determination of many such wants at the framework level is not however to accept a Marxist approach according to which they are entirely determined at the framework level (e.g. by industrial organisation) and there is no such thing as individual choice or determination at all. It is to see the social framework as a major factor in determining certain kinds of choices such as those for travel and infrastructure and to see apparently individual choices made in such matters as being channelled and directed by a social framework determined largely in the interests of private profit and advantage. See R. and V. Routley, 'Towards a social theory for ecotopia' in *Environmental Philosophy,* ANU, 1979.

Preserving Equity while Conserving Energy

Simon de Burgh

Epidemiologist
355 Riley Street,
Surry Hills NSW 2010.

Until recently, the world's energy consumption rate has tended to rise exponentially. For practical purposes, this means that fossil fuel reserves likewise have been depleted exponentially (1). If these curves ever came close to meeting, that is, if annual consumption ever approached the remaining reserves, there would be a social catastrophe in which civilisation would collapse. For civilization, fundamentally, is the application of surplus energy to tasks beyond mere survival — tasks such as building opera houses or holding conferences (2).

Energy resources, of course, are not evenly spread over the earth, and other factors would intervene before the two curves could meet. Price and simple scarcity would cause consumption to level off and then drop, and the now familiar depletion curve would result (3). All the same, with consumption still rising rapidly, oil in particular, if left to market forces, will become very expensive, very suddenly. Demand may overtake production capacity quite soon. Only the really wealthy will have access to fuel.

Diesendorf, M. (ed.) (1979). *Energy and People*. Canberra, Society for Social Responsibility in Science (A.C.T.)

Now, when scarcity prices marble or Californian redwood out of the reach of most people, the social consequences are small. If the same situation arose with energy, however, a very poor society might be unaffected, but otherwise only a very stable social order would stand. If only the rich had access to energy, a surviving society would have to be accustomed to wide social differences, or to repressive discipline, or both. Most likely, energy sources would be subjects of armed exploitation and envy between countries, and within them. It is not hard from this to imagine a scenario for World War III.

In the western democracies it is possible that we could acquire the necessary authoritarian regimes, but I think it unlikely that egalitarian traditions and expectations could be numbed sufficiently, in time to meet a really severe energy shortage in the first half of the next century. Any grossly inequitable distribution of energy would not be accepted peacefully. The survival of democracies, as well as democracy itself, will depend on equitable solutions to the energy problem.

These solutions must be implemented in time for social change and technological redirection to take place, so that a sustainable level of energy consumption can be found while fossil fuels last. The sooner world energy demand can be reduced, the better will be our chances of avoiding the grave and intractable risks of nuclear power and its Siamese twin, nuclear weaponry.

So far there is little sign of any serious-minded enterprise towards conservation from any government. There has been short term reaction to short term crisis, when the Arab nations imposed the oil boycott in 1973-74. There are plans for substitution of petrol by coal, and for American cars to be more economical in future. There is, to be sure, widespread interest and some research into solar power and plant alcohol. Nowhere is there real acceptance of the need for fundamental social changes, of the kind proposed in "A Blueprint for Survival" (1), already six years old.

The two fundamental components of total energy consumption (or, more correctly, use) are population and per capita use. Industrial nations have reached the point where they can, if they choose, reduce their populations, so the central problem here is that of per capita consumption.

Put in different terms, the most important issue in energy conservation is the pay scale. This, for two reasons:

Firstly, our enormous incomes must be disposed of somehow, and we do it be consuming energy in various ways. One's cost of living tends to expand as far as one's income. Or, put another way, energy consumption rises with wealth (4).

Secondly, expensive labor implies cheap energy, and cheap energy means that machines tend to replace labor, with increased use of energy. This is because any product is produced from the resources of labor, energy, and materials (including land). With the same material content of a given product, the money cost of the energy component must be comparatively reduced, if the money-cost of the labor component is increased. And vice versa. As the cost of labor rises, the cost of having machines to do the same work will fall, in the long run. Only lack of materials, technology or energy will interrupt this cycle.

So industrialisation is essentially the process of harnessing energy to replace labor. Technological advances allow machines to do work, so that labor of any kind, directly or indirectly, channels energy towards useful (or desired) ends. There are design advances, of course, like the stump-jump plough or electronic calculator, which are exceptions because they do more work for the same energy than their predecessors. But on the whole, technology improves the productivity of labor (production per worker) while decreasing its efficiency (production per kilowatt-hour). For example, the high productivity of industrial agriculture, per worker and per hectare of land, is matched by its inefficiency. Shifting cultivators in the Congo gain food energy from their rice, cassava and bananas an estimated 65 times the energy input. This output/input ratio is 14 to 18 for some other African rice growers, 1.29 for intensive rice growing in the USA, 0.05 for the fishing fleet in the UK. (5). As Odum put it: " . . . industrial man no longer eats potatoes made from solar energy; now he eats potatoes partly made of oil" (6).

The USA has been the leading influence in the world this century, and oil was so cheap there that new ways had to be invented to burn it (7). Throw-away society was a natural consequence, and automation only had to wait for capital investment. Since the second world war we have seen an unprecedented rise in production, in consumption, and in popular expectations. Where automation shrank the workforce in one part of industry, expansion of production in another area took up the slack. Thus, through rising energy consumption and expanding production, the western world experienced fairly full employment and low inflation with rapidly rising material wealth. It experienced this just long enough for economic expansionism to become fixed in the minds of economists and politicians, as an incontrovertable Law of Nature; and long enough for ordinary people to come to expect, as a right,that they should be materially richer this year than they were last year.

So all conventional wisdom urges us to find further fields of expansion. Yet this is to borrow more deeply from our capital funds, fossil fuels, and to shorten the time left till we must learn to live again on our income, renewable sources of energy.

Therefore it is vital to relieve this pressure to expand, to achieve economic growth. And to governments the two most important pressures for growth are unemployment and inflation: growth absorbs labor put out of work by automation, and it accommodates the expectations of the employed population for higher incomes, and so keeps down inflation.

Consider unemployment: automation puts workers off, obviously, and they must find other work. Unless jobs are opening up elsewhere (i.e. unless there is economic growth), they will not be able to, regardless of training or education. If labor is expensive, then the average factory manager, or economist, will advise that it is "economic" to install machinery which will reduce the labor necessary to produce the factory's products.

Clearly, those who claim that wage demands cause inflation and unemployment are at least partly right. But the question is, *whose* wage demands?

In fact it is the articulate upper and middle classes, people such as ourselves, who are the ones who set standards, generate popular expectations, start trends in patterns of consumption. It is unrealistic to expect the union movement to abandon its wage claims as long as substantial wage inequalities exist, and as long as unionists can observe every other group in the workforce doing its best to improve its own position. Neither is it just to expect those in lower paid and less desirable jobs to be the ones to restrain their demands.

One can say it is not wage demands so much as *salary* demands that cause inflation and unemployment. It is among the well paid that restraint must begin.

Is it possible? Wealth generates greed, and wealth in others generates envy. And yet the idea of energy conservation is well enough accepted. Outside government and the senior public service, anyone from society matrons to carpenters can be heard discussing what is to happen when the oil runs out. There is simply not the emotional acceptance to match — it is still considered eccentric not to own a car.

What is needed is a popular movement with emotion, with a sense of mission, and of belonging, as well as an ideology. I am thinking of the women's and homosexual movements, Marxism and the political left, "Black Power", even anti-censorship. The power of European societies to change under pressure of ideas is remarkable, especially when the ideas become fashionable.

So, can energy conservation become fashionable, in deeds as well as in words? Recently a group of Swedish doctors declared that they were overpaid. Can we hear from Australian doctors? A start must be made by opinion leaders, well paid people who are in a position to say publicly that they earn more than their needs. I propose that people at this conference join forces in some way, to lobby for lower salaries as a group, not just to make heroic gestures as individuals.*

When this work is begun, we can address ourselves to the question of overpaid tradesmen, and to Arbitration judges who hand down extravagant and anti-social decisions for groups like airline pilots and university professors. We can ask publicly why wage relativity always refers to other classes of work with *higher* wage rates, never lower; and we can take the idea of employment indexation seriously.

In short we must seek wage justice by lowering high wages, rather than raising the low ones. If we do not, we can expect inflation to do the job for us, but indiscriminately.

* * *

I have argued that the driving force behind extravagant energy use is popular expectation of high and rising incomes. Conventional "energy policy" can make a useful contribution to conservation, but it will be a limited one. It can reduce consumption by, for example, introducing a road speed limit (the 55 mph limit is still observed in the USA). Other measures encourage the substitution of coal for oil. But a tax on fossil fuel energy can not be very successful because its burden falls lightly on the rich and heavily on the poor, and is, at the least, inflationary for those in between. It is thus inequitable and politically difficult. Moreover its effect is diluted by inflation, since incomes tend to keep pace with increased prices.

Real gains will involve changes in work, leisure and daily life, so that energy-rich patterns are displaced by person-rich ones.

Some measures follow which could encourage these changes. The principles I have followed in suggesting them are:

1. Maintain an adequate degree of equity in distribution of energy.
2. Make energy consumption visible — remind people how much they are using.
3. Charge lightly for light consumption, and heavily for heavy consumption.
4. When taking with one hand, give with the other: e.g. offer a bounty on insulation and conversion to gas while raising the price of electricity (electricity is an inefficient form for heating and cooking, microwave ovens excepted.)
5. In general, provide inducements rather than compulsion.

Rationing

The first group of measures to consider is rationing. The very mention of the word will stop most people thinking. "We are not at war, so we don't need rationing." But it needn't mean queues and a black market. What I have in mind is not withholding of goods, but access to privileges, dispensed equitably.

For example:

1. Dispense coupons for cheap petrol with drivers' licences, 20 litres a coupon, much like the coupons allowed foreign tourists in Europe. By handing over a coupon you pay only 20 cents a litre, instead of 25, 30, or 40. Say you get 400 litres for the year, that takes you 5000 km if your car does 8 litres per 100 km, a reasonable figure. And if you share, you'll go much further.
2. Install household electricity meters which dispense cheap electricity up to a certain quota per person. Then it changes to a higher rate for the rest of the day, and at the same time it turns off the lights. The reason for this is that it is important to know that you have used your quota. It also provides motivation not to be wasteful if there is some inconvenience, because you must go to the switchboard to turn on the lights again. (Even the German industrialist in a fur collar, driving a big BMW, even he turns off his lights, because it is inconvenient to return to a flat battery.) At midnight a pulse switches the meter back to the cheap rate.
3. *Cheap travel.* When I first went travelling, young Australians left to see the world, and they didn't come back for a year or three, because the world is so far away. That was travelling, and it made a deep impression on us. Nowadays people go to Bali for a week's holiday, and probably bump into the Johnsons on Kuta Beach. Our tourism is trivial, and our mobility is frantic. One of the most trivial kinds of air travel is the cheap holiday return, so allow cheap air fares only up to a limited mileage for each five or ten years, stamped on the passport. If you must go shopping in Singapore, then go, but you will have to pay for it.

The same applies to trivial domestic air travel, though it means some form of identification would be necessary. If people, and governments, rediscovered the passenger train, there would be no need for an extra runway at Sydney airport.

Note that nowhere here is there any compulsion, there is only inducement to save.

The Energy Sales Tax

1. I would propose that nearly all sales taxes be replaced by the same dollar-worth of energy tax. Rather than attempt a full "energy economy", in which money represents energy units, we could place a tax on goods according to the amount of fossil-fuel energy required for their production and distribution. This would, incidently, give locally produced goods an advantage over imported ones, such as Gosford oranges versus Californian ones, or hand picked versus machine picked. It would give an advantage to well made articles that last, and to repairs over replacements. It would discourage the use of durable plastics for throw-away articles. Buyers could be made aware of the amount of embodied energy they use in products.

Calculation of embodied energy is complicated and difficult, but good progress is being made. Perfection is not required, and good estimates can cover numbers of similar articles. An adjustment can be made for the relative usefulness or "energy coefficients" of various fossil fuels, and their relative scarcity. With gradual introduction, the tax would have a minimal inflationary effect, because it involves no increase in overall tax.

2. For the car, of course, a progressive tax disincentive is needed to discourage the thirsty ones. Once a car is bought, the tendency is to use it, regardless of the price of petrol.

Restrictions

A further group of measures are restrictions.

Examples are:

1. The 55 mph speed limit in the USA has been mentioned. It has remained in force there since 1974, with saving of fuel and lives. We could adopt a 90 kph limit immediately without hardship to anyone except departments of main roads.
2. No balanced person needs the power supplied by our oversized cars, and they serve mainly to kill off our youth on Saturday nights. So put a limit on engine capacity per occupant for private vehicles, including, and especially, racing cars. Motor racing just rouses the passions for dangerous driving.

Conclusion

The rich no more than the rest of us can survive a real energy crisis in the democracies. Equitable measures of restraint must be accepted. The most important of these is a social movement to pressure the well-off to lower their incomes. Follow-up measures can dispense quotas of cheap energy, replace taxes on labor with taxes on energy, and put sensible restrictions on certain extravagant energy uses.

References

(1) The Ecologist, (1972). *A Blueprint for Survival.* Harmondsworth, Penguin. (See Figure 1, p. 18)
(2) Soddy, F., (1926). *Wealth, Virtual Wealth and Debt.* London, Allen and Unwin.
(3) Foley, G. (1976). *The Energy Question.* Harmondsworth, Penguin. (See Figure 10, p. 141)
(4) Foley, G. (1976). Ref. 3, p. 90.
(5) Leach, G. (1976). *Energy and Food Production.* Guilford, Surrey, IPC Science and Technology Press.
(6) Odum, H. T. (1971). *Environment, Power and Society.* New York, Wiley-Interscience. (p. 115-116)
(7) Foley, G. (1976). Ref. 3 p. 60

*Readers interested in this proposal can contact the author at the address given.

Energy Policy and the Optimal Depletion of our Fossil Fuels

R. P. Rutherford

Lecturer in Economics
University of Tasmania, Hobart.

Introduction

This paper is addressed to the specific problem of the rate at which we use our fossil fuel supplies, in particular oil. It will be shown that the rate of depletion of a fossil fuel, in common with any exhaustible resource, is critically dependent on the choice of the rate of interest used for discounting purposes. I shall argue that, because the rate of interest thrown up by the private market is likely to be too high, this will lead to over rapid exploitation of our fossil fuels and that we need to interfere with the market mechanism in the interests of future generations. A number of important problems will not be dealt with and it must be stressed that their omission is due not to a belief that they can safely be neglected, but rather to facilitate concentration on a particular aspect of the energy problem. In particular nothing will be said about our capacity to sustain a per capita growth rate in energy consumption nor about political and strategic considerations such as the desirability of energy independence. Before we proceed to examine the problem of optimal use of our fossil fuels we must first relate this issue to the question of whether we face an energy problem as a whole and to that I now turn.

Energy Crisis or Oil Crisis?

As a matter of history the current widespread perception of an energy problem and of a need to think carefully about various energy futures is the direct result of the sharp rise in oil prices of 1973 which sent shock waves through the industrialised world. This price rise was not caused in any way by an imminent shortage of oil but was rather the result of the successful formation of a cartel by the major producers with the aim of withholding supplies so as to force up the price. This assertion of monopolistic market power appears to have been politically motivated in the first instance but I think we can ascribe the durability of the arrangement to the economic self-interest of the respective oil producing nations. Despite its origin in economic sharp practice, the sudden price rise had an enormous effect on attitudes to resources. In the public mind it confirmed age old fears about running our of resources, fears which had recently been revived by the gloomy predictions of the Club of Rome study, 'The Limits to Growth' (1). We were forcibly reminded of the finite nature of the world's supply of fossil fuels and the debate over the seriousness or otherwise of this fact divided those involved into two camps: the resource pessimists, principally natural scientists and the resource optimists of whom economists were the most vociferous. In retrospect the difference between the two groups seems to have been the particular past trend they were bold enough to extrapolate, past rates of exploitation of non-renewable resources or past rates of technical progress. Differences of faith are, of course, always difficult to resolve. Perhaps the most illuminating point to be made is that each group refused to extrapolate growth rates in the very things they knew most about.

Oil and coal alone account for nearly ninety per cent of Australia's primary energy use and this can be taken as fairly typical of the situation faced by all advanced industrial nations. When taken in conjunction with past growth rates of per capita energy consumption the energy crisis appears to be much the same thing as a crisis in our supplies of fossil fuels. We seem to be led inevitably to the conclusion that we are rapidly exhausting our energy capital and that this will prove to be the brake on growth. It is hard to dismiss such worries when it can be shown that, if we extrapolate growth in consumption, Australia's recoverable reserves of crude oil would be exhausted in seven years, or even at current rates of use last only fifteen (2, p. 187). However, we must beware of lighting on such facts and drawing the inference that we face an energy crisis. There is no necessary link between running short of particular fossil fuels and running short of energy as a whole. Economists would argue that as fossil fuels became scarce their price would rise thus making alternative renewable energy sources commercially viable. Further we have to be very careful how we handle the information conveyed by figures showing known recoverable reserves. Such reserves represent the amount deemed to be economically recoverable. Thus two important determinants of known reserves are the current state of extraction technology and the market price of the fuel. To a certain extent these determinants are interdependent since, as the resource becomes scarce, the rising price makes previously uneconomic fuel bodies worth extracting and also increases the commercial attractiveness of research and development into more efficient methods of exploitation. Known reserve levels, as defined by current technology and commercial values, may bear no relationship to what ultimately usable reserves exist. Indeed as pointed out by Commoner (3), we have every reason to believe that actual physical quantities of our fossil fuels greatly exceed current estimates of accessible deposits and further, that it was the failure of the price of oil to rise in the past which led to the failure to find and develop new fields. This important point has long been known to economists. W. S. Jevons writing on the Coal Question in the nineteenth century, while concerned over 'the probable exhaustion of our coal mines', nevertheless stressed that he meant the exhaustion of deposits at a price that would be competitive with those available to nations other than Great Britain. He was scornful of the view that 'our coal seams will be found emptied to the bottom and swept clean like a coal cellar'. In a physical sense he claimed:

'our mines are literally inexhaustible. We cannot get to the bottom of them; and though we may some day have to pay dear for fuel, it will never be positively wanting' (4, p. xxx).

Nevertheless, most of us would not wish to see the cost of energy to future generations become so high as to mean their impoverishment. Ideally we would want to exploit our non-renewable sources of energy at a rate which would allow the development of alternative, renewable sources which could be obtained by future generations at a reasonable cost in terms of the other goods which must be foregone in the process. On this basis what we face is not so much an energy crisis but rather a specific set of problems pertaining to our present dependence

Diesendorf, M. (ed.) (1979). *Energy and People*. Canberra, Society for Social Responsibility in Science (A.C.T.)

on exhaustible fossil fuels, particularly oil. The choice of the optimal depletion rate is difficult because of two counteracting influences which we need to trade-off, one against the other. We do not want to use our fossil fuels so quickly that we impose a costly burden of transition to alternative energy sources entirely on future generations but, at the same time, the fuel we extract now is an important input in the creation of the capital equipment necessary for the development of alternative energy technologies. While there is consensus that the transition path to an economy based on renewable energy sources should be smooth and predictable so that the disruptive effects of sudden shortages and large jumps in the prices of fuels are avoided, there is much disagreement over the desirable speed of transition which will be determined by the rate at which we exploit out fossil fuels. Some economists adopt what might seem a rather cavalier attitude to resource exploitation perhaps taken to its logical conclusion by Kay and Mirrlees:

"There is a real danger that the world's resources are being used too slowly . . . In general we believe that the interests of future generations will be better served if we leave them production equipment rather than minerals in the ground" (5, p. 171).

This view is based on certain assumptions made within the framework of the market model typically used by economists to analyse questions of resource use. We need to examine this framework in order to understand how such a conclusion can be reached and further why there are good reasons for believing exactly the opposite — that we are using our fossil fuels too quickly.

Inter-Temporal Resource Allocation and the Market

The benchmark economists use to derive conclusions about optimizing social welfare over time is a model of the workings of perfectly competitive markets. It is a fundamental theorem of welfare economics that, assuming a fair distribution of purchasing power determined by the political process, the pursuit of self-interest under perfect competition leads to the greatest benefit for society (6).

Unfortunately, the economy is not characterized everywhere by assets owned and traded by large numbers of well informed buyers and sellers who are unable, because of the force of competition, to individually influence the market to their own advantage. In particular, there are problems posed by such things as common property resources, where use by one economic agent can impose costs on others which will not be taken into account because there is no market incentive to do so. Air and water pollution provide good examples. These 'externalities' as they are termed, and the more familiar monopolistic distortions, precisely because they lead to departures from the efficient resource use of the perfect market model, call for government intervention. Similarly, market imperfections of various kinds could lead to the rate of exploitation of fossil fuels being too rapid or too slow in comparison to the optimal rate that would result from the operation of perfect markets. Thus the approach followed is to identify distortions in the system and to suggest appropriate measures to correct its behaviour.

The market model requires that the owner of any valuable resource attempts to maximize its present value, that is the value today of all the future returns he will enjoy from ownership. The present value of a house is thus the stream of benefits it will confer on the owner, valued at the rent he would otherwise have to pay, suitably reduced to the same base by discounting. This discounting procedure is necessary because the more distant a benefit the less it is worth in present value terms. We would be foolish to regard a dollar today as being worth the same as a dollar ten years hence because if we had that dollar now we could invest it. Obviously it would rise in value earning compound interest and be worth a considerable sum after ten years.

We can apply this principle to oil extraction. Maximizing the present value of a given stock of oil implies that the contribution to present value made by a barrel of oil extracted in any period must be equalized. If it were not the owner would have an incentive to alter the dates at which he planned to extract particular quantities of oil so as to extract more in periods where the proceeds would be worth more to him in present value terms (7, ch. 1). The effect of discounting is now clear. If the present value of a barrel of oil in each period is equalized, then its money value must rise on a compounding basis determined by the rate of interest. Indeed, if this were not so, profits could be increased by extracting all of the oil now and investing the proceeds at the ruling rate of interest. How much oil is extracted in each period can therefore be seen to depend on expectations about the contribution to present value that would be made by selling an additional barrel at different times. Thus we have to estimate future changes in demand conditions and costs of supply in order to achieve the optimal allocation over time of the use of the resource. Since it is the oil industry's business to know more about these factors than anyone else, it follows that the gain to society will be maximized by relying on its judgment, provided, of course, that the system has been corrected for the sort of distortion mentioned earlier.

As the rate of depletion is optimized by the market, we must pose the question of what users of the fossil fuel will do as it becomes a scarcer and its price rises as a result. The rising price of oil will encourage on the one hand the use of the substitutes by consumers and on the other their development by producers. The eventual demise of fossil fuels, on this view, is not going to occur because in some physical sense we have run out, but simply because they will no longer be competitive with alternative energy sources. This position is succinctly summed up in a Treasury Paper on economic growth:

'It cannot be inferred from the fact that supplies of fossil fuels are finite that the prices of such fuels must necessarily rise in the decades ahead; but it can be inferred that as alternative energy forms are developed, it will become increasingly necessary for fossil fuel supplies to be cheap if they are to be used at all. It is conceivable that the great bulk of the world's store of these fuels will remain in the ground (or under the sea) for ever' (8, p. 38).

Certainly there would appear to be evidence that the market accepts this view. For instance, the overburden from open cut coal mining is often dumped in such a way as to greatly add to the cost that would have to be incurred to mine seams presently of too low a grade to be economic. The presumption must be that it is believed they will always remain so.

The market model paints quite a bright picture of resource use over time. In the long term, market signals will lead us to move over to renewable sources of energy, such as solar power. Moreover, it is difficult to argue that this is physically scarce, economists having learnt that an area equivalent to 'one quarter of 1 per cent of the Sahara receives energy from the sun equivalent to present world energy consumption' (9, p. 35). Furthermore, on the crucial question of the speed of transition, we have seen that this also would be optimal under perfect markets. It is, perhaps, rather surprising that if we introduce a monopolistic distortion to the model (under realistic assumptions), the resource will be tapped more slowly (10, p. 695). It is the acknowledgment of the significance of monopolistic distortions in the real world that accounts for the conclusion reached by Kay and Mirrles mentioned earlier. The oil cartel is a case in point, the general idea being that the wicked Arab monopolists restrict production so as to extort a higher price from consumers and this means that the resource is used too slowly.

This argument appears suspect. It is true that if a monopolistic distortion is introduced to a model of an otherwise perfect world, resource allocation will no longer be optimal. However, if there are pre-existing monopolistic elements in the system, the effect of the new 'imperfection' can be to offset these to some degree, making the situation better, not worse. Technically this is known to economists as the 'second best problem' (11). It would be difficult to argue that everyone else in the oil industry, from initial producer to the multinational manufacturers who use oil as an input and those who market it as a final good, all behave as if they were perfect competitors. Therefore the argument that the existence of an oil cartel results in the resource being used too slowly would seem to lose much of its force.

It will be recalled that a very important part was played in the earlier discussion of the optimal rate of depletion by the rate at which future benefits are discounted. In effect this was taken to be the same thing as the rate of interest established in the perfect market system. The owner of a particular resource, using his judgment of what society will want in the future and what alternatives will be available, maximizes his present value with reference to the ruling rate of interest which he uses for discounting purposes. This can be deemed optimal because this rate of interest represents the opportunity cost of his decision, what he could have earned if his assets had been invested elsewhere in the economy. Nevertheless, it is perfectly possible for the rate of interest itself to be affected by market imperfections and lead to the rate of depletion being too rapid or too slow. Further, when the resource owner forms his estimates of the future costs of supply, he is only interested in the private costs of extraction he will have to bear. If there are market imperfections of the externality type, mentioned earlier, his private judgment will understate the social costs of extraction and could critically affect the rate of depletion. For instance, he may believe that lower grade fossil fuel will be available in the future at reasonable cost only because he neglects consideration of the high external environmental costs likely to result from dealing with much larger quantities of waste material. For these reasons it can be argued that the free operation of the private market in fossil fuel will lead to too rapid a rate of use. In particular, this is likely if we have reason to believe that the market exhibits too high a rate of time preference and thus discounts future benefits too quickly.

The Social and Private Rates of Discount

One of the implications of our market analysis is that the net revenue earned on each unit of fossil fuel extracted must rise on a compounding basis with the rate of interest. If we make the simplifying assumption that the marginal cost of extraction remains constant, it follows that the price of the resource must rise at the same rate as the rate of interest (12, p. 163). This assumption allows the illustration of the effects of the discounting process and does not seriously affect the conclusions reached. The following table shows the effect of different discount rates on the price of oil assuming an initial value of $13 a barrel.

Price of oil maximizing present value under various discount rates to nearest dollar per barrel.

Year		Rate of Interest				
		5%	7%	8%	10%	15%
0	1978	13	13	13	13	13
10	1988	21	26	28	34	53
20	1998	34	50	61	87	213
30	2008	56	99	131	227	861
40	2018	92	195	282	588	3,482
50	2028	149	383	610	1,526	14,088
60	2038	243	753	1,316	3,958	56,992
70	2048	396	1,482	2,842	10,267	230,564
80	2058	644	2,915	6,135	26,629	932,761
90	2068	1,049	5,734	13,246	69,069	3,773,454
100	2078	1,710	11,280	28,957	179,148	15,266,075

Notes
(i) The rise in price is not due to inflation. All prices quoted are in constant dollars and the rates of discount shown are real not nominal rates of interest.
(ii) The table makes no comment on the purchasing power of the consumer. It must be remembered that real earnings per capita are also likely to grow over time.

Suppose our profit maximizing oil producer is using a discount rate of ten per cent. The implied expected price in twenty years time is roughly seven times the initial price. After fifty years have elapsed he expects the price to have increased more than one hundredfold and so on, as compound interest does its worst. The table reveals that the lower the discount rate used the less horrifying the expected price rise. This requires careful interpretation.

For the conservation of a barrel of oil to a given date to be justified in the eyes of the producer it must be worth at least thirteen dollars in present value terms. The higher the discount rate the higher the money price he must expect to get in order to meet this requirement. If with a particular planned output over future periods he expected that in ten years time he could sell a barrel of oil for twenty six dollars, he would be maximizing his profit provided the discount rate was seven per cent. However, if the discount rate was ten per cent, twenty six dollars a barrel would not be enough; he would need to earn at least thirty four dollars. Hence he would not be maximizing his profits since he could increase his present value by selling more barrels today at the ruling market price of thirteen dollars. It therefore pays the producer to deplete the resource at a faster rate the higher the rate of discount.

The rate of discount used depends upon the market rate of interest. In reality, because of the differing degrees of risk associated with various activities, there exists a whole spectrum of interest rates. However, this need not concern us as it does not affect the basic issue of whether interest rates on the whole are too high or not. The market interest rate is regarded as indicating private time preference. Individuals, through their decisions to save, make available part of their current income for investment. Since investment is productive, they are revealing their rate of time preference by their willingness to sacrifice present consumption in order to enjoy a higher level of consumption in the future. The time preference of the market as a whole results in a rate of interest which measures the gain required by savers to persuade them to postpone consumption. While some degree of discounting would seem to be justified by the physical productivity of capital, high discount rates are the result of pure time preference. This has been explained by such things as the uncertainty associated with future consumption and the inevitability of death.

Further, since man does not always behave like the fully rational abstraction employed in many economic models, it may be due largely to short-sightedness, to what Pigou termed a defective telescopic faculty (13, p. 25). Most of us would not want future generations to be victims of irrational pure time preference, aptly described by the philosopher and economist Frank Ramsey as 'a practice which is ethically indefensible and arises merely from the weakness of the imagination' (14, p. 543). Governments, as the guardians of the interests of future generations, should not accept pure time preference as a criterion for intertemporal resource allocation. In particular, Pigouvian myopia provides a powerful justification for intervention.

There is another serious imperfection in the operation of the market which reinforces the argument that the optimal or social rate of time preference should be lower than the private market rate. The welfare of future generations is, to an extent, a 'public good'. A public good can not be provided for efficiently by the market because an individual does not have any incentive to reveal his preferences adequately. To do so allows others to gain at his expense. Even if people were fully rational and aware of the implications of the savings rate for future generations, each person would know that alone he could not affect the total situation. An individual has no incentive to consume less for the sake of future generations if this will simply make it possible for someone else to be a spendthrift.

These arguments lead to the conclusion that the much lauded market solution to fossil fuel depletion is seriously biased against leaving future generations a fair share. The economist, Ivor Pearce, has written an amusing parable which illustrates the damaging influence of pure time preference on resource use (15). He depicts the situation in the Garden of Eden before the fall. Each person is allotted a certain amount of oil by the heavenly storemaster. To simplify matters, there is zero population growth and all inhabitants have the same tastes. However, all income is earned by fathers. If the sons want oil in order to enjoy life while young, they must incur an interest charge. Private time preference of this sort, though an imperfection, does not cause any problems in the Garden because each individual pays the debts he incurs while young when he becomes a father. Lifetime consumption of oil is exactly the same as it would be if there were not private time preference. After the fall part of the punishment is that control of the oil is handed over to Satan. This has disastrous results for inter-generational equity. Satan, perceiving the market rate of interest, sets to work applying the principles of profit maximization. Previously, private time preference could not affect future generations because of the fixed allotment of oil dispensed to each individual by the heavenly storemaster. However, Satan trying to maximize his present value is fooled by this interest rate which is not produced by higher productivity. He believes that he must so arrange depletion of the resource to ensure that the price will raise at the rate of interest. The discount rate now operates across generations. The oil he plans to sell later appears to contribute less to present value. He therefore allows a faster rate of use to the detriment of future generations. Satan is merely applying the market solution that was discussed earlier. The moral of the story is simple. The fact that the rate of interest used in the market for discounting purposes is influenced by pure time preference over and above considerations of the productivity of investment, means that market mechanism facilitates the robbery of the future by an unthinking and greedy present.

We have seen that there are good reasons for believing that the benefits the future could derive from fossil fuel left to them are not likely to be given their due weight in private profit calculations. This is a very damaging criticism of the way in which our resources are presently used. If the discount rate is wrong so is the price charged at any period. We can add to this that, even if it were right, there would be no reason to suppose that the particular mixture of monopolistic distortions found in the oil industry would lead to the socially optimal price and rate of use.

Clearly we can not rely on these emerging acçidently as the Arabs set the highest price they can charge without collapsing the industrial economies of the world.

Conservation of our Fossil Fuel

It is one thing to claim that the market rate of discount will lead to too rapid a rate of use of our fossil fuel and another thing entirely to suggest policy measures to correct this. Piecemeal intervention to lower interest rates could be disastrous. For while we have seen that lower interest rates should mean slower exploitation, when the repercussions on the rest of the economy are considered the net effect could be exactly the opposite. Lowering interest rates would increase investment and the growth rate of the economy which in turn might mean increased use of our fossil fuel stocks. This is clearly a very important empirical issue (16). The essential point is that lowering the discount rate could have untoward effects on the mix of goods we provide for the future. This would not happen if the only imperfection in the system was the discount rate itself. However, as argued earlier, the private resource owner is likely to underestimate future social costs of extraction. Correcting the discount rate does not remove this distortion and is likely to worsen its effects by making the future benefits that would stem from capital equipment look relatively more attractive than those that could be derived from fuel left in the ground. We are once again in the realm of the second best. If resources as a whole are misallocated at a point in time due to various distortions, such as failure to take into account the social costs of resource exploitation, then trying to correct the discount rate to improve the allocation of a particular resource over time could make things worse not better.

In a second best world what is required is a package of measures specifically tailored to promote conservation of particular resources and especially oil and coal. Nevertheless, I would not argue for complete government control of the industries concerned. After all the oil industry still knows more about oil than anyone else. Ideally we should be looking for ways of changing the framework in which the market operates so as to provide incentives which will lead to both lower consumption and slower depletion of Australian reserves. Whilst it would be relatively easy to ensure this by an appropriate taxation policy designed to raise the price to the consumer and to hold down the return to the producer below that available in the rest of the world, this would introduce undesirable dynamic problems associated with exploration and development. Owing to their great complexity only a few tentative comments will be offered in conclusion.

In a sense we both want to have our cake and eat it, that is for the oil companies to explore for more oil but not exploit reserves at an over rapid rate. Maintaining continued exploration is of great importance because there are long lags involved between discovery and development. In fact until recently the Federal government had a policy which encouraged the development of new fields while discouraging rapid exploitation of old established sources. The price that the producer received was the import parity price for new oil, that is for oil discovered after the 14th September 1975 and also for a given percentage of old oil. However, most of the oil from earlier discoveries earned only 22-33% of the world price (17). When taken in conjunction with the policy announced in the 1978-1979 Budget Speech of charging refineries the world price for all of their oil and hence a realistic price based on this to the consumer, this could have been an effective way of both conserving indigenous supplies and encouraging exploration. As it is, the percentage of oil from old fields which can earn the world price is to be allowed to rise rapidly to 50% by 1981 and 100% as soon as possible thereafter. We do know, however, that it is feasible to devise policies to achieve the conservation objective. There are other issues involved. The government's decision may reflect a value judgment over energy dependence, though accelerated use of our own resources would appear to lead to earlier complete dependence on imports. A rational response to strategic worries might be to stockpile Arab oil while it is being sold at what will turn out to be a bargain price. This would not preclude measures to encourage stockpiling of our own oil in the best possible way — by leaving it a little longer exactly where it is.

References

(1) Meadows, D. H., et. al. (1972). *The Limits to Growth.* London, Earth Island.
(2) Report of the Task Force on Energy of the Institution of Engineers (1977). *Conference on Energy 1977, Towards an Energy Policy for Australia.*
(3) Commoner, B. (1976). *The Poverty of Power.* New York, Alfred A. Knopf.
(4) Jevons, W. S. (1906). *The Coal Question.* New York 1965, Augustus M. Kelley.
(5) Kay, J. A. and Mirrlees, J. A. (1975). The desirability of Natural Resource Depletion, 140-176 in Pearce D. W. ed. *The Economics of Natural Resource Depletion.* London, Macmillan.
(6) Bator, F. M. (1957). The Simple Analytics of Welfare Maximization. *American Economic Review.* **47,** 22-59.
(7) Scott, A. (1973). *Natural Resources: The Economics of Conservation.* Toronto, McClelland and Stewart.
(8) *Economic Growth: Is it Worth Having?* (1973) Treasury Economic Paper No. 2. Canberra, A.G.P.S.
(9) Robinson, C. (1975). The Depletion of Energy Resources, 21-55 in Pearce, D. W. ed. *The Economics of Natural Resource Depletion.* London, Macmillan.
(10) Peterson, F. M. and Fisher, A. C. (1977). The Exploitation of Extractive Resources: A Survey. *Economic Journal.* **87,** 681-721.
(11) Lipsey, R. G. and Lancaster, K. (1956-1957). The General Theory of Second Best. *Review of Economic Studies.* **24,** 11-32.
(12) Pearce, D. W. (1976). *Environmental Economics.* London, Longman.
(13) Pigou, A. C. (1952). *The Economics of Welfare.* London, Macmillan.
(14) Ramsey, F. P. (1928). A Mathematical Theory of Saving. *Economic Journal.* **38,** 543-559.
(15) Pearce, I. (1975). Resource Conservation and the Market Mechanism, 191-203 in Pearce, D. W. ed. *The Economics of Natural Resource Depletion.* London, Macmillan.
(16) Fisher, A. C. and Krutilla, J. V. (1975). Conservation, Environment and the Rate of Discount. *Quarterly Journal of Economics.* **89,** 358-370.
(17) Pricing of Crude Oil (1978). *Petroleum Gazette.* **20,** 66-67.

Jobs, Energy and Economic Growth in Australia

John Andrews

Environmentalists for Full Employment (Vic.)
672B Glenferrie Road, Hawthorn Vic. 3122

Summary

A prime characteristic of a sustainable economy living in harmony with the environment is a fairly constant level of energy consumption.

The conventional economic wisdom is that the only way to create more jobs is by economic growth, and that economic growth necessarily calls for the use of more energy.

Over recent decades, by far the largest proportion of new jobs in Australia have come in the services sector, which accounts for a very small percentage of total energy use.

Most of the increase in energy use over this period has been consumed in primary and secondary industry, which have provided relatively few new jobs.

The argument that the greatly increased production by primary and secondary industry — which has depended on higher energy use to boost productivity — has generated the wealth required to finance the expanding services sector needs to be studied in more detail.

Recent authoritative studies show that, despite past trends, economic growth and employment growth can be decoupled from energy growth in the future.

This can be achieved by using energy savings from more efficient use of energy to fuel continued economic growth in relatively labour-intensive areas such as services.

The success of this service-sector employment growth policy would be seriously undermined by new electronic information technology aimed at automating office and other service-sector work.

Employment opportunities in a low or zero energy growth future for Australia are likely to be far better than in the high energy growth future currently planned.

But the unemployment problem will not be solved simply by changing the way we use energy. Economic and political changes will be needed as well.

Introduction

It is now becoming increasingly accepted that we cannot continue expanding energy production and consumption indefinitely, and that a prime characteristic of a sustainable society with acceptable rates of depletion of non-renewable resources, and suitably low pollution emissions and impact on other living species, is a fairly constant level of energy consumption. But what about the unemployment crisis? Don't we need to use more energy if we are going to provide more jobs? What effect would a decision to level off growth in energy consumption have on economic growth and employment growth? Such are the questions to be addressed in this paper, which investigates the complex relationship between energy use, employment opportunities and economic growth in the context of the Australian economy.

More Energy leads to More Jobs? Point Counterpoint

The conventional economic wisdom concerning the link between energy use and employment can be simply stated. The only way to create more jobs is by economic growth. Economic growth necessarily calls for the use of more energy.

If we look at the trends in total employment, total primary energy use and gross domestic product (GDP) in Australia between 1954 and 1976 (Figure 1), we can see how this view has been formed. All three factors have been growing quite steadily over the period. GDP and total primary energy have been

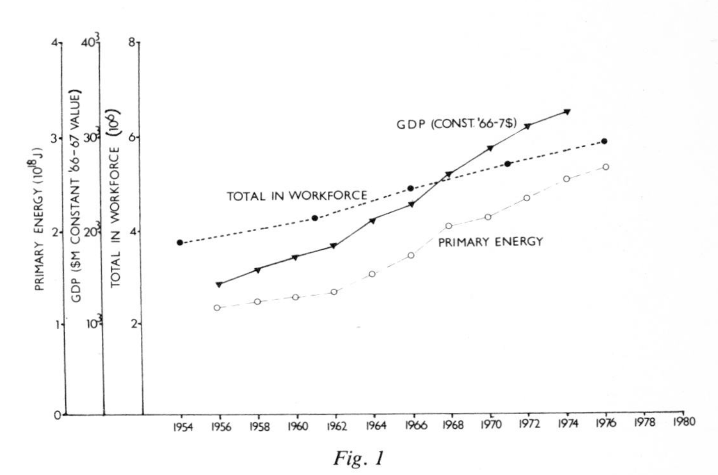

Fig. 1

Trends in total employment, gross domestic product, and total primary energy use in Australia between 1954 and 1976. Sources: Employment — Censuses 1954, 1961, 1966, 1971, Civilian Labour Force by Industry *May 1976; Energy — Kalma (1976), Department of National Development (1978); GDP — Commonwealth Year Book, 1975-76, p. 1096.*

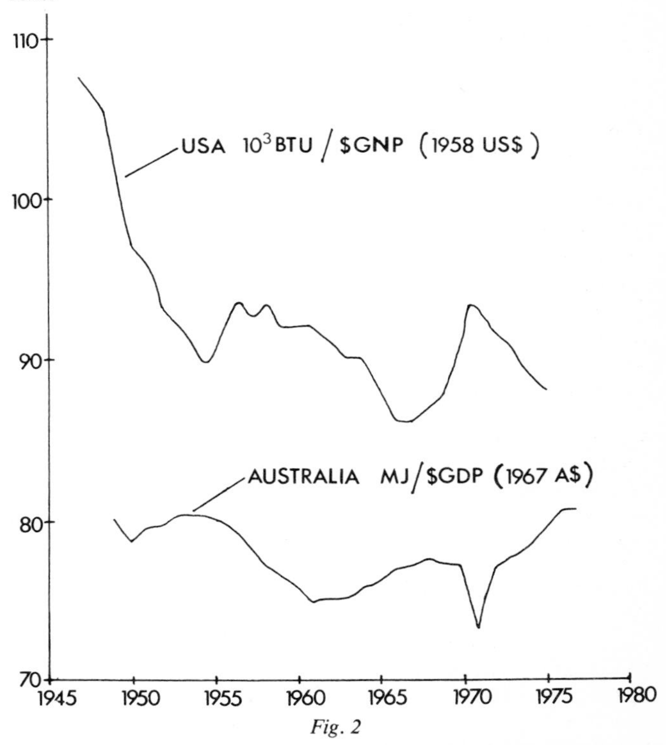

Fig. 2

The fluctuations in the ratio of primary energy consumption to gross domestic (or national) product (constant dollars) for the U.S. and Australia between 1949 and 1975. Note the declining economic efficiency of energy use in Australia since 1971.

Diesendorf, M. (ed.) (1979). *Energy and People*. Canberra, Society for Social Responsibility in Science (A.C.T.)

Table 1

Total employment, total energy use, turnover, value-added, and fixed capital expenditure, plus various computed ratios between these factors, for the Australian Standard Industrial Classification (A.S.I.C.) divisions for Victorian Manufacturing industry in the year 1974-75. Value-added gives the contribution of an industry to gross domestic product. Labour productivity is given by value-added per employee, capital productivity by value-added per unit capital invested and energy productivity by value-added per unit energy consumed. It is a serious failing that at present economic decision-making concentrates on labour productivity, while paying scant attention to energy and capital productivities (Commoner, 1978). 1974-75 dollar values are used throughout this table.

Industry Division (ASIC)	Total Persons Employed 10^3	Total Energy Use 10^9MJ	Energy/ Employee 10^3MJ = GJ	Turnover \$('74-5)10^6	Turnover/ Employee \$('74-5)10^3	Value-added \$('74-5)10^6	Value-added/ Employee \$('74-5)10^3	Fixed Capital Expenditure \$('74-5)10^6	Fixed Capital Expenditure/ Employee \$('74-5)10^3	Value-added/ Energy Use \$('74-5)/GJ	Value-added/ Capital \$/\$
Food, Beverages, Tobacco	61	21.76	355	2556	41.7	869	14.2	71	1.16	39.9	12.3
Textiles	23	5.14	223	536	23.2	209	9.1	19	0.83	40.7	11.0
Clothing, Footware	50	1.63	33	806	16.2	382	7.7	8	0.16	234.3	48.6
Wood, Wood Products, Furniture	20	1.43	71	453	22.4	215	10.6	12	0.60	150.4	17.7
Paper, Paper Products, Printing	35	12.85	369	844	24.2	434	12.5	50	1.43	33.8	8.7
Chemicals, Petroleum, Coal products	22	10.98	501	806	36.8	369	16.9	35	1.58	33.6	10.7
Non-metallic mineral products	14	22.21	1561	407	28.6	218	15.3	30	2.11	9.8	7.2
Basic metal products	13	6.68	511	550	42.1	184	14.1	28	2.13	27.6	6.6
Fabricated metal products	39	4.68	121	860	22.2	428	11.1	37	0.95	91.5	11.7
Transport equipment	62	7.75	125	1652	26.7	711	11.5	92	1.48	91.7	7.8
Other machinery equipment	70	5.65	81	1582	22.7	788	11.3	46	0.66	139.5	17.2
Misc. manufacturing	30	5.42	180	778	25.3	371	12.3	31	1.02	68.5	12.1
Total/ average	439	106.17	242	11830	27.0	5179	11.8	457	1.04	48.8	11.3

Table 2

Recent energy-jobs-output trends in Victorian manufacturing industry. Percentage changes in the factors given in Table 1 between the years 1969-70 and 1974-75 with allowance made for the effects of inflation.

Industry Division (ASIC)	Total Persons Employed	Total Energy	Energy/ Employee	Value-added	Labour Productivity = Value-added per Employee	Fixed Capital Expenditure	Fixed Capital/ Employee	Energy Productivity = Value-added per Unit Energy	Capital Productivity = Value-added per Unit Capital
Food, Beverages, Tobacco	+ 1.8	+ 17.7	+ 15.6	+ 17.7	+ 15.4	+ 1.4	− 0	0	+ 16.0
Textiles	− 20.5	+ 2.0	+ 28.2	− 21.0	− 0	− 49.5	− 36.2	− 22.6	+ 57.1
Clothing, Footware	− 22.8	+ 10.9	+ 43.5	− 7.6	+ 20.3	− 61.4	− 50.0	− 16.6	+ 139.4
Wood, Wood Products, Furniture	+ 3.2	+ 2.1	− 0	+ 23.0	+ 19.1	+ 44.5	+ 39.5	+ 20.4	− 14.5
Paper, Paper Products, Printing	+ 1.4	+ 17.5	+ 16.0	+ 13.2	+ 11.6	+ 68.8	+ 66.3	− 3.7	− 33.1
Chemicals, Petroleum, Coal products	− 4.2	+ 5.3	+ 9.9	+ 5.8	+ 10.5	− 52.7	− 50.6	+ 0.3	+ 122.9
Non-metallic mineral products	+ 1.5	+ 17.8	+ 16.1	+ 25.8	+ 23.4	+ 13.5	+ 11.6	+ 6.5	+ 10.8
Basic metal products	+ 15.2	+ 35.5	+ 17.5	+ 38.3	+ 19.5	− 25.5	− 35.3	+ 2.2	+ 83.3
Fabricated metal products	− 0.8	+ 61.4	+ 63.5	+ 14.8	+ 15.6	+ 8.2	+ 9.2	− 28.9	+ 6.4
Transport equipment	+ 10.1	+ 84.5	+ 66.7	+ 9.8	− 0	+ 19.3	+ 8.0	− 40.5	− 7.1
Other machinery equipment	+ 5.8	+ 42.3	+ 35.0	+ 16.7	+ 10.8	− 15.4	− 19.5	− 18.0	+ 37.6
Misc. manufacturing	+ 4.1	+ 22.3	+ 17.6	+ 27.2	+ 21.8	+ 15.1	+ 10.9	+ 3.9	+ 10.0
Average over all Manufacturing	− 1.6	+ 22.0	+ 24.1	+ 12.2	+ 13.5	− 7.6	− 6.3	− 8.1	+ 21.5

growing at about the same pace, while total employment has been rising at a slower rate. The Department of National Development (1978) estimate the average exponential growth rate in primary energy use between 1960-61 and 1975-76 as 5.45% p.a.

The view that economic and energy growth, and in turn job creation, all go hand in hand is at the heart of Government and business thinking in Australia. For example, in the most recent forecast of total Australian primary energy demand, by the Department of National Development (1978), an average economic growth rate of 4% is assumed for the period to 1986-87, and the average energy growth rate 1975-76 to 1986-87 is forecast as 4.08% p.a. Hence energy growth is assumed to remain tightly coupled to economic growth over the foreseeable future. To my knowledge no government publication in Australia has yet seriously challenged that notion.

However, from Figure 2 (Mardon, 1978) we can see that the ratio of total primary energy use to gross domestic product (constant dollars) in Australia — i.e. the primary energy input per dollar of GDP — has fluctuated significantly over the 1950-1977 time period. In 1953, 80.3 MJ of primary energy input were required per dollar of GDP (1967 A$), while only 73.0 MJ were needed in 1971. Since 1971 the primary energy input per $GDP has again risen steadily, so that by 1977 it had reached 80.5 MJ, i.e., above the 1953 level. Hence in 1977 we achieved 14% less value of output per unit of primary energy input than we did in 1971.

For comparison purposes, Figure 2 also shows the variation in the primary energy/gross national product ratio for the U.S.A. over the same period. The ratio for the U.S. economy fell by 20% (107.5 to 86.0 $\times$ 10^3 BTU/$GNP) between 1947 and 1966, then rose to 93.0 $\times$ 10^3 BTU/$GNP by 1971. But, in contrast to the Australian situation, the U.S. ratio has been falling steadily since 1971, signifying an improvement in the overall economic efficiency of energy use. No doubt the rising real costs of energy in the U.S. since 1971, in conjunction with the falling real costs of energy in Australia, partially account for this difference in observed behaviour of the energy/GNP ratios in the two countries.

In a climate of growing uncertainty about future energy supplies, the energy-economic growth-jobs inter-relationship has now begun to come under close scrutiny overseas, especially in the U.S.A. Questions such as the following are being asked: What is the effect on jobs of increased energy use in industry? In what areas have most new jobs been created over the recent decades of economic expansion? What effect would a slow-down in energy growth have on employment?

Environmentalists for Full Employment (EFFE) (1977) pointed out that increased energy use in most industries in the U.S. has been associated with new technology which has greatly boosted labour productivity; the same, or often greatly expanded, output can then be produced using less labour. If demand for the product doesn't rise sufficiently, some workers will be out of a job.

> *Historically, industry has sought to substitute energy for human labour. The amount each working person could produce has therefore increased steadily. But after substitution of energy for labour in each process, the total number of workers needed* decreased. *The only way the total number of workers could increase would be if there were also a rise in the demand for products. In other words, more jobs would have to be created by the increased demand than were eliminated by the energy substitution.* (EFFE, 1977)

This trend can be illustrated by a simple example. Imagine a furniture factory which employs 10 workers who, using only hand saws, chisels, drills, etc., can make 10 chairs a day. Now suppose there is a technological innovation — power tools are introduced into the workshop. With this new technology, suppose the 10 workers can make 20 chairs a day, that is, their labour productivity has been doubled at the expense of the extra non-human energy used per chair produced. Whether the introduction of power tools leads to unemployment or not depends on the market for chairs. If all 20 chairs made each day by the 10 workers could be sold, then no one need lose their job. (Generally, new labour-saving technology has cheapened the unit cost of production which has been an incentive for greater sales.) Indeed if demand for chairs was sufficiently strong, say an output of 30 chairs a day could be sold, then an extra 5 workers would have to be hired. However, in a market situation where only an output of 10 chairs a day could be sold, five workers would be out of a job — though in our ideal closed system perhaps one or two of these could find work making power tools or at the new power station built to meet the extra electricity demand. Yet not all five workers displaced are likely to find work indirectly associated with chair making in this way because generally a new technology is introduced only if it reduces the total labour inputs per unit of production — including the labour needed at all stages in the making of the new technology.

There is an intriguing third possibility in our furniture workshop. If 10 chairs a day made by this workshop and all the others like it ensure plenty enough chairs for world needs, then after the introduction of power tools, "In a sensible world, everybody concerned in the manufacture of (chairs) would take to working four hours instead of eight, and everything else would go on as before," as Bertrand Russell once pointed out. As we know, for a complex of reasons, unfortunately our present economic system doesn't allow this common-sense solution to be adopted.

The energy-jobs-production relationship in a real economy can be exemplified by recent trends in manufacturing industry in Victoria. Table 1 summarises total energy use, total employment, turnover, value-added and fixed capital expenditure, together with various computed ratios between these quantities, for all the Australian standard Industrial Classification Divisions for Victorian manufacturing industry, in the year 1974-75. The data were directly drawn, or computed, from Australian Bureau of Statistics (Vic. Office) (1974-75 a and b). Figures for total energy consumption by industry division were calculated by converting the quantity (in tonnes, litres, etc.) of each type of fuel consumed to its primary energy equivalent and then summing the resulting energy figures. In the case of gas and electricity, where consumption was given only in dollar values in the statistics, an assumption of a linear relationship between the dollar cost of gas or electricity and the actual quantity of energy consumed was made as a first approximation. Full details of the calculation of figures in Table 1, and a similar table for the year 1969-70, are given in Andrews (1978).

Table 2 shows the percentage changes in factors given in Table 1 over the 1969-70 to 1974-75 period, and, as can be seen, both of the main types of energy-jobs-output relationship exhibited in our furniture workshop example are represented. Values given in 1969-70 dollars were converted to 1974-75 dollars using an inflation factor of 1.65 (i.e. $('69-'70) 1 = $('74-'75) 1.65), which was calculated from figures given for the Australian gross domestic product at constant and current prices for these two years in the 1975-76 Commonwealth Year Book (p. 1096).

For instance, in the food, beverages and tobacco class, total energy use rose 17.7%, total value-added (the contribution of this class to GDP) also rose by 17.7%, while employment climbed by only 1.8% between 1969-70 and 1974-75. Labour Productivity (value-added per employee) rose by 15.4% and energy use per employee by 15.6%. These figures suggest that new technology introduced during this period has significantly improved labour productivity at the expense of higher energy consumption. Rising demand for food products, however, has more than offset the labour-displacing tendency of the technological change.

On the other hand, new technology in the chemicals, petroleum and coal products industry class in Victoria has allowed production to rise while total employment has dropped. Between 1969-70 and 1974-75, value-added rose by 5.8%, employment fell by 4.2%, and energy use rose by 5.3%.

But the onwards march of new technology, in Australia and overseas, has not only been associated with substituting capital plus energy for labour: it has also, in many industries, led to a change in the nature of the product as well. As Commoner (1978) points out, natural products such as leather, cotton, wool, wood, paper and soap have over time been displaced by synthetic alternatives such as plastics, synthetic fibres and synthetic detergents. In the natural-product industries, labour productivity is low while energy and capital productivities are high: in the synthetic-product industries, labour productivity is high (hence their growth), but energy and capital productivities are low.

On this basis Commoner (1978) thus argues that the environment, energy and economic crises that beset us are inextricably linked. He writes:

The same shifts in production technology that reduced the productivity of capital and energy and have cut the number of jobs usually increased the impact of production on the environment. As synthetic products replaced natural materials, more petroleum and natural gas were used both as raw materials and for fuel, polluting the environment with combustion products and toxic chemicals. The petrochemical industry demonstrates the close links among the wasteful use of energy and capital, the assault on the environment and unemployment.

Having considered the labour-substitution effect of new energy-consuming technology, let us now turn to an analysis of where most new jobs have been created over recent decades, and consider to what extent this job creation has depended on increased energy use. The short answer for advanced industrial nations such as the U.S. and Australia is the services or tertiary sector of the economy. This sector includes transport and storage, building and construction, finance, public administration, communications, insurance, wholesale and retail, health, education and welfare, and other community services.

As Daly (1972) has shown for the US economy:

The total employment within the service sector has grown from approximately 40% in 1929 to over 55% in 1967. Of the total net increment of 14 million jobs between 1947 and 1965, the service sector accounted for an increase of 4 million, but agriculture accounted for a decrease of 3 million. Services, using less energy than the other two sectors, have provided nearly the entire net increase in employment since 1947.

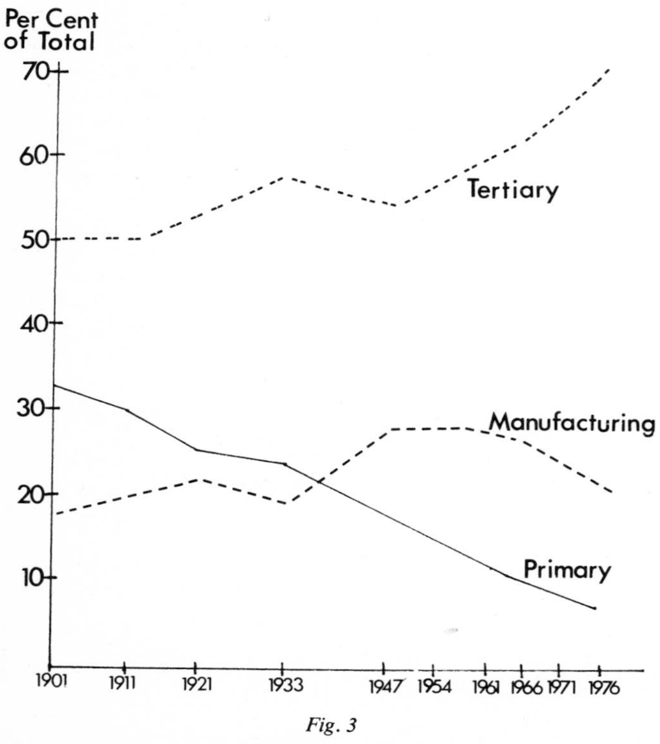

Fig. 3

Distribution of the Australian workforce between primary, manufacturing (secondary) and tertiary sectors of the economy, 1901-1976. Sources: Ford (1970); Commonwealth Year Book 1973, p. 704; Commonwealth Year Book 1975-76, p. 696.

Figure 3, showing the distribution of the workforce between the primary, manufacturing and tertiary sectors in Australia, indicates a similar trend towards expanding employment in the services sector here. Between 1947 and 1976 the percentage of the workforce in the services sector grew from 55% to 70%. The percentage of the workforce in manufacturing peaked in 1954 at 28% and by 1976 it had fallen to 22%. Since 1947 the percentage of the workforce engaged in mining, agriculture and other primary industry has dropped by more than half, from 18% in 1947 to 7% in 1976.

If we now look at employment changes in absolute rather than relative terms, we find from Table 3 that the net increase in the Australian workforce between 1961 and 1971 of 1,015,000 was made up of a mammoth 990,000 rise in service sector jobs, a relatively small 76,000 rise in manufacturing employment, and a 51,000 decline in primary sector employment. Table 4 indicates a similar pattern for the sectoral employment changes between November 1973 and November 1976.

Table 3

EMPLOYMENT CHANGES BY MAJOR SECTOR IN AUSTRALIA: 1961-1971

	Number employed (10^3)		
	1961	1971	Change
Primary	513	462	-51
Manufacturing	1140	1216	+76
Services	2572	3562	+990
Total	4225	5240	+1015

Employment changes by major sector in Australia, 1961 to 1971. Sources: Commonwealth Year Book 1962, p. 1225, Commonwealth Year Book 1973, p. 704.

Table 4

EMPLOYMENT CHANGES BY MAJOR SECTOR IN AUSTRALIA:
November 1973 to November 1976

	Number employed (10^3)		
	Nov. 1973	Nov. 1976	Change
Primary (a)	463	447	-16
Manufacturing	1352	1257	-95
Services (a)	3875	4132	+257
Total	5690	5836	+146

Note (a)
Part-time employees in forestry, fishing and hunting and mining sectors were not available separately and have been included in the Services Sector.

Employment changes by major sector in Australia, November 1973 to November 1976. Source: Australian Government White Paper on Manufacturing Industry, May 1977, p. 7.

Table 5 gives the sectoral breakdown of final end use of all primary and secondary energy, total employment and energy used per job in Australia for the year 1971-72, and enables us to look at the relationship of the employment changes shown in Table 3 to overall energy use.

As can be seen from Table 5, the services sector, excluding transport and storage, accounted for 60.5% of total employment in 1971-72, but for only 2.6% of total energy use. The energy use per job in this sector was a mere 1/33 of the energy used per job in manufacturing and 1/47 that used per job in mining. By contrast primary and secondary industry accounted for 52.4% of total energy use but only 34.0% of total employment.

Although no figures for a sectoral breakdown of Australian end use of energy are available for the years before 1971-72, it is almost certain that the large growth in service sector employment of 990,000 between 1961 and 1971 has resulted in a very small increase in total energy use. The growth in the transport and storage class over the period would have resulted in much more energy use, though relatively few extra jobs.* By far the largest number of new jobs have been created in services other than transport and storage.

Table 5

ENERGY USE (PRIMARY), TOTAL EMPLOYMENT AND ENERGY USE PER EMPLOYEE IN MAJOR SECTORS OF THE AUSTRALIAN ECONOMY, 1971-1972.

	Final end use of all primary and secondary energy		Total Employment (30 June 1971)		Energy/ employee
	10^{15}J	%Total	10^{3}	%Total	10^{12}J
Agriculture forestry fishing	47.0	2.6	386	7.8	0.12
Mining	72.0	4.0	76	1.5	0.95
Manufacturing	817.1	45.8	1216	24.7	0.67
Transport Storage	606.3	34.0	272	5.5	0.45
Other Services	46.4	2.6	2983	60.5	0.02
Domestic	196.6	11.0			
Total	1785.4	100.0	4933	100.0	

Energy use, total employment and energy use per employee by sector in the Australian economy, 1971-72. Sources: energy use — Kalma (1976); employment — Commonwealth Year Book 1973, p. 704.

From these considerations it seems reasonable to conclude that most of the new jobs created over recent decades in Australia have been in relatively labour-intensive service industries which take only a very small share of total energy use. Most of the growth in energy use has been absorbed by primary and secondary industry, where the net job increase has been very low.

In the light of the foregoing analysis it might seem at first sight that the nexus between increasing energy use and increasing employment has been broken, that there is no necessary causal relationship between the two, that the rapid growth of employment in the services sector has been unconnected to the rapid growth in energy use in the economy as a whole, especially in the primary and secondary sectors. EFFE (1977) summarised this argument as follows:

What has kept the 'more energy leads to more jobs' myth alive has been that accompanying a growing population has been a very large increase in the use of goods and services per person. There has also been a significant increase in energy use. It has thus appeared as if energy expansion had been causing economic expansion and increases in jobs. But constantly expanding demand has led to constantly expanding production and employment.

According to this view, then, it is economic growth that's the locomotive pulling along energy growth and employment as carriages behind.

However, this explanation fails to consider a very important connection between expanding primary and secondary sectors, based on increased energy use, and an expanding services sector in which most of the new jobs are generated. To finance a growing tertiary sector obviously requires capital, even though the capital investment per new job in most services is very much lower than in manufacturing or mining, say. Where does the capital come from to provide more and more services?

It can be argued that this capital has been generated by the expanding production in primary and secondary industry, the latter growth depending directly on increased energy use. Some of the wealth generated in the 'productive' sector of the economy will flow on to the Government (via taxes on company profits, for example) and into the private sector, and can then be used to finance services. In addition, an expanding 'productive' sector requires directly — and will thus finance — increasing levels of services in areas such as banking, insurance, communications, transport, building construction and public administration, not to mention other services provided by the state such as health and education.

There can be little doubt that the service sector owes its origins to the expanding surplus wealth generated by primary and secondary industry. Equally, the service sector could not have grown had it not been for the labour 'freed' from direct production by new technology which increased labour productivity. However, it should be pointed out that once a service industry† is set up, its operation creates profits which can lead to capital accumulation by this industry itself. In turn, this capital can be used to finance further expansion and hence job creation in the service sector — i.e. independently of new capital generated in primary or secondary industry. In other words, although service industries cannot be self-starting they can be self-expanding, at least in part, so far as capital investment requirements are concerned.

Furthermore, it is a common mistake (or mystification) to overestimate the extent to which wealth generated by the 'productive' sector flows via Government taxes into community services, as Fitzerald (1974) convincingly demonstrated in his report, *The Contribution of the Mineral Industry to Australian Welfare,* on the 1960's mining boom in Australia. Fitzgerald estimated that, for the period 1967-8 to 1972-3, the Australian Government paid out $55m *more* in the provision of assistance and services to the mineral industry (e.g. through the services of the Bureau of Mineral Resources) than it received from the industry in the form of royalties and tax revenues — i.e. the Australian Government was a net loser in financial terms during our great minerals' boom. Meanwhile, State Governments received a gross $260m income from mining royalties between 1967-8 to 1972-3, though Fitzgerald gives no estimate of State financial assistance to the mineral industry over this period. However, even the total take of $260m is barely one quarter of the $1024m which accrued over the same period to foreign direct shareholders of the principal mining companies involved. In short, mining companies enjoyed remarkably low income-tax rates and royalty payments during the sixties mineral boom. With its recent decision not to impose a resources tax on mineral exploitation, the present Federal Government has indicated clearly its intention not to change this situation. Under these conditions only a very small proportion of earnings from mineral developments find their way into needed community-service expenditure.

In the U.S., Hazel Henderson (1978) has strongly attacked the argument that the private productive sector is the 'Golden Goose' from which all our wealth derives. Rather, she argues, "the dawning view (is) that the public sector is becoming the private sector's 'Milking Cow' "

Today, the Golden Goose model of our economy conceals the extent to which private profits are won by incurring public costs, which are mortgaging our future.
. . . Therefore, it is not surprising that the second paradigm now rapidly emerging is that of the state as the private

*Employment in transport, storage and communication increased by approximately 12,000 people between 1961 and 1971, in contrast to an employment rise of 990,000 in the services sector as a whole over the same period.

†The following argument applies to all service industries which receive payment directly from the consumers of their product. Of course, it does not apply to Government services such as education which are paid for out of state funds since there is no 'profit' involved which can be directed in part or whole to capital accumulation.

sector's Milking Cow, where the politically and economically powerful raid the tax revenues and channel federal, state and local budget funds to enrich themselves, while the path of technological innovation follows producer priorities and capital cities rather than consumer choices and public needs.

She concludes that the Golden Goose is still certainly excreting, but not necessarily laying Golden Eggs!

Yet whether or not there has been a necessary connection between energy growth and more jobs *in the past,* there is no need to keep these two factors coupled in the future, as we shall see in the next part of this paper.

Is Full Employment Possible under Zero Energy Growth?

Generate less energy?
Sure and generate galloping unemployment!

So read one ad put out by a U.S. electricity utility. Is there any truth in it? The Energy Policy Project of the Ford Foundation (1974) considered the connection between energy use and employment in the U.S. in detail and concluded as follows:

There is much talk — and considerable anxiety — about the supposedly close and unbreakable relationship between energy consumption and employment. Both the econometric model (of the U.S. economy) and the analytical work of the Project Staff reveal that such commonly held fears are unfounded. While it is true that a sudden and unexpected energy shortage can cause, and has caused, major unemployment our conclusion is that a long term slowing of energy growth signalled by clear policy commitments, slowly rising (energy) prices, and appropriate compensatory policies could actually increase employment.

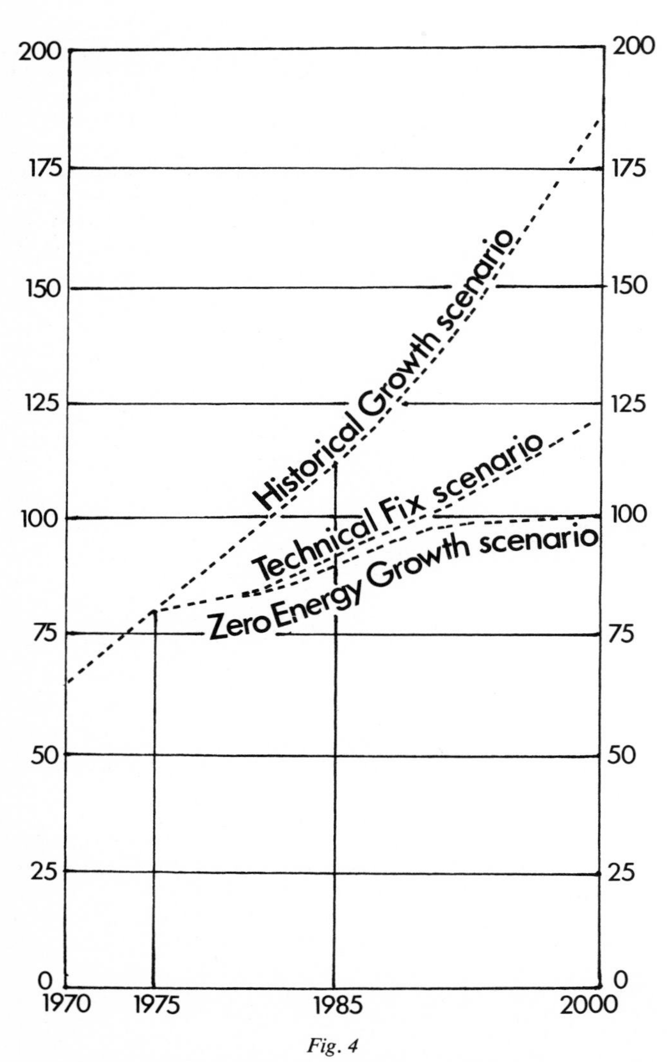

Fig. 4

Primary energy use in the U.S. in the three alternative energy scenarios investigated by the Ford Foundation's (1974) Energy Policy Project.

The Ford Foundation (1974) looked at three possible future energy scenarios for the U.S. up to the year 2000, *historical growth* in which energy use continued to rise according to past trends, *technical fix* in which conservation measures slowed energy growth, and *zero energy growth* in which strong conservation policies and other measures reduced growth in energy use to zero from 1985 onwards (see Figure 4). One of the most important findings of this study was that under the zero energy growth (ZEG) scenario, economic growth could continue at much the same rate as in the higher energy growth scenarios; that is, energy growth and economic growth could be decoupled.

The basic strategy to achieve this decoupling was to use the energy saved by tough conservation measures to fuel continued growth in the economy without increasing total energy use. Suggested energy conservation measures included: better fuel economy for cars; tighter building codes to ensure more energy-efficient housing; expansion of public transport; financial incentives for recycling metals, glass and paper; and more efficient industrial energy use. The Ford Foundation report recommended an energy tax which would have the double-edged effect of discouraging energy consumption and raising money for Government to subsidize energy conservation measures and development of renewable sources of energy.

But importantly, future economic growth under ZEG would be mainly in provision of goods and services with low energy inputs and high labour inputs per dollar value of output. For example, strong economic growth and employment growth areas would be services such as education, health care, day care, cultural activities, urban amenities such as parks; and in industries involved with energy conservation and renewable energy technology.

As can be seen from Figure 5 (Ross and Williams, 1977) the energy content per dollar of product varies considerably across the spectrum of goods and services. (These figures refer to the total energy input to a particular product, calculated by summing all the energy used at each stage in its production, going right back to raw materials.) At one extreme, the energy needed to produce and distribute a dollar's worth of aluminium is about equal to the energy content of a dollar's worth of petrol. At the other extreme, the energy required in the provision of a dollar's worth of medical service is only one-twentyfifth as much as the energy in a dollar's worth of petrol. Generally, following a zero energy growth path would mean a steady shift in the make-up of the economy towards labour-intensive goods and services, i.e. towards the left of Figure 5. The present trend, both in the U.S. and Australia, towards more energy-intensive leisure activities — such as off-road vehicles and caravan touring — would also need to be reversed.

In industry, the aim of energy conservation would be to improve the efficiency with which energy is used, i.e. to produce a dollar's worth of output using less energy input. Ross and Williams (1977) have shown that there has been a steady fall in the energy use per dollar of value-added in US manufacturing industry since 1945 — in short, energy productivity has been rising (See Figure 6). In a ZEG future this trend would have to continue; for instance, the Ford Foundation (1974) claimed that "some energy-intensive industries such as steel could maintain current levels of output with one-third less energy than now used."

The changes in energy productivity in (value added per unit energy input) given for Victorian manufacturing industry in Table 2, however, show clearly that we are shaping up very poorly compared to the U.S. in improving the efficiency of our energy use. Between 1969-70 and 1974-75 energy productivity in Victorian manufacturing actually *fell* by 8.1%. The largest falls were in the transport equipment (-40.5%) and fabricated metal products (-28.9%) divisions, both of which greatly increased their consumption of natural gas over the period considered.

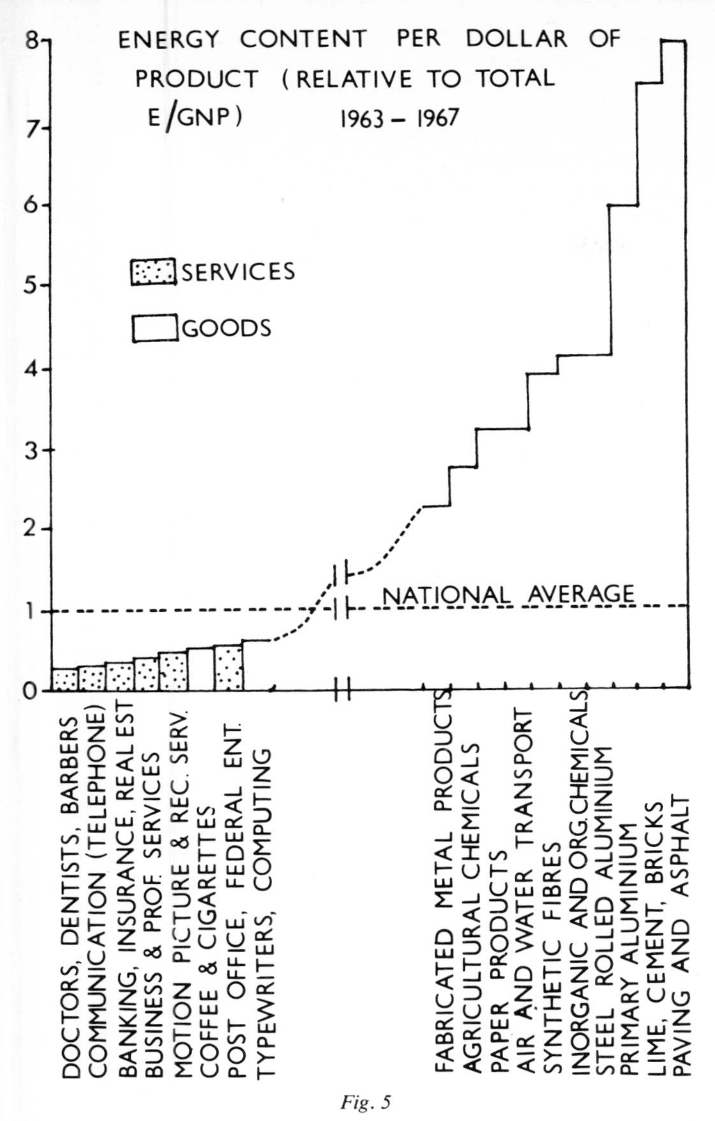

Fig. 5

Total energy consumption (including inputs at every stage of production going right back to raw materials) associated with each dollar of product purchased in the U.S. for various product categories. (1963 — 67 average values.) Source: Ross and Williams, 1977, Figure 11.

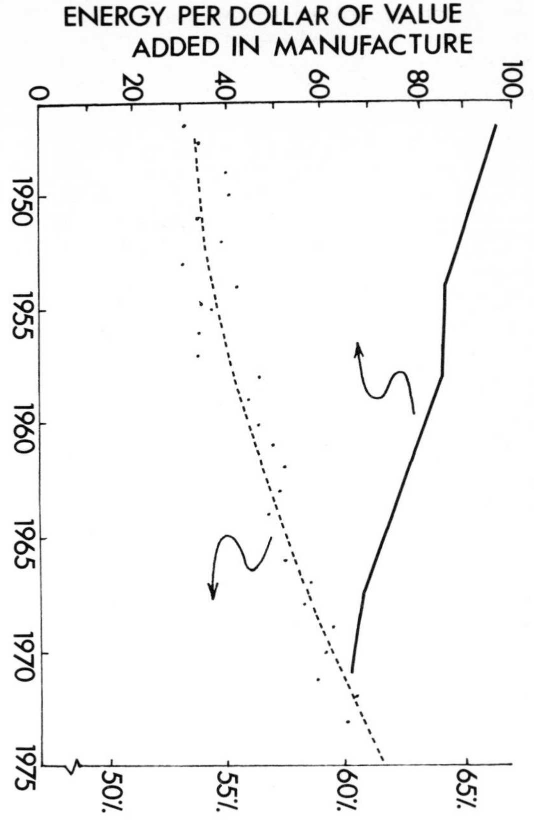

Fig. 6

Energy used per dollar of value-added (adjusted to compensate for inflation) by U.S. manufacturing industry 1950-1975. The steady fall indicates that the energy productivity has been improving, in marked contrast to the situation over recent years in Victorian manufacturing industry where energy productivity has been falling (see Table 2). Source: Ross and Williams, 1977, Figure 1.

Interestingly, the three most energy-intensive industries — chemicals, petroleum and coal products, basic metal products, and non-metallic mineral products — all managed to improve their energy productivities over the period, by 0.3%, 2.2% and 6.5% respectively. In these industries, no doubt energy costs are a significant proportion of total costs so that economies in energy use become important.

Affirmation of the Ford Foundation's findings regarding the viability of a ZEG future has come recently from the already quoted study by Ross and Williams (1977), which was prepared for the Sub-Committee on Energy of the Joint Economic Committee of the U.S. Congress. Ross and Williams show that a slower labour force growth and continuation of the shift towards less energy-intensive goods and services could cut U.S. energy growth from its average level of 4% p.a. between 1960-1973, to less than 2.5% p.a. from 1985 to 2000. They further say that improvement in the efficiency of energy use — especially in the areas of space heating and cooling, water heating, car travel, and cogeneration of steam and electricity at industrial sites — could cut this 2.5% growth rate back to near zero from 1985 onwards. All the while, economic growth would continue.

Ross and Williams also stress the job-creating potential of energy conservation policies:

> *A determined national effort aimed at reducing energy inefficiency would also create many new job opportunities. New businesses and industries would be needed to produce, market, install, maintain, and repair energy conservation technology; new building insulation materials, heat pumps, electronic controls for regulating energy use in buildings, new types of batteries and other local energy storage systems, new coal burning cogeneration services, retrofit equipment for large air-conditioning systems, communications systems that substitute for transportation, and so on.*
>
> *Producing the new equipment needed for this conservation program would constitute a major economic effort — the investment of hundreds of billions of dollars over the next decade or so. But this program would be less costly than developing the corresponding energy supply capacity.*

They conclude, like the Ford Foundation's EPP, that "there are substantial opportunities for uncoupling energy growth from economic growth in the present (U.S.) economy." Even U.S. Energy Secretary, James Schlesinger, now appears to have embraced this new wisdom. He said recently:

> *If we begin to moderate our energy growth rate now — and decouple the growth in energy consumption from growth in G.N.P. — the U.S., the industrialised West and Japan may be able to move through the 1980's with minimal long-term disruption . . . growth rates may slow somewhat, but they should remain healthy and reasonably predictable. (EFFE, 1978).*

In Sweden, where the Government decided in 1975 to cut back energy growth from its then level of 4.5% p.a. to 2% up to 1985, a report on energy strategy by the Secretariat of Future Studies (1977a) concluded ". . . we do not think there are any long-term conflicts between the goal of holding down energy usage and the goal of upholding full employment . . . the energy-thrifty consumption pattern that is based on public services (care sector etc.) is also employment intensive."

In Australia we face an immense unemployment problem over the next decade whatever energy future we move towards. This arises because of the large number of expected entries into the workforce over the coming period and continued immigration. Assuming a net immigration of 50,000 p.a. as presently planned by the Federal Government, Birrell (1977) estimates:

> *. . . that if unemployment is to be reduced to a minimum of 10,000 (i.e. allowing for inevitable 'frictional' movement in the labour force) some 800,000 new jobs would have to be created between 1976 and 1981.*
> *The magnitude of the job creation problem can best be comprehended by considering the fact that between 1971 and 1976 only 427.3 thousand new jobs were created, and two thirds of these were in part-time work; unless there is a great improvement on this we can expect unemployment to be at least around the half million mark by 1981.*

As in the Ford Foundation's ZEG scenario, Birrell also sees the main hope for job creation in Australia as continued expansion in services, but he draws attention to two significant problems in sustaining past rates of employment growth in this sector:

> *Labour-saving innovations in office work, via the exploitation of electronic devices and computers (e.g. word processors), could begin to eat into the potential labour force growth. Perhaps more important, in the rapidly growing government services area, future expansion may be limited by tax payer opposition . . . most of the 300,000 new jobs created between February 1973 and February 1977 were in the Government-funded community services area (health, education, welfare etc.) (Birrell, 1977).*

New electronic information storage, transmitting and processing technology such as microprocessors and word processors is likely to reduce substantially the need for clerical workers, typists and stenographers. It has been estimated, for instance, that in West Germany, which has the largest office population in Western Europe, that 40% of all office work will be automated by 1990 (ABC, 1978). No equivalent figures are yet available for Australia. See, however, Trans National Co-op, (1978). This is an area the Federal Government's Committee of Enquiry into technological change, scheduled to deliver its report towards the end of 1979, will no doubt investigate. We have in fact already witnessed the ramifications on employment of new communications technology to be introduced into the telephone system by Telecom. The conclusion is inescapable that new technology could severely limit growth of job opportunities in services such as banking, insurance, communication, government administration and the wholesale-retail trade over the coming decade.

And this imminent technological change signals a new age in automation in which the historical link described earlier in this paper between increasing labour productivity and increasing energy consumption will be severed. The new microelectronic technology which is the hallmark of the coming revolution in office work is remarkably energy and materials *non*-intensive in both its production and usage. Word processing systems are likely to demand in total less energy then comparable systems based on standard electronic typewriters.

Services such as health, education and welfare, on the other hand, are relatively immune to the job-eliminating tendency of technological change. An improvement in both the quantity and quality of these services almost always depends on increasing staffing levels, for example, increasing the number of teachers to lower the pupil-staff ratio. The problem comes in raising the money necessary to expand these services in the face of what Birrell calls "taxpayer opposition". One way around this problem might be to channel some of the revenue from resource taxes, levied on all non-renewable energy and mineral resources, into more Government-financed community services.

Another possibility which would make some contribution to meeting the 800,000 new jobs by 1986 target is the expansion in industries associated with energy conservation materials and equipment, and renewable energy technology, which implementation of a ZEG future would urgently require. One of the few Australian studies of the job-creating potential of solar energy, by the Total Environment Centre in Sydney, estimated that a solar collector industry to supply NSW's low-temperature heat needs would employ about 10,000 people and have a turnover of about $100 million a year (Nicholls, 1976). An expanding solar industry would further call for growth in the basic metal products, glass and fibre-glass insulation industries, as well as giving a boost to the plumbing industry.

California has taken the lead in actually implementing a high-employment low-energy-growth future. The State's Governor, Jerry Brown, has identified the development of renewable energy sources and the creation of new jobs in energy as the major priority of his administration (Environmentalists for Full Employment, 1978).

California's goal in solar water and space heating is 20% of all residences and commercial buildings by 1985. Fourteen acts to stimulate job creation plus increased solar heat usage have passed the State's legislature this year. The California Public Policy Center (1978) estimates that a solar industry meeting feasible Californian space and water heating needs between 1981 and 1990 could generate over 376,000 jobs over this period, with 36% directly related to solar employment and 64% generated indirectly. Of the direct solar jobs, 21% would be in manufacturing and 57% in installation, which are at present the two areas of greatest structural weakness in the State's economy. This level of job creation in 1977 would have halved California's soaring 7.8% unemployment rate.

If we look to changes that could be implemented in the short-term, it is encouraging to find that practically all of the proposed small changes in the way we use energy — e.g. a marginal shift from cars to public transport — are likely to increase net employment opportunities as well as reduce total primary energy consumption. Hannon (1977) has investigated the labour impacts of a wide range of energy-conserving shifts in the U.S. economy and his estimates of the new jobs generated per 1% saving in total U.S. primary energy use are listed in Table 6. Table 6 also includes a crude transformation of these figures to the Australian economy, assuming the U.S. ratios of net employment change to net energy change apply also to the Australian situation. The final column estimates the likely job creation in Australia resulting from a feasible level of energy saving due to implementation of the conservation measures considered. For example, a 5% shift from car to bicycle for urban travel would save 0.7% of our total primary energy use while creating 3,500 new jobs, although these figures should be regarded as highly speculative.

There is obviously a need for a much more thorough analysis of the employment impacts of energy conservation measures in Australia. The econometric models of the Australian economy being devised by the IMPACT Project of the Industries Assistance Commission would probably be the most appropriate tools for performing the necessary simulations and calculations.

Conclusion

Australia compares very poorly with other developed countries in the amount of research done to investigate and publicise information on the range of alternative energy strategies which we could pursue — each based on different values, different priorities. As yet, we have no studies conducted here which clearly articulate our energy options in the manner that, for example, the Energy Policy Project of the Ford Foundation (1974) has done for the U.S.A., Chapman (1975) has done for Britain, the Science Council of Canada (1976) and Valaskakis (1976) for Canada, and the Secretariat for Future Studies (1977b) for Sweden.

I believe State and Federal Governments in Australia should collaborate in setting up a research team to analyse and articulate our possible range of energy futures and their social,

Table 6

THE EMPLOYMENT OF ENERGY CONSERVATION MEASURES IN THE U.S. AND AUSTRALIA (U.S. Figures from Hannon 1977)

Energy-saving measure — change from:	New jobs created per 1% saving of total energy use in USA	New jobs created per 1% saving of total energy use in Aust. (a)	Actual saving likely in Australia (as % of total primary energy)	This would require in Australia (i.e. to implement previous column) (b)	Jobs created in Australia
Plane to train (intercity)	744,000	25,000	0.22%	10% of air travel shifted to trains	5,500
Throwaway to refillable drink containers	600,000	20,000	0.23%	50% shift to returnable containers for all soft drinks, beer, milk, fruit juice	4,600
Car to train (intercity)	560,000	19,000	0.34%	10% shift from car to train for non urban travel	6,460
Expenditure on new freeway construction to health insurance	512,000	17,000			
Car to bicycle	160,000	5,000	0.7%	5% shift from car to bicycle travel in cities	3,500
Electric to gas cooker	128,000	4,000	0.03%	10% of electric cookers replaced by gas cookers	120
New freeway construction to new public transport construction	24,000	800			
Present to increased home insulation	12,000	400	0.7%	insulate 50% of houses currently uninsulated	280

(a) Assumes U.S. ratios of net employment change to net energy change also apply to Australia and that total Australian primary energy use is 1/30 that in U.S.
(b) Shifts are measured in energy terms, though these would probably correlate closely to units of service (e.g. passenger kms, bottles sold, etc.)

Computer estimates of the net total employment change brought about by implementation of various energy-conserving shifts. The first column gives Hannon (1977)'s figures for the new jobs created per 1% saving in total U.S. primary energy use obtained as a result of implementing the respective conservation measure. Hannon calculated these figures as follows. For each unit of service, for example a passenger-mile, the differences of the direct and indirect energy and employment demands of an activity and its alternative are calculated. To these differences are added the direct and indirect energy and employment caused by the expenditure of any dollar savings on average personal consumption. A ratio of the net employment of the net energy change is then formed and normalised to 1% of total U.S. energy use. The second column in the table attempts to relate these figures, in a very rough way, to Australia, assuming the U.S. ratios of net employment change to net energy change apply also to the Australian economy. The final column estimates the likely job creation in Australia resulting from a feasible level of energy saving due to implementation of the conservation measures considered.

political and economic ramifications. An important part of this investigation would be to analyse the different employment impacts of the energy paths considered. A preliminary report of this research team could be used as the basis for full public debate on which strategy to adopt (Crossley 1979). Possibly an appropriate procedure to enable this public participation would be to set up a Federal Government Enquiry into energy policy in Australia.

In conclusion, it should be made clear that adoption of a low-energy growth future, *by itself,* is not going to lead us back to full employment (i.e. a level of 1-2% frictional unemployment) in Australia over the coming decade. Obviously, the unemployment crisis we face is not simply a function of the way we use energy. Primary causes of unemployment are economic and political factors upon which energy use has some bearing but it is by no means the determining influence. Moreover, the new electronic technology which is likely to have the greatest impact on unemployment levels over the next few years is unlikely to increase our energy consumption. The way back towards full employment must therefore involve political and economic changes, and in addition changes in the way we use energy. However, the arguments and information given in this paper strongly suggest that adoption of a low or zero energy growth strategy in Australia would considerably improve the unemployment situation both in the short term and especially later this century.

Further research is called for to prove this contention in the Australian context. Yet I would be very surprised if our findings here were not similar to the following conclusion of Henderson (1978) for the U.S. economy:

> *. . . our economy has overshot the mark in its substitution of capital for labour. In fact, I contend that in hundreds of production and service processes, labour has now become the more efficient factor of production and as natural resources become increasingly scarce we must employ our human resources more fully.*

References

ABC Radio "Broadband" (1978). August 10, programme on technology and office work.
Australian Bureau of Statistics (Vic. Office) (1974-75a), *Manufacturing Establishments: Details of Operations 1974-75.*
Australian Bureau of Statistics (Vic. Office) (1974-75b), *Manufacturing Establishments: Usage of Electricity and Fuels 1974-75.*
Andrews, J. (1978). *Energy Use, Employment and Economic Growth in Victorian Manufacturing Industry 1969-70 to 1974-75,* Environmentalists for Full Employment (Australia) Working Paper 2.
Birrell, R. (1977). Immigration and Unemployment: The Implications of the Green Paper on Immigration Policies and Australia's Population, *The Australian Quarterly,* December 1977, 36-49.
California Public Policy Center (1978). *Jobs from the Sun,* Los Angeles.
Chapman, P. (1975). *Fuel's Paradise,* Chapter 6 (Penguin, Harmondsworth).
Commoner, B. (1978). *Energy and Labour: Job Implication of Energy Development or Shortage,* address to Conference on Jobs and Environment sponsored by Canadian Labor Congress, Ottawa, February 20, (reprinted by Environmentalists for Full Employment (Aust.), Melbourne).
Crossley, D. J., (1979). Energy Policy and People, in M. Diesendorf (ed.) *Energy and People,* (Society for Social Responsibility in Science, Canberra).
Daly, H.E. (1972). Electric Power, Employment and Economic Growth; A Case Study in Growthmania, in *Toward a Steady State Economy,* Ed. H. E. Daly, (W. H. Freeman, San Francisco).
Department of National Development (1978) *Demand for Primary Fuels: Australia 1976-77 to 1986-87* (Australian Government Publishing Service, Canberra).
Environmentalists for Full Employment (U.S.) (1977) *Jobs and Energy* (EFFE, Washington).
Environmentalists for Full Employment (U.S.) (1978). *Jobs and Energy Update* (EFFE, Washington).
Fitzgerald, T. M. (1974). *The Contribution of the Mineral Industry to Australian Welfare* (Australian Government Publishing Service, Canberra).
Ford, G. W. (1970). Work, in *Australian Society: A Sociological Introduction* 2nd Edition, Eds. A. F. Davies and S. Encel.
Ford Foundation, (1974) Energy Policy Project, *A Time to Choose* (Ballinger, Cambridge, Mass.).
Hannon, B. (1977). Energy, Labour, and the Conserver Society, *Technology Review,* March/April, 47-53.
Henderson, H. (1978). *Creating Alternative Futures: The End of Economics* (Berkley Publishing Co., New York).
Kalma, J.D. (1976). *Sectoral Use of Energy in Australia,* CSIRO Division of Land Use Research, Technical Memorandum 76/4,
Mardon, C. (1978). Conservation of Urban Energy Group, private communication.
Nicholls, J. (1976). *An Economic Case for Solar Energy,* (Total Environment Centre, Sydney).
Ross, M. H. and Williams, R. H. (1977). *Energy and Economic Growth,* study prepared for Sub-committee on Energy of the Joint Economic Committee of the U.S. Congress (U.S. Government Printing Office, Washington).
Science Council of Canada (1976). *Human Goals and Science Policy,* by R. Jackson.
Secretariat for Future Studies (1977a), *Energy in Transition; A report on Energy, Policy and Future Options* (Secretariat for Future Studies, Sweden).
Secretariat for Future Studies (1977b), *Solar Sweden,* (Secretariat for Future Studies, Sweden).
Transnational Co-operative (Communications Project Collective) (1978). *White Collar Technology,* (Transnational Co-operative, Sydney).
Valaskakis, K., (1976). *The Conserver Society: A Blueprint for the Future? (GAMMA, University of Montreal, Montreal).*

Planning Criteria to cope with the Entropy Crisis

Terry L. Lustig

School of Civil Engineering
University of N.S.W.
P.O. Box 1, Kensington, N.S.W., 2033

Introduction

Since Georgescu-Roegen (1) pointed out that "the economic process consists of a continuous transformation of low entropy into high entropy, that is *irrevocable waste*", there has been a growing realisation that energy use (2, 3) and environmental change are related (4), and that both can be understood in terms of a change in entropy production (5, 6). That is, the so-called energy crisis and environmental crisis are two aspects of what might be called an entropy crisis.

Although economic evaluations of projects need not be in terms of money, almost all are. Lately however, most project evaluations have required environmental aspects to be included, and as this cannot be done in terms of money (7) they have tended to be subjectively tacked on to the economic studies. Some of the worst examples of this are the Environmental Impact Statements in the U.S.A., where often the letter of the law is complied with, but not the spirit (8). Even multiobjective planning which tries to avoid this problem (9), is limited by its final stage, in which the economic and environmental goods are traded off against each other. Since economic goods are often considered as "real" or measurable, in comparison with environmental goods which are regarded as subject to non-market evaluation, the trade-off too often favours the economic goods.

There is a need to develop criteria which consider economic processes and environmental effects together in a more objective manner. Without adequate measures of effectiveness, the most thorough environmental study may not affect a decision. For example, consider a question of whether to build a nuclear power plant. Even with the same data, if one person's main objective were to increase economic benefits to the present community, the decision might differ from another's whose main objective was to increase them over the next two centuries, during which time the resulting environmental effects could become more important.

I shall suggest two criteria for evaluating the economic, environmental and energy aspects of a project. Both are based on time and entropy.

2. Entropy

I make no apology for using a technical term like entropy in this paper. If you are going to continue your interest in the problems we are all discussing here, you will hear that word more often.

Although mathematically somewhat complex, the physical significance of entropy is fairly simple. It is often described as a measure of the availability of energy for doing work: hence it is a measure of the *quality* of energy.

For example, a heavy body which is let go (or water in a stream), falls downhill. The energy in the body, when it reaches the bottom, is less than when it started. The difference is not destroyed, merely converted into heat. The first law of thermodynamics says that energy cannot be created or destroyed. However, if we rigged up a suitable device, the energy in the falling body could turn a machine, or even generate electricity (as in hydroelectric power). That is, the

Diesendorf, M. (ed.) (1979). *Energy and People.* Canberra, Society for Social Responsibility in Science (A.C.T.)

energy in the falling body (or water) is available for doing work.

We roughly describe the energy in this state as having low entropy. Once the energy is dissipated as heat, we say it has higher entropy. The second law of thermodynamics says that the entropy of the universe always increases, that is, all processes involve an increase of entropy.

In other words, all changes are in some way irreversible, and many common sayings are virtual restatements of the second law. For example:

You can't unscramble an egg.
It's no use crying over spilt milk.
You're a long time dead.
You can't turn the clock back.
You can't make a silk purse out of a sow's ear.

Till recently, technical efficiencies have been calculated in terms of the first law:

$$\text{efficiency} = \frac{\text{useful work done}}{\text{energy input}}$$

However, Berg (2) has suggested a more useful measure in terms of the second law:

$$\text{efficiency} = \frac{\text{useful work done}}{\text{maximum possible work done}}$$

Table 1 demonstrates that second-law efficiencies can often be of the order of one tenth the first-law values, and can show much more clearly where there is scope for energy savings.

For example, instead of using electrical energy to heat water directly, we could use the energy in a highly efficient heat pump which would supply about 7 times as much heat. Again room airconditioners can be quite inefficient — 5% of theoretical maximum. Because of the poor air distribution system, it may be necessary to heat air to over 50° C to deliver it at 20° C (2).

Table 1
Comparing 1st and 2nd Law Efficiencies (2,3)

Energy Use	1st Law	2nd Law
Room Heating (gas)	0.6	0.03
Electrically Heater Water	0.75	0.015
including power station losses	0.25	0.015
Gas Heated Water	0.5	0.03
Boiler	0.8	0.225
Airconditioning	2 (coefficient of performance)	0.045
Automobile	0.1	0.1
Electric Motor	0.3	0.3

I hope you will now see that the so-called energy crisis is strictly-speaking an entropy crisis. That is, we should be concerned with the quality as well as the quantity of energy supplied (2). But this is not all. Another aspect of the entropy crisis is the environmental crisis.

3. The Environmental Crisis

Bormann (4) has shown that our rapid use of stored energy affects our environment, and that our natural areas must be fostered if we are to reduce adverse impact.

Citing the case of forest ecosystems, he shows that activities such as urbanisation and transportation, harvesting and extraction of natural resources, and regional air pollution, cause stress to forests.

Forests stabilise landscapes, moderate temperature, filter air and water, grow forest products and provide aesthetic and recreational enjoyment. But to the extent that these functions, all powered by solar energy, are no longer carried out, "they must be replaced by extensive and continuing investments of fossil fuel energy, and other natural resources if the quality of life is to be maintained" (4).

Our natural regions not only store energy, they slow the rate of entropy increase (Sec. 3.1). By contrast, adverse environmental impacts speed up entropy production (Sec. 3.2).

3.1. Nature slows entropy production

Consider a forest which is cleared for farming (Fig. 1). While it is there, up to 5% of the incoming solar energy is retained in the forest system, mostly as (low entropy) cellulose. If left alone, the trees containing this cellulose last for decades or centuries. As each tree dies, the wood decays and the stored energy is dissipated in a state of high entropy.

If the land is cleared, solar energy input is reradiated back into space within a day or so, again in high entropy form. Finally, if the cleared land were left alone, a forest like the earlier one could eventually grow back.

It would appear that nature tends to evolve so as to slow down the *rate* of entropy increase of the incoming energy (Appendix 1). In thermodynamic terms, such phenomena are described by Prigogine's Principle of Minimum Entropy Production. Where there is more rapid change (e.g. seasonal variations, egg fertilisation, convection, human invention), it is more correct to use Glansdorff and Prigogine's (10) General Evolution Criterion (equation A3 of Appendix 2).

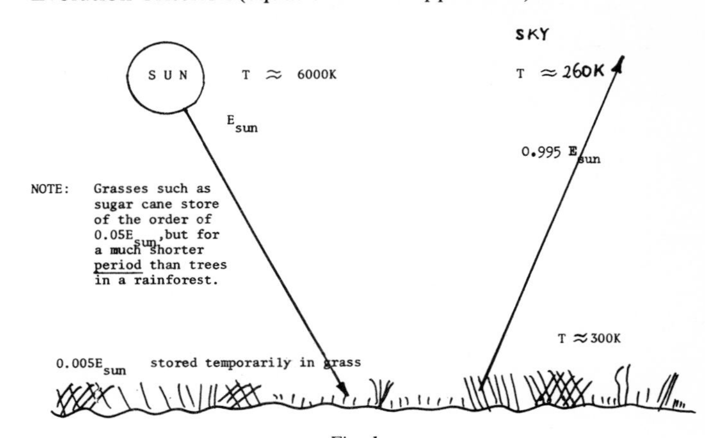

Fig. 1a
Grass slows down entropy increase of 0.5% of the incoming energy. A bare field slows almost nothing.

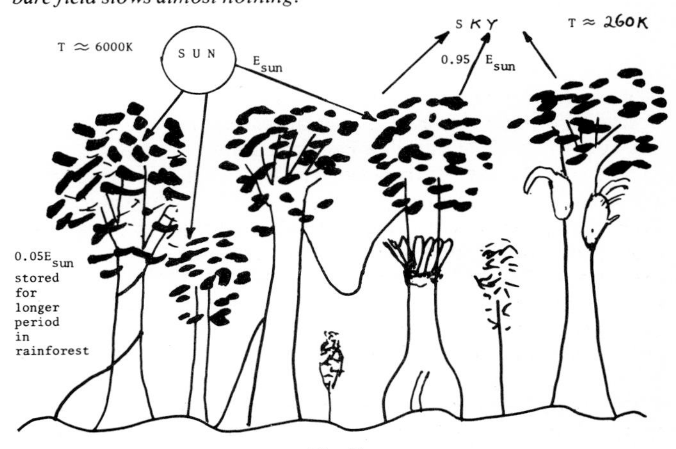

Fig. 1b.
Rainforest slows down entropy increase of 5% of the incoming energy.

Some other environmental examples of this tendency towards a lower entropy production rate are (Appendix 1) *biological growth of an individual, species evolution, river meandering, soil stabilisation.*

3.2. Negative change

I denote the overall tendencies of natural evolutionary processes as "positive change" and equate "negative change" with adverse environmental impact. That is, adverse impact is defined as an effect which runs counter to the long-term trend of natural processes (i.e. ignoring short-term catastrophes).

If, as discussed in Section 3.1, natural processes generally reduce entropy production, measures of environmental impact would then be functions of how much the entropy production increased.

For systems which tend to follow the Minimum Entropy Production Principle, a suitable measure of impact is the change in the time-averaged entropy (Appendix 2) (11). However, where the system had, before disturbance, normally been changing relatively quickly, I suggest, as a measure, the change in the local development δD (Appendix 2) which is a generalisation of the above time-averaged entropy.

4. Economic Evaluation

Attempts at economic evaluation of environmental problems tend to use money as the basic measure of effectiveness (12). However, money is a social convention (13). Man's economy "is submerged in his social relationships" (14), and contrary to Adam Smith's assertion (15), a society's economic system need not depend on the existence of markets (16).

I shall indicate below (Section 4.1) why we should rely less on monetary units for public projects when trying to evaluate their economic worths. In Section 4.2 I shall put forward an alternative, thermodynamic, economic evaluation; and since private projects are much more constrained by market prices, I shall suggest (Section 4.3) how governments could use resource taxation to make the market act to reduce environmental damage and unemployment.

4.1 Prices

Economic value is not identical with price, which is "a parochial reflection of values. (Prices) depend . . . firstly, on whether the objects . . . can be denied to some members of the community . . . (and secondly, on) the fiscal power of public administration. By contrast, (economic) value is a category which can change only with the advance of knowledge" (17).

Nevertheless, many economists advocate making decisions using cost-benefit analysis (which is based on prices) so as to move towards a potential Pareto optimum (18). A Pareto improvement results when at least one person is made better off without making someone else worse off. However, a *potential* Pareto improvement is a situation where net gains *can* be redistributed so that at least one person is better off with no-one being worse off (18).

4.1.1 Pareto optimality

The difference is crucial and leads to a paradox. A situation could exist where an economic change from say, state *A* to state *B* could be regarded as a potential Pareto improvement and *vice-versa* (19). The paradox can sometimes be resolved if the gainers in a change compensate the losers (20). This also raises the question of whether a *prior* redistribution would have resulted in a better improvement. All in all, without redistribution, a potential Pareto improvement may give us no guidance on how to move to Pareto optimality (21).

Mishan (22) discounts the probability of such a contradiction occurring; that is, it is unlikely that after deciding to move from state *A* to state *B*, it would appear desirable to go back to state *A*. He argues that the Paretian changes resulting from any project are relatively small, and that the distributional and price changes will not be significant. However, this would depend on community values also remaining constant, and attendances at conferences such as "Energy and People" must show that attitudes are changing — not least because of past efforts at making potential Pareto improvements, usually as indicated by cost-benefit analyses.

As well as there being difficulties with *potential* Pareto-optimal changes, the Pareto optimal criterion itself ignores income distribution (23). In fact improvements in Paretian efficiency (utility) and distribution (equity) are often incompatible (24). It has been suggested that projects be weighted to show utility and equity effects. The weighting might be deduced from past decisions (24). This approach however, would not allow for changes in values.

Furthermore, the Pareto criterion in use does not normally allow for envy or altruism (25), so such an improvement may not necessarily result in increased happiness. A study of self-rated happiness in two similar Israeli settlements, which differed mainly in their inequality of income distribution, showed that a greater number of people would be less happy than would be happier with a Pareto-optimal change that resulted in greater inequality. The people in the more equal settlement were generally more concerned than those in the other group, that equality should be maintained (26).

There are also difficulties in trying to use Pareto-optimality in a monetary economy, as it then becomes necessary that marginal prices equal marginal costs (27). (Marginal cost is the cost of producing one more unit of the commodity, and marginal price is the price to buy one more unit of the commodity.) Because of imperfect markets, this rarely happens, and marginalism, far from being "the cornerstone of economics" (28), forms "an unnecessary adornment on the apparatus of economic analysis" (29). While marginalist theories seem to provide plausible explanations of economic behaviour, they rely so heavily on simplifying assumptions and preservation of the existing distribution of wealth, that economists "would be no worse off accepting the obviously indispensable Law of Demand on Trust" (29).

Georgescu-Roegen (30), while reinforcing these points, lays the finger on the essential flaw of Pareto's economic theory: it is mechanistic rather than evolutionary, and is not applicable to societies which are changing. People do not merely grope "with given means towards given ends according to given rules" (30) as Pareto claimed. Economics is also concerned with how "new economic means, new economic ends, and new economic relations are created" (30).

To summarise, Pareto-optimality, marginalism and market prices usually provide measures of only the utilitarian aspect of economic reality. But, since public works should be concerned with all aspects of economic worth, we should not measure the benefits and costs only in terms of prices. We need methods which bring out the equity aspect, and the thermodynamic approach suggested below could be one.

4.2 Thermodynamic economic evaluation

As a start to avoiding the problems of prices, marginalism and interest rates (Sec. 4.2.3), and to also allow for environmental changes, the change δD of the local development (Appendix 2) could be used for public projects as a supplement to normal microeconomic evaluations. Project inputs consist of natural resources and labour, and these could be measured in terms of their entropy and time changes. Resource entropy measurements should in principle be relatively straightforward (though not necessarily simple). However, entropy measures of personal time use would depend on the social structure.

4.2.1 Social value of time

We need to know how much a society values time relative to resources use (i.e. entropy change). That is, to maintain the activities of a person for a given period, we want to gain some idea of how much entropy change results. We can denote this as the "social value of time" (SVT) and measure this in units of say *MJ/K. a.* (i.e. megajoules/Kelvin-year) (31).

The essential determinant of this SVT is how much different societies are constrained by space and time. Societies which do not value personal time very highly do not use up much resources. We, on the other hand, consider time very important and are happy to try to save it by greatly increasing entropy.

It seems that SVT should be a function of social structure. The more we depend on outside interdependent sources to satisfy our needs, the less we tend to rely on our neighbours. Interdependent sources are powered by large amounts of low entropy, while equivalent neighbourhood resources and activities mostly need smaller inputs.*

*Some examples are food production, water heating and travel to work. Growing food in one's backyard requires little energy subsidy, whereas food systems of industrialised countries nowadays require energy subsidies of the order of a factor of ten (32). Solar water heating pays for its capital cost in a few years, relative to the cost of centralised energy (32a). Finally, travel to work is often by car. If the time worked to earn the money to acquire and run a car were added to the time to drive and maintain it, the average speed would be less than 10 km/hr (33). If the work is done locally, one can go by foot or bicycle as well as by car. For all modes, the aggregated average speed would not differ much.

Furthermore, our interdependent systems have led to somewhat unstable socio-economic structures. Nowadays our cities can be disrupted by relatively few people — of the order of one hundredth of one per cent of the population.

This trend towards economic interdependence has come about through our excessive reliance on money as a measure of effectiveness. Since the market cannot measure all aspects of our economic activity, a market economy will favour those aspects which can be priced, and discount those which can't. Economically interdependent activities can be more easily priced than neighbourliness, and the logic of the market will tend to promote centralised social structures at the expense of local ones (34).

This in turn encourages our social relationships to become measured in terms of money (35) and much of our national production, particularly unpaid domestic work, is underrated (36).

Till now, most studies of our energy uses have considered only energy obtained from mainly centralised sources, and then categorised them according to the task which they are used for (e.g. cooking, transport etc), and the socio-economic characteristics of the users (e.g. age, income, location).

The corresponding measure of energy use has usually been the per caput annual throughput (incorrectly called the annual per caput energy consumption — energy cannot be consumed). However, this measure reflects the *status quo* and is not very informative about the changes going on in our evolving society.

An alternative measure, the marginal rate of substitution of entropy change for man-hours which I recently suggested (11), is also not very satisfactory, as it too is influenced by the distribution of income. Furthermore, it could fluctuate greatly, given that marginal changes in society are discontinuous rather than continuous.

Finally, while both measures are functions of the social structure, they do not greatly help us understand how that structure influences energy use.

It is suggested that social structures could be better understood if we considered how a society was divided into social units. There would be hierarchies of social units, starting with the unit at the lowest level, the domestic unit. By studying all energy, resource and personal time flows between and within social units, we should gain an appreciation of how centralised or self-sufficient a society was, and how social structures determine the social value of time.

There will be some difficulties in determining what the boundaries of a social unit are, and thermodynamic theory should help us. One criterion for the existence of a social unit would be its stability, for if it were not stable, it would soon cease to exist. The sufficient conditions for thermodynamic stability are (37) (Appendix 3)

$$\delta^2 s < 0$$

and

$$\frac{\partial}{\partial t}(\delta^2 s) \geq 0$$

where s is the entropy per unit mass, and t is the time, δ^2 symbolises the 2nd differential.

That is, by studying the entropy changes of a group of individuals over time, we should be able to determine who belonged to the social unit (stable), who did not (unstable), and who was indeterminate. Other, stronger criteria based on Liapounoff functions (38) would need to be developed to refine this technique, to reduce the area of indeterminacy (39) and thus define the boundaries of the social units more precisely.

As it would take some time to develop these improved Liapounoff functions, the indeterminate regions would have to be subjectively decided on.

Assuming we have somehow defined the boundaries, we must find the SVT for each configuration of social unit. This would be a function of the resource and time flows within and between the different social units. They would depend on the divisions of labour and hence also on the degrees of alienation which may result from these divisions (40). By knowing the SVT for the various social units we should gain an understanding of a society's structure, the relationships within and between the different levels of social units, and the degree of centralisation.

This SVT could be a function of the unit's local development D or time-averaged low entropy storage $\bar{S}$ (Appendix 2), and the resources used (entropy increased) over a given interval. In lay language, we look at how much resources are being used up, and what the social unit is creating in the process. That is, we regard a human as one more animal whose social system forms an important component of the local ecosystem — but whose activities can cause positive or negative change (section 3.2) — and measure its social units as we would a biological individual within its ecosystem.

We could calculate D or $\bar{S}$ for a social unit from the entropy changes caused by the unit's activities — its production and consumption. However, we should not consider production only in terms of work output, nor consumption just as leisure use. There is no definite distinction between work and leisure (41), and a better categorisation of activity appears to be whether on balance it is creative or destructive. (This would also get around the problem (42) of how to value leisure in countries with high unemployment — where leisure is enforced.)

Over time we could find out if D (or $\bar{S}$) was increasing or decreasing, or staying the same,

i.e. if $\frac{dD}{dt} < 0$ (creation)

$\frac{dD}{dt} > 0$ (destruction)

or $\frac{dD}{dt} = 0$ (maintenance, e.g. eating)

The relationships between one social unit and the others is a function of the transport and communication links, which are also entropy flows. (Many of these links result from individuals belonging to more than one social unit, such as the link associated with commuting between one's domestic unit and one's work unit.)

The way social units could be combined to form units at higher levels (such as community units, which are formed mainly from domestic units) would depend on these links. Again, these higher level units could be classified according to their rates of change of local development ($\frac{dD}{dt}$) and their resource usages.

In some cases the problems of determining SVT might be simplified. For example, in assessing the effects of activities such as education and training, we could value the man-hours of training plus the associated teacher man-hours amortised over the period these skills are used. Because the teacher man-hours also require training and we could be led to an infinite regress, we need to calculate these relative values iteratively.

The overall method of determining SVT needs further study. But initially it might be approximated by the average or marginal substitution rate of entropy increase for personal time. These rough values could be varied to check sensitivity for the particular project.

Once we have some idea of the SVT for the different social units, and can combine them to determine the SVT of the region which is affected by the project, we can start the thermodynamic evaluation.

4.2.2 Thermodynamic evaluation

As an example, in an airport study, one would measure the changes in resource entropy (energy and materials) and man-hour entropy associated with passenger and cargo movements, access, operation and construction, neighbourhood disruption, and environmental impacts.

That is, the activities associated with say, passenger movements would be added (in thermodynamic units) as follows:

(a) Transporting passenger from A to B
 — entropy increase (i.e. fuel used etc.)
 — (hours going from A to B) × SVT

(b) Transport between say home and airport
- — entropy increase
- — (time taken) × SVT

(c) Pro rata construction costs for one passenger
- — entropy increase (energy and resources used in construction)
- — (labour i.e. equivalent unskilled man-hours) × SVT

(d) Proportionate share of operation costs
- — entropy increase
- — labour × SVT

(e) Neighbourhood disruption
- — entropy increased by altering the urban layout (e.g. resources used in demolishing and constructing houses)
- — (personal time lost by families needing relocation) × SVT
- — (time lost through access traffic and resulting accidents) × SVT

(f) Environmental impacts
- — time-averaged entropy increase or *D* (cf. Sec 3.2)

These entropy costs could be compared with those of other alternatives as is done with cost-benefit analyses.

4.2.3 Interest rates

The social time preference rate (S.T.P.), which is the ultimate justification of discount rates for public projects, derives basically from the individual's fear of death (43).

However, some economists argue that the State should be "the trustee for unborn generations as well as its present citizens" (44). Furthermore, the Second Law of Thermodynamics tells us that we must prepare not for a steady-state economy but a declining one (45). Since neither interest rates nor market mechanisms can allow for the claims of future generations (46), our economics will have to change so as to encourage us to rely mainly on solar inputs (which future generations will find effectively undiminished) together with amounts of terrestrial resources which deplete the existing stock by only a small amount (47).

Therefore, if we are to consider posterity, public project evaluations must first of all have zero discount rates. If any positive rate is used, it will mean that the interest of generations after the second or third are effectively being ignored.

There is another question (48) — whether we *should* be concerned with posterity.

The opposing argument states that firstly, technological substitution and change will enable future generations to cope with reduced resources (49). This has already been dealt with above: the Second Law of Thermodynamics ensures that eventually we shall run out of available, accessible (low entropy) resources and will *not* be able to cope, if we desire to maintain the present standard of living.

The second argument points out that historically, succeeding generations have often regarded the preceding well-intentioned efforts as wrong, and that men's vices and greed have frequently "done more to advance civilisation, however unintentionally" (49). This does not, however, excuse us from trying to improve things albeit using imperfect knowledge, and we should at the least seek to leave a better world for the next generation than the one we received (50).

I consider therefore that we should have regard for posterity. It is *discounted* by the social time preference which therefore should be ignored. Interest rates could then be dispensed with when evaluating the economic and environmental effects of public projects.

Instead, the influence of time should be worked out using parameters such as the local development and the time-averaged entropy. These are evaluations over time, since their effects are integrated over the project's period of existence (Appendix 2), rather than being discounted relative to one instant of time, the present.

This thermodynamic approach still does encourage benefits coming sooner rather than later. If low entropy is stored early on in the project, it is integrated over the whole period of storage. By contrast, the same amount of low entropy stored someway through the project for say, half the above period, will be deemed half as valuable (cf. Fig. A2.1).

4.3 Private projects

Two items having the same stored low entropy (e.g. bread and cake), require the same resources for production, but may be quite differently valued by people. How should we choose between them? Or was Marie Antoinette right?

If the project is a private one, there seems no alternative to evaluating in terms of current prices. However, these prices would depend partly on the costs of entropy resources. In order that resources were used effectively they could be taxed in proportion to their associated entropy increases. In other words, energy, materials and environmental impacts would be charged for according to how much the entropy increased. This would be the correct basis for a resources tax, and prices could then reflect these charges.

Interest rates would still be charged, as long as money was used for transactions. Nevertheless, as the social value of time reduced, and our society operated more in harmony with the natural environment (Sec. 4.2.1 above), so interest rates would generally fall. That is, time would no longer be so highly valued, and the social time preference rate would decrease.

5. Qualifications

5.1 Indeterminacy of thermodynamic processes

The Entropy Law determines neither *when* nor exactly *what* will happen as entropy increases (1), so these thermodynamic planning criteria cannot predict. A positive change (Sec. 3.2) such as biological evolution can only be evaluated by these criteria after it has occurred.

In the same way, if a species is extinguished, we may destroy something which might have proved to be of use in the future, or which may have been the basis of significant additions to the gene pool. Similarly, an invention, which is also a positive change, may alter the relative efficiencies of the alternatives.

5.2 Ethics

Government actions — public projects and economic regulation — should aim to benefit the community, and their effectiveness could be valued in terms of thermodynamics (Sec. 4.2). However, a proper thermodynamic evaluation needs to regard social systems as components of the local ecosystem, rather than as independent entities. We need to adopt an ecological point of view which acknowledges mankind's "biological roots" and rejects the anthropocentric "heroic Western ethic" (7) which seems to derive from Christianity whose God adopted human shape (51).

In other words, if we are concerned with benefits to people, we must study them as a species within their environment, so that we have to take account of their surroundings in our evaluations.

Some would go further. They suggest that even if a species is of no use to us now or in the future (say smallpox which is now almost extinct) we need to reflect whether it should be allowed to live in its own right (7).

5.3 Implementing an ecological ethic

As such an ethic became adopted (and this would require much educational and political effort) governments could start to manipulate the social value of time.

If the SVT was lowered, employment would increase (replacing machines by labour) and environmental impact should further reduce. At the same time, there would be increasing public emphasis on thermodynamic economic evaluations rather than monetary economic ones.

Nevertheless, it would not be easy to lower the SVT, since the inequality of income distribution would have to be reduced. Richer people tend to use up more energy per head than the less rich, and so have an above-average SVT. Moreover, the greater the inequality, the more the not-so-rich are tempted to seek to emulate their wealthier acquaintances. Since they can become very rich normally only by exploitation, the more unequal the wealth distribution, the more it will tend to reinforce itself (52). Perhaps redistribution along the lines suggested by Stretton (53) may work.

6. Conclusion

Economic evaluations need not be only in terms of money and prices, which signify different values from society to society. Since we must nowadays also consider the environmental as well as the money costs of our projects, it will be useful to use thermodynamic measures which apply to both.

Thermodynamic evaluation techniques would still use economic analysis, but the answers could often be quite different from those using money. A thermodynamic approach would also indicate the way governments could influence market prices to reduce unemployment and environmental damage.

Government projects on the other hand, which are intended to benefit the community as a whole, could be evaluated in terms of market prices to indicate efficiency and at the same time equity could be measured by finding out the impact on people and their environment in terms of the time-averaged entropy change or the change in local development δD.

Finally, these thermodynamic measures help eliminate the perennial problem of what interest rate to use for government projects. Once we recognise the claims of future generations and note that the Second Law of Thermodynamics necessitates a declining economy, we cannot do other than ignore interest rates altogether.

Acknowledgment

I wish to record the help and ideas given by people at this Conference. David Howell, David Gallagher, Geoff Fishburn and Zula Nittim at the University of N.S.W. and Nicholas Georgescu-Roegen also gave useful criticisms.

Appendix 1

Examples of Slowing Rates of Entropy Increase

(see also Appendix 2).

1. Biological growth

After an initial period (following birth etc.) a growing individual's entropy production per unit mass starts to decrease (54) and throughout adulthood (when no energy need be used for growth) asymptotically approaches a minimum.

2. Species evolution

If a new, evolved species leaves more descendants than its ancestors, it has greater fitness in its environment (55) since it uses the resources of its environment more effectively. That is, it "feeds on" less "low entropy" (56) per unit mass. So the unit entropy production is lowered.

3. River meanders

Wilson (57) and Yang (58) have observed that the meandering of perennial rivers conforms with the Minimum Entropy Production Principle. The water as it flows downhill, gains entropy. However, once a river starts to meander, the river water loses potential energy more and more slowly, so that the rate of entropy increase per unit mass approaches a minimum.

4. Soil stabilisation

Since soil stabilisation slows the entropy gain per area of the surface soil particles, the entropy production must approach a minimum.

Appendix 2

Thermodynamic Measures of Environmental and Social Organisation

(Note: Some relations given below are dimensionally different from those given in an earlier paper (Ref. 11) which can lead to inconsistencies.)

A2.1 *Definition of the local development D*

Define D as

$$D = \sum_{i=1}^{n} D_i \tag{A.1}$$

n is the number of individuals in the system

$$D_i = \frac{\int_0^{t_i^*} \int_0^{t} \Phi_i \, dt \, dt}{t_i^2} \tag{A.2}$$

Here t is the time, t_i^* is the longest time that any of the material comprising the individual i exists (i.e. the maximum time between initial growth and final decay), $\Phi_i(t)$ is the Glansdorff-Prigogine (10) local potential defined by

$$\frac{d\Phi_i}{dt} = \sum_j J_j \cdot \frac{dX_j}{dt} \leq 0 \tag{A.3}$$

J_j is the jth thermodynamic flow (e.g. energy, matter, chemical changes), and X_j is the corresponding "force" or gradient (e.g. functions of temperature, or gravitational or chemical potential).

Equation (A.3) is the General Evolution Criterion: for constant boundary conditions the local potential approaches a minimum value, the local reference state, where the system is steady and locally stable.

If however, the boundary conditions are not fixed, the system may evolve to a new reference state. Each such state may have different kinds of physical and chemical inhomogeneities which produce a "dissipative structure" (10). A smoke ring, for example, has its own stable structure which stores kinetic energy until it is dissipated.

The local potential might be understood in lay terms by interpreting the differential $d\Phi$ as the increase in low entropy (i.e. extra high quality energy or resource) inputs which are required to keep the energy flow rates the same as for the reference state, after the system has been disturbed from that state. Environmental examples of such resources are fertilizers to maintain nutrients in farm land, water treatment for eutrophied waterways, and air-conditioning to compensate for microclimatic change caused by local tree felling.

A2.2 *Time-averaged entropy*

When an entity or individual is not evolving, the local development D can be simplified to the time-averaged entropy — the average entropy from the start of growth to the end of decay (Fig. A2.1).

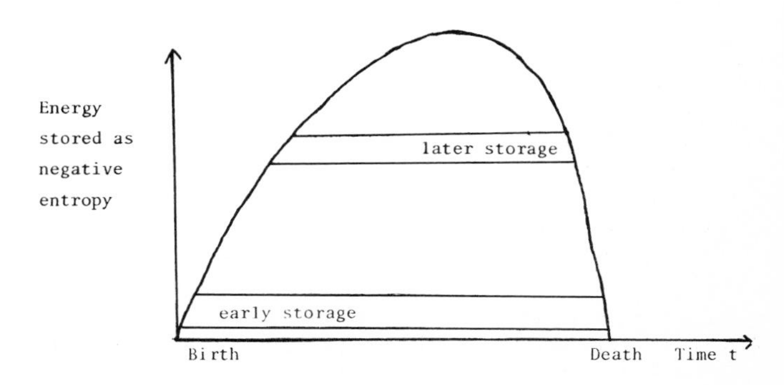

Fig. A2.1
Energy stored by the entity or individual over its period of existence.

As an example, consider the life of an individual. At around birth, organic differentiation ceases — Prigogine's Principle of Minimum Entropy Production then holds (as in Appendix 1) and the entropy production per unit weight of the individual decreases until death (57).

This principle can be written as (10)

$$\frac{d^2S}{dt^2} \leq 0 \qquad \text{(A.4)}$$

$$\text{i.e.} \quad \lim_{t \to \infty} \frac{dS}{dt} = minimum \qquad \text{(A.5)}$$

S is the entropy, and t is the time.

If the stored (low entropy is plotted against time (Fig. A2.1), we can characterise the individual by its average storage $\bar{S}$, *say, i.e.*

$$S = \frac{\int_0^{t^*} S dt}{t^*} \qquad \text{(A.6)}$$

That is, S equals the area under the curve, divided by the period of existence.

Appendix 3

Stability in the Sense of Liapounoff (Lyapunov) (38,39)

If the system under consideration behaves according to the differential equations

$$\dot{\underset{\sim}{x}} = f(\underset{\sim}{x}, t) \qquad \text{(A3.1)}$$

where $\underset{\sim}{x}$ is the set of position co-ordinates, $\dot{\underset{\sim}{x}}$ the time differential, and t is the time, then let $L(\underset{\sim}{x}, t)$ be a Liapounoff function.

If $L(\underset{\sim}{x}, t) > 0$ (A3.2)

and $\dot{L}(\underset{\sim}{x}, t) \leq 0$ (A3.3)

then the system described by (A3.1) is stable in the sense of Liapounoff. If instead of (A3.3) we have $\dot{L}(\underset{\sim}{x}, t) < 0$ (A3.4)

then the system is asymptotically stable.

If however,

$L(\underset{\sim}{x}, t) < 0$ (A3.5)

$\dot{L}(\underset{\sim}{x}, t) < 0$ (A3.6)

the system is unstable.

Finally, if with $L(x, t) < 0$

either $\dot{L}(\underset{\sim}{x}, t) \leq 0$ (A3.7)

or $\dot{L}(\underset{\sim}{x}, t) \gtreqless 0$ (A3.8)

the stability is indeterminate.

All the above relations hold if L changes sign.

References and Notes

(1) Georgescu-Roegen, N. (1966). *Analytical Economics: Issues and Problems.* Harvard U.P.; (1971) *The Entropy Law and the Economic Process.* Harvard U.P.

(2) Berg, C. A. (1974). A Technical Basis for Energy Conservation. *Technol. Rev.,* Feb., 15-23.

(3) American Institute of Physics (1975). Efficient Use of Energy. AIP Conf. Proc. No. 25.; summarised in *Physics Today* (1975) Aug., 23-33.

(4) Bormann, F. H. (1976). An Inseparable Linkage: Conservation of Natural Ecosystems and the Conservation of Fossil Energy. *Bioscience* **26**, 12, 754-60.

(5) Georgescu-Roegen, N. (1975). Energy and Economic Myths. *The Ecologist,* 164-74, 242-52; (1977) *Energy and Economic Myths.* Pergamon.

(6) Commoner, B. (1976). *The Poverty of Power.* Jonathan Cape.

(7) Ehrenfeld, D. W. (1976). The Conservation of Non-Resources. *Am. Scientist,* **64**, 648-56.

(8) McIntyre, I. School of Civil Engineering, University of N.S.W. (1977) pers. comm.

(9) de Neufville, R. and Marks, D. H. (eds) (1974). *Systems Planning and Design,* Prentice-Hall, N. J. Chaps. 20-22.

(10) Glansdorff, P. and Prigogine, I. (1964). On a General Evolution Criterion in Macroscopic Physics. *Physica,* **30**, 351-74; (1971). *Thermodynamics of Structure, Stability and Fluctuations.* Wiley-Interscience.

(11) Lustig, T. L. (1978). *Project Evaluation with Thermodynamics?* Conference on Environmental Engineering, Sydney. Inst. of Eng., Aust., Nat. Conf. publ. No. 78/5.

(12) Dorfman, R. and Dorfman, N. S. (1972) (Ed.). *Economics of the Environment.* W. W. Norton, for example.

(13) Samuelson, P. A., Hancock, K. and Wallace, R. (1970). *Economics.* Australian Edition, McGraw-Hill; Aristotle, *Nichomachean Ethics,* 1113a.

(14) Polanyi, K. (1944). *The Great Transformation.* Beacon, p. 46.

(15) Smith, A. (1826). *Wealth of Nations.* Reprinted in Newman, P. C. et al (1954). *Source Readings in Economic Thought,* Norton, p. 126.

(16) Polanyi, K. op. cit. p. 43 (ref. 14).

(17) Georgescu-Roegen (1971). op. cit. p. 287 (ref. 1).

(18) Mishan, E. J. (1971). *Cost-Benefit Analysis.* George Allen and Unwin, p. 159.

(19) Scitovsky, T. (1941). A Note on Welfare Propositions in Economics. *Review of Economic Studies,* Nov.

(20) Little, I. M. D. (1957). *A Critique of Welfare Economics.* Oxford, p. 98.

(21) *Ibid,* p. 100 ff.

(22) Mishan (1971). op. cit., p. 321 (ref. 18).

(23) Sen, A. (1973). *On Economic Inequality.* Clarendon, p. 6 ff; Little, op. cit. Chap. 6.

(24) Weisbrod, B. A. (1968). Income Redistribution Effects and Benefit-cost Analysis, in S. B. Chase (Ed.). *Problems in Public Expenditure Analysis.* The Brookings Inst. p. 177-208; reprinted in Layard (1972) *Cost-Benefit Analysis.* Penguin; Mishan (1971). op. cit. p. 311 ff (ref. 18).

(25) Mishan (1971). op. cit. p. 312, but see p. 311.

(26) Morawetz, D. et al. (1977). Income Distribution and Self-Rated Happiness: Some Empirical Evidence. *The Econ. Journ.* **87**, 511-22.

(27) Bōhm, P. (1973). *Social Efficiency: A Concise Introduction to Welfare Economics.* Macmillan.

(28) Ruff, L. E. (1970). The Economic Common Sense of Pollution. *The Public Interest,* No. 19, 69-85, Spring; Reprinted in Ref. 12.

(29) Mishan, E. J. (1961). Theories of Consumer's Behaviour: A Cynical View. *Economica,* **28**, 1-11, Feb.; (1967). *The Costs of Economic Growth.* Staples Press. p. 45 ff.

(30) Georgescu-Roegen (1971). op. cit. (Ref. 1), p. 318-20, 343.

(31) A similar idea, but which uses money as an intermediary, is put forward by Hannon, B. M. (1975). Energy, Growth and Altruism. Limits to Growth '75 Conference, Houston.

(32) Odum, E. P. (1975). *Ecology.* Holt, Rinehart and Winston, p. 209; compare also the intermediate case of Hong Kong, Newcombe, K. and Bowman, K. (1978). Intensive Animal Production at the Crossroads: The Potential of the Urban Fringe. *Search,* **9**, 10, 351-9.

(32a) But see Georgescu-Roegen, N. (1978). Technology assessment: the case of the direct use of solar energy, *Atlantic Econ. J.,* Dec.

(33) Illich, I. (1974). *Energy and Equity.* Calder and Boyars, Lond. p. 31.

(34) Stretton, H. (1976). *Capitalism, Socialism and the Environment.* Cambridge U.P. p. 197; Apps P. (1975). Child Care Policy in the Production-Consumption Economy. Victorian Council of Social Services. Chap. 1.

(35) Marx, K. (1964). *Economic and Philosophic Manuscripts of 1844.* International, N.Y., p. 148 ff; Polanyi, op. cit. (Ref. 14) Chap. 4.

(36) Stretton, op. cit. pp. 184, 188-9, 196 ff.

(37) Glansdorff and Prigogine (1971). op. cit. (Ref. 10).

(38) Hahn, W. (1967). *Stability of Motion.* Springer; Pars, L.A. (1965) *A Treatise on Analytical Dynamics.* Heinemann, Lond.; La Salle, J. and Lefschetz, S. (1961). *Stability by Liapounov's Direct Method.* Academic.

(39) de Sobrino, L. (1975). The Glansdorff-Prigogine Thermodynamic Stability Criterion in the Light of Lyapunov's Theory. *J. Theor. Biol.* **54**, 323-33.

(40) Marx, K. and Engels, F. (1970). *The German Ideology.* International Publishers, N.Y., pp. 19, 53-56.

(41) Young, M. and Willmott, P. (1973). *The Symmetrical Family: A Study of Work and Leisure in the London Region.* Routledge and Kegan Paul. Chap. 8.

(42) Georgescu-Roegen (1967, 1971). op. cit. (Ref. 1).

(43) Feldstein, M. S. (1964). The Social Time Preference Discount Rate in Cost-Benefit Analysis. *Econ. J.,* **74**, 360-79; Reprinted in Layard, op. cit. (Ref. 24), p. 252-3.

(44) Pigou, A. C. (1920). *Economics of Welfare.* Macmillan.

(45) Georgescu-Roegen (1975). op. cit. (Ref. 5), p. 245.

(46) Ibid, p. 249-50.

(47) Ibid, p. 251.

(48) Passmore, J. (1974). *Man's Responsibility for Nature.* Duckworth. Ch. 4.

(49) Ibid, p. 81-4.

(50) Ibid, p. 86-91.

(51) Ibid, p. 12.

(52) Stretton, op. cit. (Ref. 37), p. 269 ff.

(53) Ibid, Chap. 10.

(54) Hiernaux, J. and Babloyantz, A. (1976). Dissipation in Embryogenesis, *J. Non-Equilibrium Thermodynamics, 1,* 33-41; Nicolis, G. and Prigogine, I. (1977). *Self-Organisation in Non-Equilibrium Systems: From Dissipative Structures to Order through Fluctuations.* Wiley-interscience; Odum, H. T. (1971). *Environment, Power and Society.* Wiley-Interscience, p. 75; Zeuthen, K. E. (1953) Oxygen Uptake and Body Size in Organisms. *Q. Rev. Biol.,* **28**, 1-12; Zotin, A. I. and Zotina, R. S. (1967). Thermodynamic Aspects of Developmental Biology, *J. Theoret. Biol.,* **17**, 57-75; Trintscher, K. S. (1965). *Biology and Information: Elements of Biological Thermodynamics.* Consultants Bureau, N.Y.

(55) Williams, M. B. (1970). Deducing the Consequences of Evolution: A Mathematical Model. *J. Theoret. Biol.,* **29**, 343-85; Peters, R. H. (1976). Tautology in Evolution and Ecology. *Am. Nat.,* **110**, 971, 1-12.

(56) Schrodinger, E. (1944). *What is Life?* Cambridge U.P.

(57) Wilson, K. C. (1956). Derivation of the Regime Equations from Relationships for Pressurized Flow by the Use of the Principle of Minimum Energy-Degradation Rate. *Civil Eng. Res. Rep.* **51**, Queen's University, Ontario.

(58) Yang, C. T. (1971). On River Meanders. *J. Hydrology,* **13**, 31-53.

A Two-Part Tariff to encourage Energy Conservation

Gerhard Weissmann

c/- PO Hahndorf, South Australia 5245

Mr Roger Morse of CSIRO, Melbourne, said in Adelaide recently that Australia's energy demand growth was 5.4% in 1972, 2% forecast for 2000 and expected to reach zero growth in 2025. He did not mention the mechanism by which this contraction in growth is to be induced.

At the same symposium representatives of the S.A. Gas-Company, the Electricity Trust and others agreed that conservation is necessary, but the main thrust was directed at 'Energy Management', to get more out of resources to increase profits. Mr Tostevin, a Consulting Engineer, extolled industry to optimise energy use in order to minimise costs.

The fact that optimization is deemed possible, implies that there exists an elastic system which is amenable to feedback control. This feedback loop is therefore held to offer the possibility to induce optimization by loading the cost of central supplies of energy and other services.

Two counter-acting factors can thus be recognised: on the one hand the *demand for central supplies of energy* and other services and on the other hand the *need for conservation* in these areas, which has been established beyond doubt in minds not closed by dogmatic profit motives.

Now, as long as the monetary system still functions, the proposal is to use the price-elasticity to influence the demand in the direction desired. Highgrade energy can either be supplied in practically any degree of refinement from central generating plants, using non-renewable fuels, or, with some lesser but more appropriate degree of refinement, decentralised and by individual effort from sun, wind etc. in various ways: solar heaters, wind-wheels, vegetable or fuel growing, fish-ponds, methane-producers or simply use what is there, such as the warm air under the roofs of houses to heat the living areas.

Conservation of centrally supplied energy should therefore by way of its price structure offer the incentive for people to make individual efforts to supply their own needs where possible, matched in refinement to the use.

Specifically in energy supply there are two parts:

(a) The level of *Demand,* i.e. at which rate of flow energy is expected, meaning whether energy is expected to be supplied 'in large lumps' and how large these are to be.

(b) The total quantity of *Energy* used in a given time, that is the lake filled by the trickle or stream or the amount of fuel used.

In the case of electricity the level of demand is known as the *power,* is measured in Kilowatts (kW) and is determined by the size of load connected to the supply: a 2 kW heater or cooker or air-conditioner needs to be matched with a generator and a supply system capable of providing and carrying this size of flow.

The energy consumed is measured in Kilowatt-hours (kWh): if the heater were 'on' for one hour it would use 2 Kilowatt-hours (kWh) of electric energy. Coal to the equivalent of 2 kWh plus conversion and transmission losses, which in the case of electricity amount to another 4 kWh, would have to be burnt in the powerhouse.

To conserve energy and limit further fouling of the environment, both demand and energy would have to be influenced: the time a heater is 'on' should be shortened to what is physically and financially tolerable, but also it has to be made clear that where a 1 kW heater or a ½ kW element would suffice, then this should be used to save in the aggregate the building of another powerstation and transmission lines. Conservation should therefore be encouraged by a price structure effective in both ways, hence having two parts:

Part D. Demand

A 'hire'-charge for the size of system required to meet the demand. To make the hire charge obvious, it should be shown as a separate entry on any account to bring to the user's notice that he is contributing to the perceived 'need' for a new powerstation or refinery or other central supply installation and associated distribution system.

Part E. Energy

This is the charge for energy or fuel or services used during the accounting interval, month, quarter etc., and should be level from the first unit used to the last. Its price should particularly not vary with time of the day or the use it is put to, because if it did, it would be a quantity discount or subsidy to those users who can avail themselves of the facility.

This type of Two-Part Tariff should be applied wherever something is supplied by publicly owned systems. Its application ranges over such services as watersupply, transport generally, post, electricity, gas, local council services and many others and could usefully be employed to influence the preferences of people away from socially costly systems towards more benign and environmentally softer solutions people may think of themselves.

The price levels of both the D and the E parts could be so fixed as to reflect the primary input required for service relative to alternatives as well as the cost for individual self-help, substitution or doing-without. In this way the demand for a kW of electricity should cost three times that for an equivalent demand of gas, which in turn should be a multiple of the same demand met in coal (where appropriate). This then would bring, say, solar heaters into favour by the individual as soon as the price of a kW of demand in electricity is higher than that of a kW equivalent of solar collector. The latter then would present a financially as well as energetically more attractive alternative with the added advantage of increased employment thrown in.

The social consequences of imminent innovations in electronics, such as the Viewdata system to transmit information over the television screen and replace daily newspapers, could be reflected in including in the demand charge for such services the cost of disruption to the present systems of employment and give proper weight to these consequences in their price structure. Another aspect is contained in the choice given to the consumer in selecting or rejecting levels of demand by the balance of price vs. service offered. Hopefully, some people may refuse to subscribe to 'improved' service if this means increased unemployment.

Diesendorf, M. (ed.) (1979). *Energy and People*. Canberra, Society for Social Responsibility in Science (A.C.T.)

Since the first repercussions of the misinterpretation of Keynes' economics are now coming into the open with the recently ended Telecom dispute and the rumblings in the banking world, the introduction of the principle of the two-part tariff should be a matter of urgency in order to get the awareness of social cost indoctrinated before there will be more serious disruptions. To planners in public utilities the elements of Security of Supply, the continued Occupation of Employees, Environmental Disruption etc. have been unquantifiable entries in the evaluation of projects because there was no way in which these could be reflected in the charge for energy or services alone, since the benefits accrued to a different section of the community from those who bore the costs. The present system in effect often penalises those who limit their own demand on principle, by making a minimum charge or else forcing them to resort to extraordinary lengths or capital expenditures to escape 'the system'.

To illustrate the workings of the proposed two-part tariff some examples:

Telephone

Here, a two-part tariff already exists with 'rental' paid regularly, yet not diversified where it should be: for special services available to the subscriber. Telephones with STD or ISTD facility cost the same as those connected to a mechanical or even manual exchange, whereas the provision of these higher service levels should attract higher rentals than telephones without them. On the top of the list of charges should be the computer data-link service for which level the present rental seems to be levied anyway. If a choice or trade-off between level of service and demand charge were possible, then its availability might have reduced the pressure for introduction of computer exchange service and the present dilemma of jobs lost to technology might be solved by people's preference for a lower priced, low speed human exchange. If individual lines might not be treated differently, then the majority of subscribers on an exchange might determine its character.

Water

Demand is given by the size of supply orifice: a house with a 6 mm diameter connection makes one quarter the demand of a house with a 12 mm diameter orifice and the latter should bear a rental four times higher. All water consumed would cost the same per litre from the first to the last.

Gas

Central gas supply may be treated in the same way as water, the demand level being determined by the supply aperture.

Electricity

Industrial maximum demand tariffs exist, but often with decreasing slope, and their application is an exception rather than the rule, while their application universally would be quite simple and less involved than the present system of double metering with time-clocks and central switching control: the demand level in kW to be selected by the customer and charged for, can be controlled by an automatic overload switch which trips when demand exceeds the agreed level. This would have the effect of inducing people to stagger the use of appliances to fit into as low a demand limit as possible and use either cooker or water-heater or radiator, but not all simultaneously and it would make people look for lower rated appliances to avoid excess demand. In industry it would hopefully have the effect that replacement of people by machines would be halted if every additional kW installed and replacing workers would cost as much as the same kW supplied manually.

Another aspect which could be reflected in the demand charge would be the level of security of supply, as already mentioned. Supply via a single line protected by a fuse will be out of service for as long as it takes to replace the fuse. This forms the lowest level of security. The next higher level would be some automatic reclosing device, with the highest level provided by central supervision at the highest level of technology and via multiple paths and stand-by arrangements, the principle being that the shorter and less frequent the outages the higher the demand charge. Logically similar treatment could be given to environmentally sensitive installations necessitated by consumers demanding certain levels of supply in a specified locality which makes undesirable demands on the system, the extra costs of meeting it to be reflected in the demand charge.

The price of energy consumed should be level from the first to the last kWh, since the amount of fuel used to produce it does not vary with time of day or kind of use it is put to. In particular modern generators are capable of daily load cycling and in a situation of fuel scarcity there is no point in transferring the time incidence of consumption from one hour of the day to another.

If one wanted to accommodate some low-demand consumers of the 'pensioner'category, it might be possible to start the demand charge at some small level greater than zero to permit the use of basic comfort such as lights and radio or television without prejudice. Generally, there is no reason why the demand charge should be uniform geographically or over the range of demand. It would seem quite logical to reflect all sorts of social costs in its level and it would seem to be the ideal medium to bring that cost to the notice of the customer in a cause and effect statement.

Public and general transport

The type of vehicle or craft used is already sometimes reflected in the price charged, but more consistent application of the principle would be desirable. For instance the Hovercraft ferry on Sydney harbour attracts a higher fare than the Manly Ferry, but a 707 service costs the same as a 747 flight, where it should be more. The alternative of bus or railway does attract a lower rate, but relatively the plane trip should cost ten to twenty times the surface trip if it were priced correctly on a Joule per kg.m effort expended by the vehicle or craft. For low cost transport and where suitable, sea-transport may again come into its own with the result of renewed employment in shipbuilding and allied services.

Summary

People expect to be supplied with *high-grade energy* via a *central supply* system.

The *energy* derived generally from fossil fuel is becoming scarcer and its consumption leads to environmental fouling.

The *provision of central supply systems* is often of an excessively high standard without giving the user any chance of opting for a less demanding level or softer alternative.

Both factors should be conserved. As long as money is still scarce, the price structure could be used to do this, by charging separately for the two components:

a 'Hire-' or 'Demand-Charge' for requiring the system

a 'Use-Charge' proportional to the quantity consumed.

The Hire-Charge may also be made to reflect the degree of refinement of the energy or service supplied, opportunities for employment provided or lost and the social and amenity costs involved.

The proposed Two-Part Tariff could ultimately be the closing link in the planning loop providing the planner with information on the lowest level of service tolerated by the consumer.

Money, Energy and the Future

D. G. Evans and A. S. Atkins

Centre for Environmental Studies, University of Melbourne
Parkville 3052, Victoria.

The Economic System

Because of the obvious power of money to influence our personal lives we often tend to view the activities of nations in monetary terms, and in so doing we forget that money is merely a convenient device for arranging the flows of raw materials, services and finished products which make up the patterns of economic activity. However this feeling is so strong that the very term *economic* has come to mean something that can be expressed in money terms. A better view of economics, of course, is that it is concerned with the allocation of scarce resources between virtually limitless competing uses, which focuses attention more properly on the physical quantities of resources rather than the money paid for them.

But how well does the economic system perform this allocation task? Is it likely to keep working adequately into the foreseeable future, or should we look for radical modifications of it? The present system of western capitalism has evolved over a long period of technological, social and political change. Adam Smith's concept of *laissez-faire* (1) (i.e. the idea that the owners of resources and means of production should be free to manufacture and trade in pursuit of their personal gains, and that this is the best way to ensure the welfare and economic progress of the nation as a whole) is still prominent. However it apparently leads to an unstable economy (cycles of over-investment followed by recession) and the great world-wide depression of the 1930's finally led to the abandonment of the belief that there was a self-adjusting balance between supply and demand, and persuaded governments that stable operation of their economies required some government intervention. The twenty-five year period 1945-1970 has been characterised by relatively strong government action which has ensured unprecedented stability, with close-to-full employment accompanied by moderate but fairly steady inflation rates of a few percent per year (see Figure 1).

However since about 1970 western capitalism has been undergoing a period of high inflation accompanied by high unemployment, and currently accepted economic doctrines and relationships are therefore under further challenge, perhaps of a more fundamental kind than hitherto. It appears that the community now identifies social progress and equity as being as important as economic progress, and that one of the major tasks of the economic system in the future will be to ensure social and political stability, with due recognition of *international* and *intergenerational* equity.

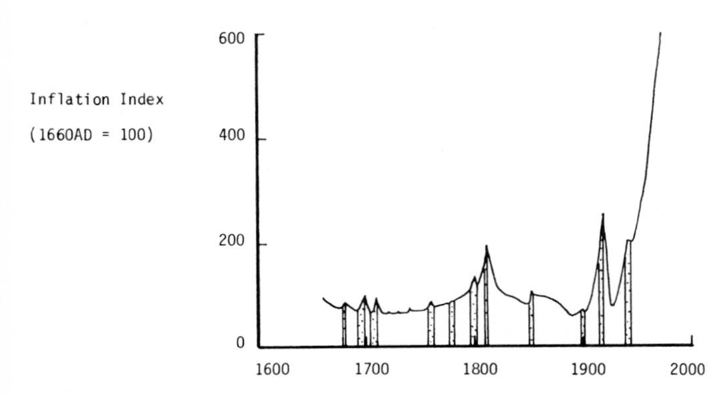

Figure 1: History of world inflation (2). Periods of war are shaded.

Resources

A nation has certain natural resources such as climate, land, minerals and people, and certain resources it has built up over past generations into a capital stock which includes institutional structures such as political and social organisations, physical structures such as cities, roads, bridges, factories and farms, and a system of educational and professional skills (3). If this capital stock is not maintained by a continuous fresh input of natural resources it will degenerate and eventually disappear. Most natural resources (e.g. land, water and air) have developed over a long period in such a way that they are continually regenerated by natural systems powered by solar energy. However some natural resources (such as minerals), once used, can only be re-generated by reworking processes involving the use of energy resources, normally the particular mineral resources we call fossil fuels, and the fossil fuel resources themselves cannot be regenerated (except in a time scale of millions of years).

It follows that an important national goal should be to make the optimum use of the nation's natural resources in order to confer the greatest benefit on the nation as a whole, including future generations. This will involve, amongst other things, nurturing and increasing the nation's capital stock of resources, as it is evident that only poor use can be made of natural resources without input from such capital resources (take, for example, South America, rich in natural resources, yet one of the poorest continents economically and socially). It will certainly call for full employment and financial stability, and, if the benefit is to continue into the future, the lowest possible consumption of natural resources consistent with these other aims.

The economic system should be such as to facilitate these aims, not frustrate them. However during the Great Depression of the 1930's the operation of the economic system was such that resources of labour and machinery were idle; in some nations food was overproduced (in terms of the money available to buy it) and was physically dumped to avoid depression of prices. This Depression was ended not by economic management but by the impetus to production given by World War II, which however did demonstrate the correctness of Keynes' analysis of the economic system of his day (4). As already noted, since that time governments have increasingly adopted Keynesian methods of economic management and extensions of his ideas by his followers to cope with non-depression situations. This has resulted in good use being made of the existing resources of capital and labour, and, as we shall see shortly, an expansion in the economies of

Diesendorf, M. (ed.) (1979). *Energy and People*. Canberra, Society for Social Responsibility in Science (A.C.T.)

most countries (this expansion was assisted, no doubt, by advances in technology also catalysed by the war).

However during the 1960's it became obvious that the continued growth of material production and the rise in capital stock were being achieved at the expense of rapid depletion of natural resources. The environmental movement identified the destruction of natural ecosystems and the loss of air and water quality, sometimes irreversible, and more recently the so-called energy crisis has emphasised for us the ever-quickening depletion of the great mineral resources of the world. The pricing and allocation mechanisms of the conventional economic system do not appear to be able to deal adequately with these special problems, because of the short time-scale of economic decision making and because such resources are external to human endeavours as measured by the monetary system; nevertheless when they are gone some of these endeavours will no longer be possible (5).

The thesis worked out in the rest of this paper is that the economics of the future must take cognizance of this whole question of resource depletion, and that since resources are really stores of high-grade energy in particular forms, this will necessitate an examination of the energy requirements of all human activities.

The Energy/Money Cycle

A simple example

Consider a primitive village surrounded by farms, as analysed by Odum (6). The farms produce enough food for themselves and an excess which is sold to the villagers. The villagers in turn provide services as labouring, milling and blacksmithing to the farmers. Assuming negligible interchange with the world outside, the demands for food and services will be such as to fix automatically, by the market mechanism, the prices to be paid for each, and the allocation of labour to the farms and village. The use of money is a more flexible way of arranging for these than is bartering.

Such an analysis, however, ignores the input of resources from nature: sunshine, soil nutrients, wind and water (7). We may regard the farm/village complex as a self-maintaining unit which takes in these resources at the farm and uses them up (degrades them) at both farm and village. It might appear that the farm is the real producer and the village a parasite, but the presence of the village improves the whole system by feeding back specialist skills to permit more effective utilisation of nature's resources. Figure 2 shows Odum's illustration of this energy cycle (6).

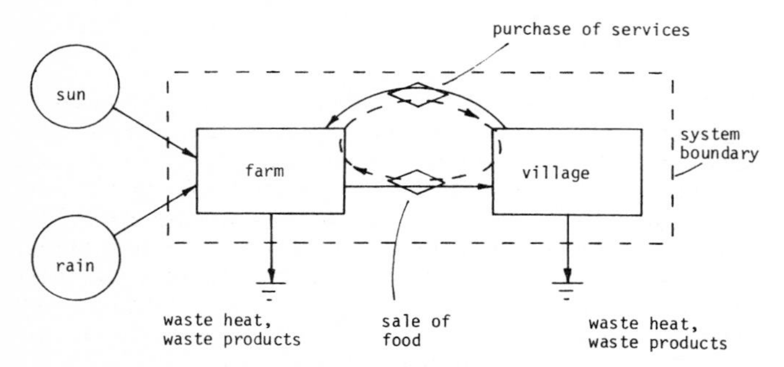

Figure 2: Energy and money flows in a primitive farm/village complex (6).

In this diagram energy in the form of sun and rain flows into the farm. Some of it leaves the farm in the particular high-grade form of energy we call food, and the remainder leaves in forms such as waste heat and waste products which are of such low quality that the farmer can make no further use of them. The flow of food energy into the village likewise leaves it partly in high-grade forms, in this case services and manufactured goods, and partly as useless low-grade energy. The overall result for the system as a whole is that all the energy flowing into the system leaves it in a degraded, useless form, as indicated by the flow-to-sink symbol. The broken lines indicate flows of money in the opposite directions from the flows of energy, with the diamonds symbolising interchange of money with goods or services (secondary energy). Note that the energy flows right through the system, dropping in quality as it goes, but with portion of it circulating round the system. It is this portion, and only this portion, which is balanced against money flows.

The national economy

We can generalise this simple picture to the economy of a nation as a whole, as shown in Figure 3. In principle the diagram is the same; modifications include:

(i) the addition of external energy from fossil fuels as well as the sun (here the covered-tank symbol is used to indicate a store that we are drawing from or depositing into).

(ii) replacement of the farm by an array of energy-conversion industries, of which the farm is but one (but note that each has the same characteristic of taking in external energy and working on it with feed-back energy from elsewhere in the system to convert some of it into the desired form — food, petrol, electricity etc., and discharging the rest as degraded energy).

(iii) replacement of the village by another storage called national assets, which will include all those capital assets mentioned earlier, from which resources can be drawn as required to control the energy-conversion industries.

(iv) addition of a money store to the money cycle.

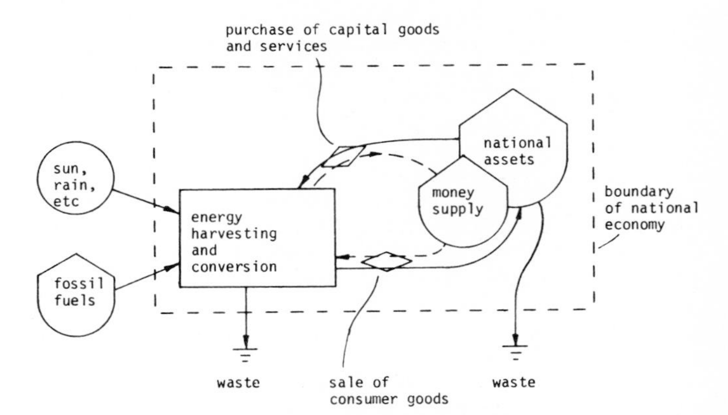

Figure 3: Energy and money flows at a national level (6).

If we ignore international imports and exports of energy and money we again see that inside the economy these two must be in balance, but only if the rate of storing energy is balanced by a corresponding rate of storing money, and vice versa. Within limits (which we will discuss below) the government can use the money store to stimulate the rate of circulation of money and hence resources. For example, by increasing the interest rate the government can persuade companies and individuals to buy shares in the energy-conversion industries, which will then build conversion devices and hire men to draw more energy in from outside the system. This will not only increase the rate of energy conversion and the build-up of capital, but will also increase activity elsewhere in the economy because of the money paid to new employees and the money flowing to suppliers of the conversion devices.

The above is a fairly conventional description of current economic management methods, but approached from the energy side rather than the money side.

Ratio of money to energy

It is commonly stated that there is a strong correlation between gross domestic product (money circulation rate in Figure 3 per year) and energy use (fossil fuel input in Figure 3 per year) for various nations, irrespective of their populations, their technological development (energy harvesting and conversion industries in Figure 3), their climates, etc., the implication being that greater development can be achieved only by greater use of fossil fuels.

Figure 4, which has been adapted from the work of Smil and Kuz (8), shows that in a very general way this is true for a range of European economics over the 21-year period 1950-1971. However when this diagram is examined in detail it is seen that the correlation is not strong. Not only does the money/energy ratio vary greatly from nation to nation, but also it changes greatly with time for many of the nations (Table 1 gives values of this ratio read off Figure 4).

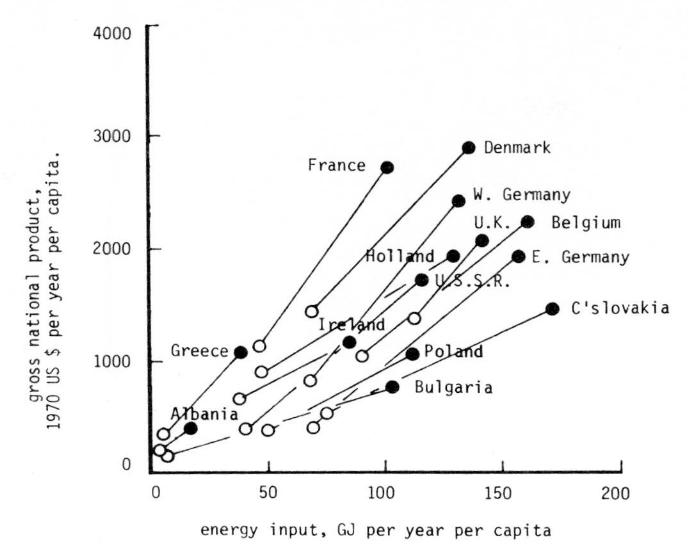

Figure 4: The relationship between energy use and the production of goods and services for various European economies for 1950-1971 (8). Economies with an appreciable use of hydroelectricity have been excluded: ○ = Use in 1950, ● = Use in 1971. All gross national products are expressed as 1970 US dollars.
(A gigajoule, GJ, is 10^9 joule; it is the energy which can be obtained from about 30 litres of petrol or 300 kWh of electricity.)

Table 1

Nation	1970 US $ per GJ 1950	1971
Greece	65	29
Albania	54	25
Bulgaria	22	8
France	25	27
Denmark	22	21
Holland	20	15
Ireland	18	13
Belgium	11	14
West Germany	12	18
U.K.	12	15
U.S.S.R.	10	15
East Germany	6	12
Poland	8	9
Czechoslavakia	7	8

Table 1: Money/energy ratios for various European economies for 1950 and 1971. These values have been calculated as the ratio of ordinate to abscissa of points on Figure 4. Note the relatively low figures for centrally-planned-economy countries, due to concentration on energy-intensive heavy industry (see also Figure 5).

The variation with time is demonstrated even more strikingly by Figure 5, which shows the ratio for the U.S.A. over a hundred-year period. The Resources for the Future team, from whose work (9) this figure has been adapted, interpret it as follows: over the period 1880-1920 the ratio fell as the U.S. economy changed from an agrarian one to a heavy industrial one. After 1920 it began to rise again, probably because of electrification of industry and mechanization of agriculture (10). Since World War 2 the ratio has been fairly stable. The ratio for Australia over the past ten years or so has also been fairly stable, but at a higher figure of about sixteen 1970 US $ per GJ.

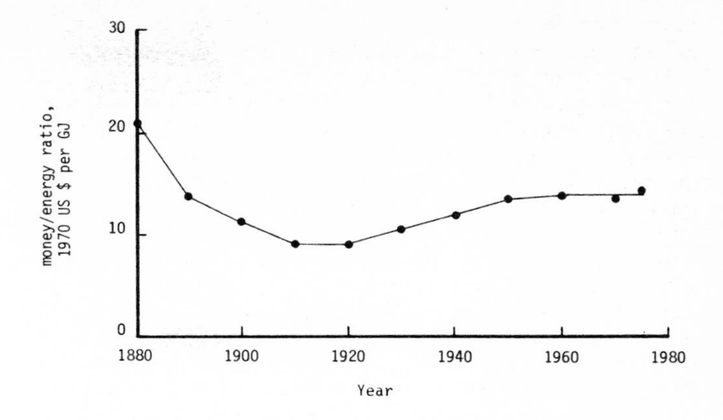

Figure 5: The change of money/energy ratio in the U.S.A. with time (9).

The question implied by comparisons of this type is: can we raise our standard of living without putting more energy in? Behind that question is the further question: does GNP measure standard of living? Figure 4 demonstrates that since World War II European nations have consistently increased their per capita energy consumption, presumably in the belief that this would benefit their citizens. However, the scattered nature of the correlation indicates that this belief is not soundly based.*

Growth

As seen from the above (and from our daily newspapers) governments since 1950 have believed in growth, as measured by GNP. Why? Are they right to do so? Where will it all end?

We will defer responses to these last two questions till later in the paper, and consider here only why nations, at least since World War II, have been devoted to the idea of growth as a good thing in itself. There is no single reason; but perhaps the following three threads wind through all arguments:

(i) the notion that more consumer goods and services per person per year is the same as a better life. To someone who works 12 hours a day to obtain barely enough food to live, this argument is irresistible,

(ii) the notion that a steady-state economy may be more difficult to manage than a growing one. This is an important feature of conventional economic thinking, but has little impact on the thinking of ordinary people.

(iii) the notion that if one section of the community, or one nation amongst the community of nations does not grow, those growing around it will eventually engulf it.

Resource externalities

As noted above, nations with a high income per capita and a high rate of economic growth have so far been able to achieve these by drawing in more external energy resources. However, as outlined earlier, this ability is now slipping away. Unacceptably high inflation rates are being accompanied by unacceptably high unemployment rates and recession. Odum (6) argues that this is due to increasing pressure on these external resources. Stimulation of the economy by conventional monetary and fiscal measures can no longer stimulate the energy-conversion industries to bring in more high-grade resources because they are not there to be brought in (or the owners, in anticipation of future scarcity, have already raised prices, which has the same effect).

As an example of this we could consider whaling: a whole ecosystem of nutrients/krill/whales exists in equilibrium, with young whales growing up to replace the old ones which die, the size of the population being controlled by the size of the food

*Compare, for example, Bulgaria and France: in 1950 both nations had about the same money/energy ratio, but the Bulgarians had much lower incomes than the French; by 1971 Bulgaria had increased its per capita energy use to just above France's level, but Bulgarians were still amongst the poorest people in Europe, while the French were still amongst the richest.

supply and the random fluctuations in it. People come along and initially take a few older whales, releasing food for younger ones and not appreciably disturbing the whole ecology. The business looks profitable, so whaling companies stimulate the harvesting of whales by investing money in ships, whaling stations, etc. which in other times might have been invested in other development activity such as rubber or coffee. Soon the limited whale population becomes depleted and the catches per ship smaller; in order to keep their existing whaling stations operating profitably the companies might then invest in more ships. A greater and greater stimulation is occurring, but with progressively smaller yields. Eventually the supply of whales has virtually disappeared and the companies are out of business or bankrupt. Quite apart from the destruction of a species, the whole enterprise foundered because the whaling companies mistakenly believed that *they* were managing the supply of whales, whereas in reality they were merely able to harvest from a stock whose size was controlled entirely by factors external to their own activities.

The Price of Energy

A comment one commonly hears on the energy situation today is 'The trouble is that energy has been underpriced'. The unspoken implication is that there are proper ways of pricing energy that governments and companies know about, but perversely refuse to use. The problem is one of time scales: in the short term (the time frame for market economics) the market sets energy prices appropriate to company profitability, government revenue, etc., but this says nothing about pricing of a fixed resource in the long term interests of the nation as a whole.

In reality, all natural resources like the sun shining on the ground or the coal buried in it cannot be valued by the short-term outlook of the economic system of each generation, as they exist independently of it. The energy to be won from these resources starts to cost something only when people build devices for appropriating them to their purposes, such as coal mines. The *cost* of energy as coal in this example is then merely the arbitrary cost of the mining right (if any), plus the cost of mining, plus any royalty the community may see fit to apply. The *price* will be what people are willing to pay, and the return to the mining enterprise will be a function of the difference between extraction costs and selling price. If the company considers the expected returns are adequate it will proceed; if not it will invest in something else, e.g. coffee or rubber. In a monopoly situation the owner might reduce output, and charge far higher prices, constrained only by the costs of substitutes, and unless the government intervenes (e.g. by raising the royalty) the profits would be higher than normal.

Before 1973 the competitive market system resulted in oil being relatively low in price, largely because there were many competing oilfields in the world where the cost of winning the oil (prospecting plus drilling) was less than 5 cents per gigajoule (compare Table 1). As a result nations which already had the wealth and a large capital stock to serve their economic system could bring as much oil as they wished into their energy conversion devices without too much concern for efficiency of conversion, depletion of resources, energy/money ratio, etc., and use this energy to build up further capital assets. In this situation user nations draw on the cheapest oil first, which means, for example, that expansions in the use of oil in the U.S.A. since about 1960 involved Middle Eastern oil from rich new fields which were being tapped, rather than U.S. oil, which was almost half gone (the cheaper half) by that date.

Not surprisingly, in 1973, as soon as they were politically able to do so, the Middle Eastern countries set up a producers' cartel to edge towards a monopoly situation, and quickly raised the price close to that of possible substitutes. Consuming nations could not afford to buy so much oil, and their own energy-conversion industries became source-limited; overall production decreased and recession set in. Stimulation of their economies in the Keynesian manner (i.e. to achieve full utilisation of capacity), instead of reversing this process, merely contributed further to inflation, as Odum's analysis predicted.

Energy Crisis

The Shorter Oxford Dictionary defines a crisis as 'the point in the progress of a disease when a change takes place which is decisive of recovery or death'. Our disease is addiction to fossil fuels: we need ever-increasing doses to keep us going. The disease has now reached its crisis; what is to be done?

To pursue the metaphor just a little further, consider the following extreme possibilities:

1. Let the disease run its course: as each preferred form of fossil fuel becomes scarce, devise ways of substituting the next form, using ever-increasing quantities, until at last nothing is left outside our system and even the store inside the system runs down,
2. Find another energy source bigger and stronger than all those before, to act as the final fix,
3. Attempt a cure. Dry the poor addict out.

We must now consider the likelihood that each of these would result in 'recovery or death'. If we believe a crisis exists, by definition we believe that what is done *now* will *decide* what the outcome will be. The earlier part of this paper attempted a diagnosis in terms of money/energy relationships. Let us now apply this diagnosis to the problem of predicting the course of the disease under the effects of these three extreme treatments.

(1) *'Do nothing'*

This option presumes that the present market system will continue to operate with rates of use growing exponentially, and available energy going preferentially to those who can best afford to buy it. The resources from which these supplies are to be drawn are limited, although we do not know yet what the ultimate limits are. We will find these out, however, soon enough, as exponential growth eats into any resource very quickly. Hubbert (11) estimated the extent of the various resources in 1968, and analysed the question of depletion under the forces of exponential growth followed by depletion under scarcity control. His estimates of ultimately recoverable resources of oil, gas and coal and his analyses of their depletion rates, as shown in Figure 6, have not been seriously challenged (see Table 2).

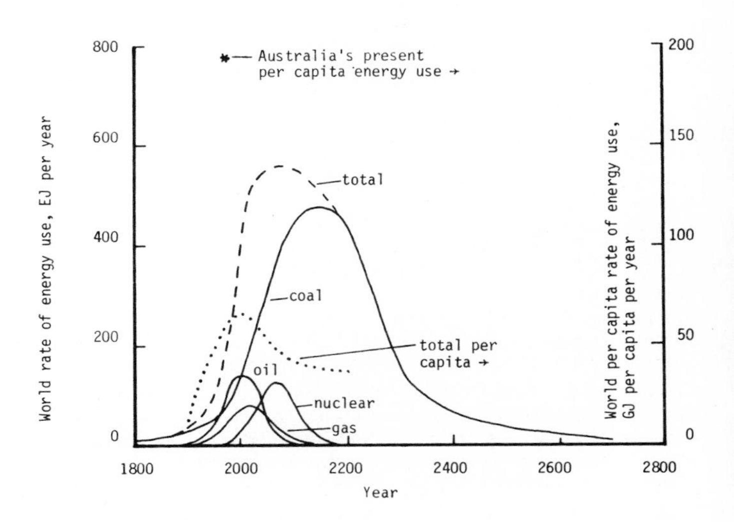

Figure 6: Past and future depletion of various fuels under market control conditions. Since the rate of use of each fuel is plotted against time the area under each curve represents the total energy resource obtained from that particular fuel (12). The curves for coal, oil and gas are due to Hubbert (11). The curve for nuclear fuel assumes that only thermal reactors will be used, but that further uranium resources will be discovered and thorium resources will be utilized, to give five times the present estimated resources shown in Table 2. The broken line for the rate of total energy use has been constructed by adding the rate of use of the individual fuels at any given time. The dotted line showing total rate of energy use per capita was calculated by dividing the total rate of energy use at any given time by the world population at the same time, as estimated by the U.N. Population Conference, Bucharest, 1974 (13). (EJ = 10^{18} joule, GJ = 10^{9} joule).

Table 2

fuel	Present known reserves, EJ (14)	ultimately recoverable resources, EJ (14)	(11)
coal	15000	165000	130000
oil	4400	12000	12000
gas	2500	7000	7000
uranium	1100	2000	—

Table 2: Known world reserves and estimated ultimately recoverable resources of mineral fuels (for uranium it is assumed that only thermal reactors will be used) (14, 11).

There is doubtless a good deal more uranium available than Hubbert estimated; after a further decade of vigorous exploration his estimates have been proved low by a factor of about two. However his conclusion would still stand: if we use only thermal reactors, uranium (and thorium) would ultimately supply far less energy than coal (Table 2). In Hubbert's words 'Failure to make the transition to a complete breeder-reactor program before the initial supply of uranium — 235 is exhausted . . . would constitute one of the major disasters in human history'. But note *why* he is for Option (2): he believes that without a major 'fix' the patient will die. (He might not of course die, but merely be left a simpleton, i.e. mankind — perhaps a few per cent of the present population — reverts to a primitive existence based on daily intake of solar energy, without any store of capital to permit the harvesting of solar energy more effectively than could be managed by primitive people.

The dotted line in Figure 6, which gives the *per capita* energy consumption of the world, shows how fragile the human condition already is. Despite exponential growth in world energy use, the *per capita* energy use has nearly reached its peak, and after 2000 A.D. will start to decline, having reached a figure only one third of Australia's present per capita use.

(2) Super technology

We have used the term 'Super technology' for this option to emphasise that it is not just a logical extension of Option (1). In Option (1) governments leave the question of the survival of mankind to short term market forces, whereas in Option (2) they try to rescue mankind by government intervention.

We have already noted that in the absence of monopolistic selling before 1973 these market forces resulted in an extremely low price for oil which encouraged a vast increase in the level and range of its uses, displacing some alternative fuels and discouraging investment in alternative resource harvesting systems. As an example of the effect of such short-term market forces on the long-term outlook of energy planning authorities, in 1972 the Central Electricity Generating Board in England announced that because of the rising cost of coal it would build no more coal-fired power stations. Within a year the OPEC cartel started the oil price hike on its way, and within two years the CEGB announced plans for new coal-fired power stations based on opening up completely new coal mines.

It would appear that the events of 1973 have been educative for governments and producers alike, and most governments have now considered that intervention of some kind is necessary. The question is: what kind? Three types seem possible:

(i) relatively minor interference with pricing (as in the recent Australian budget) to try to give smooth transitions between one energy use and the next (i.e. to eliminate fluctuations in the total energy use, as shown in fact in Figure 6). This is merely a gloss on Option (1).

(ii) major investment in super technology for harvesting new forms of energy (15, 16).

(iii) intervention with a view to achieving a stable, no-growth situation (this might also involve the development of new technology, but super technology is unlikely). We will discuss this shortly as Option (3).

It has already become obvious that any form of super technology, even the use of thermal fission reactors, will require massive government intervention. What is not yet clear is what it will cost in terms of other opportunities foregone. This issue is raised in the concluding words of a provocative paper by Kay and Mirrlees (17) on 'The Desirability of Natural Resource Depletion', viz. 'in general we believe that the interests of future generations will be better served if we leave them production equipment rather than minerals in the ground. In the currently topical case of oil, the arguments that the world is using too little rather than too much seem irresistible'. The currently topical case is North Sea Oil. A U.K. White Paper suggested that this oil field be depleted fast, over 20 years, to permit, amongst other things, appropriate investment in new energy technology. In the end the decision was made to retain the time scale but not the investment type: the oil was to be used to revitalise the whole of the British economy, and make it competitive on world markets. The developing of new energy technology for the next generation was too costly a burden for the present one. Back to Option (1).

One could suggest other examples: the small effort going into fusion research; the moratorium on breeder fission development in the U.S.A.; the out-right refusal all over the world to put money into super solar technology (Bockris in 1977 (16) suggested that to have a real chance of success here an expenditure equivalent to that required to put men on the moon would be required).†

Even the development of thermal fission reactor technology has fallen far short of targets set a few years ago, and is now seen to require massive government intervention. This would suggest that, despite much posturing, Option (2) will not occur. This is not to say that nuclear fission may not advance much further than it has already. What is being suggested is that government intervention, including the autonomous generation of the required funds, at the level required to 'go for broke', is unlikely. In capitalist economies at least (see note (18)), the present generation is unwilling to make such a problematical investment for the succeeding ones (see also de Carmoy (19) on 'The U.S.A. faces the Energy Challenge').

(3) No growth

The above hypothesis on super-technology is based solely on what already seems to be emerging, and much careful investigation will be required before we can be sure of it one way or another. In the meantime the operation of market forces, paradoxically, will probably ensure that we do not embark on Option (2) prematurely, and will probably also ensure that we don't even commit Hubbert's ultimate sin of thoughtlessly burning all the uranium-235 before we start the breeders.*

A period of economic and energy modelling may next be expected (indeed has already started — e.g. Refs (20-25)) in which attempts will be made to delineate the feasible options more carefully. We are optimistic enough to hope that this will result in a determination to deal with the disease of *exponentialitis*. This will no doubt involve an intermediate period of adaptation while the world's stock of motorcars, cities, employment opportunities and ideas, etc., is turned over, and we learn to build up a capital stock more suitable for the

†*Editor's note:* There may be good reasons additional to those of cost for not making super investments in super technologies: for instance, it is very unlikely that any of these technologies could make a significant contribution to energy supply within the next 30 years, even if funds were available. Nevertheless, the total investment in super technologies in the USA runs into tens of billions of dollars; it is far greater than the total investment in the simpler, renewable-energy technologies (e.g. solar thermal, solar photovoltaic, windpower) which together could make a substantial contribution before the turn of the century.

**Editor's note:* Hubbert's "sin" is based on a misconception of the physics. "Burning" up all the uranium-235 would actually *produce* sufficient plutonium-239 to start up the breeders. So, provided that an economically viable industry for the reprocessing of spent oxide fuel could be developed, a breeder age would not necessarily be foreclosed by delaying its introduction and continuing with burners.

steady-state society. Above all we will need to develop a new economic system *not* based on the notion that money can be used to create resources. Perhaps we will also discover that the stable world population is smaller than what we have today.

How is this to be achieved? We don't know. What we are sure of, however, is that it cannot be achieved by the actions of individuals working as individuals. Any one of us can opt out, and try to live the simple life, but in so doing he must realise that he takes with him his share (actually far more than his share, when regarded as a single *world* citizen) of the world's capital assets (the money required to buy the land, the scientific skills and knowledge built up over centuries and stored in libraries and universities), and he could dissipate that in a generation or two to become the primitive that Option (1) required a century or two to produce. The challenge is rather to the whole institutional system called society, using the mechanism we call the economy, and the changes have to be made within the institution by changes to the mechanism.

To get some feeling for the shape of this challenge let us return briefly to the money-energy system sketched out in Figure 3:

(i) Ultimately the system runs on brought-in energy, not on circulating goods or money. We could obtain greater benefit per person by bringing in energy at a higher rate or by degrading it at a lower rate. Since we suspect that the rate at which we can bring in *renewable* energy resources is lower than the present rate of use of fossil fuels we should devise methods of operating our society that degrade brought-in energy more slowly,

(ii) The money store is merely a proxy for the built-up resources store, but we have been using it as a method of manipulating the whole system. In the new economics it must return to its proxy role. Marx correctly identified the way that the use of capital was distorting the use of labour resources, but centrally-planned economies based on his ideas seem to distort the use of energy resources even more drastically than does the capitalist system (refer again to Table 1, which shows the low money/energy ratios of the centrally-planned socialist economies, which are almost certainly due to the heavy reliance on massive centrally-planned and operated energy harvesting and conversion industries).

(iii) It would be possible for us to invest a greater part of our present built-up resources in energy harvesting devices and methods for tapping renewable resources, or even use the remaining fossil fuel resources to do this, so long as we had ensured that the energy harvested was of higher value than the energy expended. We should note, however, that this was precisely the scheme rejected by the British when considering use of the North Sea Oil. The lesson here, no doubt, is that the British saw the problem in terms of capital-intensive super-technology. We could consider instead advanced energy-conversion systems far less energy-intensive than the present ones, such as miniaturised mechanical and electronic devices; housing settlements designed to minimise transport costs, minimise waste, maximise recycling, and maximise the harvesting and retention of solar energy; farms using biological control of pests, biological harvesting of nitrogen and reuse of wastes by symbiotic relationship with towns (garbage and sewage); and so on.

By and large such energy-conversion systems are not favoured by an economic and political structure which has taken shape during a period of cheap and virtually unlimited energy. Thus the problem for governments is not, as is often supposed, that of finding funds for massive investment in research and development directed towards large-scale, centralised, government-controlled systems for bringing new forms of energy in, but rather the more subtle problem of facilitating decentralised, ecologically sound, energy-efficient conversion systems which may be uneconomic by present criteria, but which will still be tenable when our stocks of fossil fuels begin to run out.

References and Notes

(1) Smith, Adam, 'Wealth of Nations'.

(2) Figure 1 has been adapted from a similar diagram given in *The Economist*, July 13, 1974.

(3) We have expressed the analysis of resource use in national rather than international terms, because people still see resources as belonging to nations rather than to the world.

(4) Keynes, J. M. (1936). *The General Theory of Employment, Interest and Money*, McMillan.

(5) Some progress has been made on the economics of environmental degration (see for example A.V. Kneese (1977), *Economics and the Environment*, Penguin). The most usual method is to make the polluter pay for returning the degraded environmental element acceptably close to its original state, but how does one work out the cost of replenishing a coal deposit?

(6) Odum, H. T. and Odum, E. C. (1976) *Energy Basis for Man and Nature*, McGraw Hill.

(7) In the analysis given in the Odums' book, ref. (6) above, it is taken that all resources may be thought of as energy resources, since they require energy to create and maintain them: for example, pure water has a higher energy value than sea water because it requires the earth's solar-powered climatic cycle to create fresh water from salt, but left to itself the fresh water spontaneously degrades to salt water.

(8) Smil, V. and Kuz, T. (1976) *Energy Policy* **4**, 171.

(9) Schurr, S. H. and Darmstadter, J. (1976), *Resources* No. 53, 1.

(10) Analyses of agricultural activity (see for example the paper by R. M. Gifford, Chapter 9 in *Energy, Agriculture and the Built Environment*, Ed. R. J. King, Centre for Environmental Studies, University of Melbourne, 1978) show that mechanisation of farming leads to a *decrease* in the output of energy as food per unit input of fossil-fuel energy. However it would appear that the effect on the economy as a whole is an *increase* in the dollar value of goods and services produced per unit input of fossil fuel energy.

(11) Hubbert, M. K., Chapter 8 in *Resources and Man*, Freeman, 1969.

(12) The equations for the mathematical construction of curves for depletion of resources under various constraints are given as an appendix in the book by the Odums (Ref. (6) above). Hubbert's curves for coal, oil and gas given in Figure 6 are a simplified version of Odum's model 5 in which depletion curves follow historically established depletion curves for time which has passed, the overall depletion curve is assumed to be symmetrical, and the ultimately-recoverable resource is estimated from other data.

(13) Report by Secretary General to U. N. Population Conference, Bucharest, (1974).

(14) Urie, R. W. et al. (1977). Chapter 1 in 'Towards an Energy Policy for Australia', Institution of Engineers Australia, Canberra.

(15) Hafele, W. (1978) *J. Inst. Nuc. Eng.* **19**, 12.

(16) Bockris, J. O'M. (1977). *Energy — The Solar Hydrogen Alternative*, ANZ.

(17) Kay, J. A. and Mirrlees, J. A. (1975). in *The Economics of Natural Resource Depletion*, McMillan.

(18) As Figure 4 and Table 1 have already hinted, this might not be true for centrally-planned economies.

(19) de Carmoy, G. (1978) *Energy Policy* **6**, 36.

(20) Meadows, D. H. et al. (1972) *The Limits to Growth*, Potomac Associates.

(21) Mesarovic, M. and Pestel, E. (1975) *Mankind at the Turning Point*, Hutchinson.

(22) Charpentier, J. P., IIASA, RR-74-010 (1974), RR-75-035 (1975), RR-76-018 (1976).

(23) IIASA, Proceedings of IIASA Working Seminar on Energy Modelling, CP-74-003 (1974).

(24) Thomas, J. A. G. (Ed) (1977) *Energy Analysis*, IPC.

(25) Leontief, W. (1976) *The Future of the World Economy*, UN.

Reflections on the Social Impact of the Rising Atmospheric Carbon Dioxide Concentration

Roger M. Gifford

CSIRO, Division of Plant Industry
Canberra

Introduction

The global reserves of fossil fuels, mostly coal, are enormous — enough to sustain the energy needs of civilization for several hundred years (1). However, exploitation of these resources at a pace consistent with both smooth extrapolation of past trends and available fossil fuel reserves may lead to unacceptable changes in the global carbon cycle. These changes could have substantial social effects and create severe strains in international politics in coming decades. In this paper I examine the facts and uncertainties of the changing features of the global carbon cycle, and of the possible agricultural and climatic effects. These facts and uncertainties are explored in relation to implications for society and energy policy.

Atmospheric Carbon Dioxide; Facts and Uncertainties

Continual monitoring of atmospheric carbon dioxide in Hawaii and at the South Pole since 1958 (2) reveals that the annual average concentration has been rising steadily. In the 1960s the annual increment was about 0.7 parts per million by volume (vpm) per year. This has now increased to about 1.2 vpm per year. Monitoring over the North Atlantic (3), over the Tasman Sea (4) and elsewhere confirms this trend. The global average is currently 334 vpm having risen from an estimated 290-295 vpm at the turn of the century (5).

Global fossil fuel consumption has risen exponentially over the last century (1, Figure 1). The constant annual growth rate of 4.3% per annum was interrupted only temporarily during the World Wars and the Great Depression. The rate of fossil carbon dioxide release has been sufficient to account for two times the observed rise in atmospheric CO_2, were that release to remain entirely airborne (2).

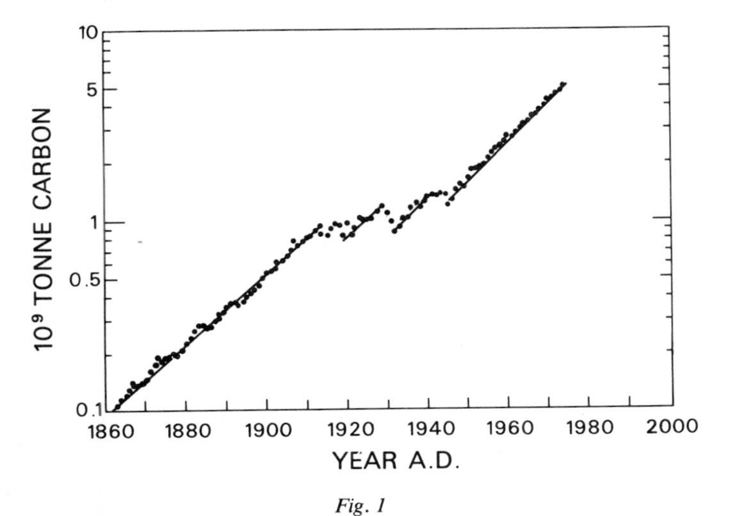

Fig. 1

Weight of carbon released from fossil fuel burning and cement manufacture (log scale). From Baes et al. (1).

It is assumed that the released CO_2 which does not remain airborne can only be absorbed by the oceans by dissolution, and by the biosphere through photosynthetic carbon fixation. Modellers of the equilibration of carbon dioxide between atmosphere and ocean have found that they can just explain this approximate 50:50 partitioning of released CO_2 between atmosphere and 'elsewhere' if the biosphere can be assumed to be merely a small sink for the excess CO_2 (6) or at least not a net source (7). The mixing parameters for the ocean-atmosphere equilibration are based on experimental measurement of the distribution of isotopes such as radioactive ^{14}Carbon and ^{222}Radon (6).

Despite this apparently nice correspondence between observation and theory there is a major uncertainty relating to deforestation, especially in the Third World. Besides the clearing of land for agriculture there is considerable informal removal of wood for cooking and heating by the Third World rural poor who have no alternative fuel (8). Preliminary assessments by some suggest that the net release of CO_2 from this source, together with the associated oxidation to CO_2 of soil organic matter, may range from as little as 10% to as much as 350% of that released from fossil fuel (9, 10, 11, 12). If this is correct, the models of ocean-atmosphere equilibration are wide of the mark and the problem of atmospheric CO_2 build-up may be primarily related to activities of the World's poor rather than the World's rich. This, of course, would have a profound effect on any interpretation of social implications.

Impact of Rising Atmospheric Carbon Dioxide

Before the uncertainties relating to the role of deforestation were raised, it seemed probable that the atmospheric CO_2 concentration would double to about 600vpm by AD2025-2035 (13). This estimate assumed that the growth in fossil fuel use would start at the long term value of 4.3% p.a. but gradually decline, leading to a peaking of the fossil fuel era between 2050 and 2100AD followed by a decline in annual use culminating in its final disappearance between 2200 and 2400AD. One might question this assumption on the grounds that high fuel price is now with us to stay and will lead to more rapid flattening off of the fossil fuel use curve. But equally there are great pressures building up to convert coal to oil as crude petroleum becomes less accessible. The ratio of energy in synthetic liquid fuel to energy in the parent coal is in the region of .35 (14) in contrast to the ratio of .89 for refined petroleum products from crude oil (15). This means that for each unit of liquid fuel energy consumed by the final consumer there will be at least twice as much CO_2 released. So if there is a major movement in direction of coal liquefaction it is unlikely that growth in fossil CO_2 released will be able to be reduced rapidly, even if the rate of consumption of secondary energy by final consumers is static.

It is not out of the question therefore that atmospheric CO_2 may at least double before stabilizing or declining, even if the existence of an impending global problem is recognised and concerted action taken. If action is not taken, atmospheric CO_2

Diesendorf, M. (ed.) (1979). *Energy and People*. Canberra, Society for Social Responsibility in Science (A.C.T.)

is likely to go much higher than twice the pre-industrial level before peaking. What might be the impact of a CO_2 doubling?

Impact on agriculture

It is well known that the rate of photosynthetic carbon fixation by plant leaves is dependent on the concentration of its substrate — carbon dioxide — in the surrounding air. Glasshouse crop producers use this fact to stimulate crop growth by enriching the glasshouse atmosphere with CO_2. But despite the large body of literature on horticultural use of CO_2, it is hard to use it to interpret what would happen to field crop and pasture production. Horticulturalists use much higher levels of enrichment than we expect for the atmosphere in the next 50-80 years. Also temperature, water and nutrient supply are usually optimal. This rarely pertains in the field.

Few CO_2 enrichment experiments have been performed throughout the life-cycle of field crops. It has been suggested that because most field crops of the world have their yield limited by other factors like low nutrient, water or light supply, or non-optimal temperature, most crops will not respond to higher atmospheric carbon dioxide. This is not so. For example, in an experiment on wheat, I reduced the water available to a crop throughout the season to a level which reduced grain yield to that of the average Australian yield. A 73% increase of CO_2 concentration doubled grain yield under these conditions, whereas grain yield increased only 37% in response to the same CO_2 enrichment when water supply was abundant.

So, although much more data is needed, we do not have a basis for assuming that agricultural yield will not increase somewhat as atmospheric CO_2 rises, if other conditions stay constant. This presumably is a socially advantageous response; but it assumes that all else remains equal. In fact increased atmospheric CO_2 is likely to have other effects such as modification of climatic patterns.

Impact on climate

Carbon dioxide absorbs infra-red radiation in the long wavelengths which are emitted from Earth, but does not absorb solar radiation of shorter wavelengths. Thus, as atmospheric CO_2 builds up, the lower atmosphere tends to get warmed and the upper atmosphere tends to cool down. It is hard to estimate the magnitude because there are so many positive and negative feedbacks involved in the global circulation of atmosphere and oceans. But the consensus seems to be that a CO_2 doubling would cause about a 1.5-3°C rise in global mean surface temperature (16), although nobody would be able to state categorically that the range of plausible values may not be from little more than 0 up to +6°C. There is one detailed three-dimensional general circulation atmospheric model by Manabe and Wetherald (17) which has been widely regarded as the best model so far and has been used to predict the effect, at equilibrium, of a doubling of atmospheric CO_2. Figure 2 shows the results. The overall average surface warming of 2.9°C which was predicted is not evenly distributed, being minimal at the equator and large (about 10°C) at the poles. The model also responded with a 7% increase in global average precipitation, but again this was not evenly distributed, being maximal over the permanent ice-caps. In some latitudes rainfall actually decreased following a CO_2 doubling. Although one cannot put any real credence on the latitudinal distribution of rainfall, it is sobering to note that regions getting less rainfall according to the idealized model included areas which have mediterranean climates such as the S.E. and S.W. corners of Australia.

The warming in the temperate regions may benefit agriculture by extending the growing season. But in the hot regions it may reduce growing seasons by lengthening the period when potential evaporation exceeds rainfall too greatly to permit crop growth. This would be especially so in those regions where rainfall decreased, but detailed predictions are quite impossible at present. The general weakening of the intensity of atmospheric circulation as the temperature differential between poles and tropics diminished, could have

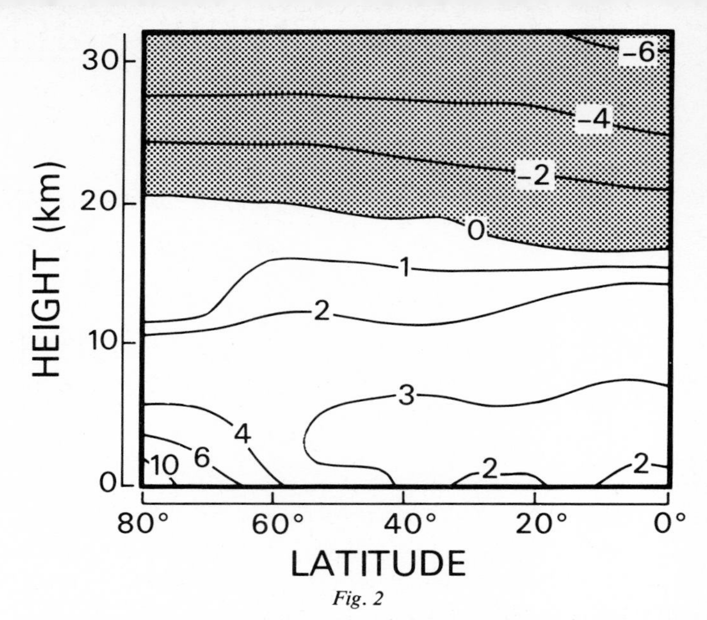

Fig. 2

Predicted temperature (C) change associated with a doubling of atmospheric CO_2 concentration from 300 vpm to 600 vpm. From Manabe and Wetherald (17).

unpredictable effects like altering the typical paths that tropical storms follow.

The greatest feature of concern is the melting of polar ice. If the ice-caps were to melt entirely, then sea-level would rise by 50-60 metres (2) an effect which, though gradual, would have immense social impact. Fortunately the temperatures shown in Figure 2 are not sufficient to cause any melting of the land based ice-caps of Greenland and Antarctica. They are sufficient to cause the melting of floating sea-ice however (17). Indeed it is the retreat of the highly reflective sea-ice which is partially responsible in the Manabe and Wetherald model for the amplified temperature response in polar regions. Since sea-ice is already floating on water, its melting would have little effect on sea-level. Moreover, there is an effect which would tend to reduce sea-level despite global warming. Much of the increased precipitation predicted by Manabe and Wetherald was in the polar regions falling on the ice-caps. So concurrent with the melting of floating sea-ice, there would be gradual build up of the depth of the land-based ice-caps.

There is one area of polar-ice instability which is of considerable concern. This is the West Antarctica region. The West Antarctic ice-cap is largely resting on rock which is below sea-level. It is buttressed on either side against the land-based East Antarctic ice-cap by ice-shelves of floating ice — the Ross Ice-shelf and the Filchner-Ronne Ice-shelf. According to Mercer (18), a 5°C rise in annual average temperature in that region would lead to the melting of the ice-shelves accelerated by a self-sustaining process called the 'calving bay mechanism'. Were the protective buttresses of the ice-shelves removed it is suggested that the West Antarctic would deglaciate relatively rapidly within, say, a century leading to a sea-level rise of about 5m. If such were to happen the social consequences for all coastal cities and for low lying agricultural land, such as in the Netherlands, Bangladesh and the Mekong and Nile Deltas, would be great. It is believed by some that the inherent instability of West Antarctic may make its response irreversible once started, even if temperature dropped back to normal.

As for all other aspects of the atmospheric CO_2 problem, West Antarctic deglaciation is uncertain and partly speculative. The possibility of a 5m sea-level rise being triggered is beyond the scale of impact which economists would normally bring into cost-benefit analyses of technological options. Costs more appropriate to cost-benefit analyses would be such matters as the costs of coping with hotter more humid environments in some regions, hotter drier environments elsewhere and the costs of coping with more frequent flooding in urban areas with inadequately sized drains. But the uncertainties involved in all aspects at the moment make it quite impossible to make any assessments of that type. Moreover, the possibility that the effects will not be simply and reversibly related to the absolute level of CO_2 in the atmosphere, renders assessment even more difficult. Besides the possible triggering and irreversibility of West Antarctic deglaciation, there is the uncertainty as to whether the climatic pattern, once shifted by an anthropogenic CO_2 build up, would return to 'normal' once the CO_2 were allowed to re-equilibrate with the oceans, or whether it would evolve along a new path (19).

The Longer Term Future

So far I have addressed myself to the next half century and responses appropriate to a CO_2 doubling. If, however, the known reserves of fossil fuel are burned on a cycle leading to their ultimate exhaustion in about 200 to 300 years time, as discussed above, then atmospheric CO_2 is predicted to peak at 6 to 8 times the pre-industrial level by 2100-2200AD (13). I do not think there are any workers in the field who would doubt that such a rise would be climatically catastrophic. The average surface temperature difference between full glacial and full interglacial periods of the ice ages was only about 5°C. So we must not generate excess CO_2 at such a rate. Hence international agreements are needed. The uncertainty over whether in the immediate future the CO_2 enrichment of the atmospheric will make things worse or marginally better for mankind is a serious impediment to such agreements. If it were generally assumed that things would be marginally better at first because of improved agricultural yields and warmer climate in the cool temperature regions, where up to now most of the political power is now located, then we would soon arrive at a situation where movement in either direction would be detrimental. The situation would be that man could either carry on enjoying the fruits of increased fossil fuel consumption and suffer catastrophe later due to CO_2 effects, or he could suffer considerable organisational trauma by hastily downplaying fossil fuels after a CO_2 problem was positively identified and in due course suffer both reduced agricultural production potential and a cooler climate. In that situation the optimal course would be the awesome political task of a centrally determined global policy of fossil fuel use and forest management to hold CO_2 concentration constant and in the end gradually to reduce its concentration to the original level, or to its new steady value when net CO_2 release has ceased.

In any event, what is indicated is the immediate need for some international institution to be set up to start drawing guidelines to (a) agree on criteria to define the acceptable limits of anthropogenic change in atmospheric CO_2 level, (b) establish the 'outer limit' goals for global fossil fuel use and deforestation/reafforestation needed to keep the CO_2 level within these guidelines and, (c) formulate policies to manage the equitable distribution between nations of the responsibility for control of fossil fuel use, deforestation/re-afforestation practices and any other practices to control the atmospheric CO_2 level.

An International Institution for Formulating Agreements Relating to Atmospheric Carbon Dioxide

The power that such an institution would need would be to set quotas of net CO_2 release permitted by each country each year. Immediately, the near impossibility of implementing this idea is evident. First, the time period over which the problem will emerge and over which a commitment to a solution must be sustained is much longer than the life of elected governments. So the temptation to procrastinate and avoid far-sighted decisions will be great. Secondly, the institution would need to rely on each country to present its own figures for the quantity of fuels burned. International inspection would not be possible, unlike international inspection of, say, the number of whales caught. Carbon dioxide molecules cannot be counted. Thirdly, even if all countries could be relied upon to return reliable statistics of the commercial fuels burned, it would be very difficult to estimate accurately the level of informal collection of non-commercial fuelwood which is not replaced by regrowth. Satellite imagery might help although changes from year to year may be hard to resolve in that way.

Moreover, even if reliable annual returns could be collected, the depressing record of the International Whaling Commission since its inception (20) suggests that it is most unlikely that adherance to restrictive quotas would be achieved. The more so since restriction of energy use, unlike restriction of whales caught, hits at the very heart of the economic system. Given these administrative hurdles, we are thrown back onto the question as to how long we can wait before taking action. What criteria could be used to decide when undesirable changes are commencing? And once the first signs of undesirable changes have been detected, would it be too late to avert serious consequences?

It looks as if it will be about 20-30 years before we will be able to assert with fair certainty that global annual average temperature is rising beyond the bound of expected natural variability. Natural long term fluctuations of the mean global annual-average temperature are in the range of about ±0.5°C of the long term average (21). We are currently in a cooling phase, at least in the Northern Hemisphere (22), but we can expect this to bottom out in the mid-1980's (23). Then any CO_2 engendered warming will reinforce the natural warming (23). If the CO_2 effect on temperature is a reality, then we may expect the warming phase to pass through the long term ceiling for fluctuations in 20-30 years. If that were the pattern which pertained, the rate of temperature increase would be likely to rise rapidly as the natural rise was reinforced by the CO_2 induced rise. A rapid retreat from fossil fuel use would be called for, but might not be achievable in practice.

This observation should have its impact on present energy policy. The oil consuming nations are poised on the brink of committing massive investment into alternatives to crude petroleum as they enter the downward trend of oil's depletion cycle in the 1990's (24,25). Coal liquefaction, shale oil and tar sands are strong contenders. If that is the direction followed it will take 20-25 years to build up a substantial contribution. Moreover, the environmental, social and technical difficulties being faced by the nuclear industry may lead to the expansion of coal-based electricity for some time yet. If so and a CO_2/temperature problem is later confirmed, it will not in practice be possible to shift rapidly to some other alternative, even if the research investment has been made into these alternatives. The commitment would have been to decommission substantial proportions of fossil fuel-based plant.

But these comments are based on the presumption that it is fossil fuels which are the main source of the extra CO_2. What if the highest estimate of CO_2 release via deforestation (11) turns out to be correct? In that case total release would be 23×10^9 t (carbon) yr^{-1} of which only 22% would be from fossil fuels. A large proportion of the remainder would be due to deforestation following informal collection of non-commercial fuelwood, and to the concomitant oxidation of soil organic matter in the Third World. Then the situation would differ in several ways. The rate of equilibration of CO_2 into the oceans would have to be much faster than can currently be accounted for; this would suggest that more rapid reversal of the atmospheric CO_2 situation would be possible than is currently expected. The potential for build up of atmospheric CO_2 would be less than we now think, since the size of the whole pool of living plus dead biomass is only $1{,}800\text{-}3{,}800 \times 10^9$ t (carbon) (11) of which only a minor fraction could conceivably be oxidised, compared with the known pool of *recoverable* fossil fuel carbon of about $10{,}000 \times 10^9$ t (1). If the quest for Third World development were successful and those countries were to move from fuel wood to fossil fuel dependence, the CO_2 released for the same level of useful energy would be much less. First because the wood is burned very inefficiently for heating and cooking (8) and fossil fuel burning devices are more efficient. Second because (for the hypothetical extreme situation relating to CO_2 sources to which we are here referring) much of the carbon released is from the soil organic matter following loss of forest rather than from the direct burning of wood. However, it is unlikely that the bulk of the Third World poor could move into the fossil fuel economy without a concomitent increase in the level of industrialization, standard of living and hence per capita use of useful energy.

Although it would be a very different social and political problem if deforestation really were the major cause, it would be inappropriate for the rich world to use that as a basis for avoiding looking at its own behaviour. The rate of fossil fuel burning is accurately known and readily accounts for the observations. The wide range of judgements on the rate of deforestation-related CO_2 release hardly represents compelling evidence for that hypothesis.

Conclusions

The implications of the observed steady rise in atmospheric carbon dioxide are plagued with uncertainty. The cause of the problem is probably, but may not be solely or (according to a few) even predominantly, fossil fuel burning. The increase, if sustained, may lead to considerable socio-economic trauma within a century through rise in sea-level and regional shifts in the margins of some agricultural zones whose edges are determined by high temperature and low rainfall or which are determined by the reliability of tropical storms. On the other hand, it may be that the associated temperature rise is modest, that West Antarctica remains stable and does not disintegrate, and that a rise of CO_2 is largely beneficial through a rise in agricultural productivity and even through a countering effect on the rise of stratospheric ozone due to anthropogenic chlorofluoromethane and nitrogen oxides (26). This latter set of alternatives would postpone social trauma until the period after fossil fuels have been superseded. Then we could expect the population size which had been fostered by the enhanced agricultural production to drop back as production declined again.

In the face of so much uncertainty, I must confess that society's response is likely to be 'no action' for the time being. That is certainly the response of those in the fossil fuel business to whom I have presented the problem. Moreover, scepticism of the reality of the problem has been enhanced by the willingness of those who are interested in fostering nuclear energy to discuss the problem (e.g. note the publisher of reference 2). The most serious policy issue which is raised is whether or not the world can risk the enormous investment in widespread coal-to-oil conversion technology and other advanced liquid fossil fuel technologies to keep the developed world moving along the same track that it is on now, and indeed extending it into the developing countries. The answer one gives depends on one's taste for risk, especially risk to others in different places and times. The considerations discussed here suggest that perhaps we should forego the possible medium-term benefits of a predominant thrust into technologies to use the less convenient fossil fuels; instead to concentrate our efforts on diversifying energy sources, on achieving structural changes in society which are conducive to intense thermodynamic thrift, and on developing economic systems which ensure equitable distribution of wealth in a 'no energy growth' society.

References

(1) Baes, C. F. Jr., Goeller, H. E., Olson, J. S., and Rotty, R. M. (1976) *The global carbon dioxide problem.* Oak Ridge National Laboratory Report ORNL-5194. 72p.

(2) Eckdahl, C. A. Jr., and Keeling, C. D. (1973). Atmospheric carbon dioxide and radiocarbon in the natural carbon cycle I. Quantitative deductions from the records at Mauna Loa observatory and at the South Pole. In *Carbon and the Biosphere* Eds. G. M. Woodwell and E. V. Pecan, Publ. U. S. Atomic Energy Commission.

(3) Bolin, B. and Bischof, W. (1970). Variations of the carbon dioxide content of the atmosphere in the norther hemisphere. *Tellus* **22**, 431-42.

(4) Pearman, G. I. and Garratt, J. R. (1972). Global aspects of carbon dioxide. *Search* **3**, 67-73.

(5) Callendar, G. S. (1958). On the amount of carbon dioxide in the atmosphere. *Tellus* **10**, 243-8.

(6) Siegenthaler, U. and Oeschger, H. (1978). Predicting future atmospheric carbon dioxide levels. *Science* **199**, 388-95.

(7) Stuiver, M. (1978). Atmospheric carbon dioxide and carbon reservoir changes. *Science* **199**, 253-8.

(8) Makhijani, A. and Poole, A. (1975). *Energy and Agriculture in the Third World.* Cambridge, Mass., Ballinger 168p.

(9) Bolin, B. (1977). Changes of land biota and their importance for the carbon cycle. *Science* **196**, 613-5.

(10) Adams, J. A. S., Mantovani, M. S. M., and Lundell, L. L. (1977). Wood versus fossil fuel as a source of excess carbon dioxide in the atmosphere: a preliminary report. *Science* **196**, 54-6.

(11) Woodwell, G. M., Whittaker, R. H., Reiners, W. A., Likens, G. E., Delwiche, C. C. and Botkin, D. B. (1978). The biota and the world carbon budget. *Science* **199**, 141-6.

(12) Wong, C. S. (1978). Atmospheric input of carbon dioxide from wood burning. *Science* **200**, 197-200.

(13) Keeling, C. D. and Bacastow, R. B. (1977). Impact of industrial gases on climate. In *Energy and Climate* National Academy of Sciences, Washington, D. C.

(14) Herman, S. W., Cannon, J. S. and Malefatto, A. J. (1977). Coal liquefaction. In *Energy Futures: Industry and the New Technologies.* Cambridge, Mass., Ballinger 661p.

(15) Chapman, P. F., Leach, G. and Slesser, M. (1974). The energy cost of fuels. *Energy Policy* Sept. 1974, 231-43.

(16) Schneider, S. H. (1975). On the carbon dioxide-climate confusion. *J. atmos. Sci.* **32**, 2060.

(17) Manabe, S. and Wetherald, R. T. (1975). The effects of doubling the CO_2 concentration on the climate of a general circulation model. *J. atmos. Sci.* **32**, 3-15.

(18) Mercer, J. H. (1978). West Antarctic ice-sheet and CO_2 greenhouse effect: a threat of disaster. *Nature* **271**, 321-5.

(19) Smagorinsky, J. (1977). Modelling and predictability. In *Energy and Climate.* National Academy of Sciences, Washington D. C.

(20) Ehrlich, P. R. and Ehrlich, A. H. (1970). *Population, Resources, Environment.* San Francisco, W. H. Freeman, 383p.

(21) Pearman, G. I. (1977). The carbon dioxide-climate problem: recent developments. *Clean Air,* May 1977, 21-6.

(22) Kukla, G. J., Angell, J. K., Korshover, J., Dronia, H., Hoshiai, M., Namias, J., Rodewald, M., Yamamoto, R. and Iwashima, T. (1977). New data on climatic trends. *Nature* **270**, 573-80.

(23) Broecker, W. S. (1975). Climatic change: are we on the brink of a pronounced global warming. *Science* **189**, 460-3.

(24) Gifford, R. M. (1978). Energy flow in Australia: Matching needs and supplies. *Aust. J. Public. Admin.* **37**, 69-83.

(25) Australia. National Energy Advisory Committee (1978). Report No. 3. A Research and Development Program for Energy. Australian Government Publishing Service. 20p.

(26) Groves, K. S., Mattingly, S. R., Tuck, A. F. (1978). Increased atmospheric carbon dioxide and stratospheric ozone. *Nature* **273**, 711-5.

Note

Any social interpretations expressed in this paper are those of the author and do not necessarily reflect the views of CSIRO.

Killer-Joules — Health Implications of Energy Use

Diana and Ian Maddocks

215 Brougham Place, North Adelaide 5006.

The cruder population indices of human 'health' — life expectancy or infant mortality — roughly parallel the affluence of the groups measured, and the amount of energy which they habitually consume. Any reduction in these indices (improvement in 'health') seems to depend more upon heightened economic prosperity than upon medical effort, and influences which are clearly energy-related, for example, improved transport, may be just as important as (say) a higher level of education in the community.

No universally-agreed definition of 'health' is available, and one can argue on the one hand that the greater use of energy resources by human populations improves their health, while on the other hand giving rise to harmful effects and deterioration in health. In this paper our intention is not to attempt some sort of ledger balance of the gain or loss of health consequent upon energy exploitation, but to emphasize the subtle but significant ways in which dependence upon slave energy is having unfavourable effects upon human well-being. Our consideration has been greatly facilitated by a recent paper by Newcombe (1), who devised a comprehensive classification for the relationship between "Energy Use and Human Wellbeing". Our approach seeks to be simpler, and is influenced by our medical background. We have used six broad headings:

1. Inimical Environmental Change
2. Violence
3. Alienation
4. Rapidity of Change
5. Loss of Control
6. Disease

Inimical Environmental Change

Particularly in Cities, but potentially anywhere on the globe, changes continue which increase pollution of all forms — chemical, particulate, radiation, noise etc. — jeopardizing normal physical and physiological function, and initiating pre-disease states. The aesthetic potential of human surroundings deteriorates, the size of connurbations stretches patterns of human movement and promotes anxiety through repeated exposure to the unfamiliar.

Violence

Human bodies are caught up in the mismanaged motion of motor vehicles and other machines, or exposed to the explosive potential of many kinds of fuel. Violence is also deliberately initiated by man against man because of frustrations which stem partly from uses of and attitudes to energy.

Alienation

In seeking to cocoon himself within arbitrarily-defined limits of comfort, man separates himself from earth rhythms and cycles, and minimizes or ignores the regularity and alternation of the natural world. "Jet-lag" is an extreme form of the disharmony induced by our imposed urgency. We are also separated from each other, boxed individually in motor cars, turned away from face-to-face contact to experience vicarious electronic or animated images, lacking immediacy and physical confrontation.

Rapidity of Change

"Future Shock" is always with us. The continuing evolution of techniques brings a steady discounting of experience, and a reversal of older hierarchies in which mature individuals had the greater measure of authority, and everyone 'knew their place'. The replacement of people by machines, and the repeated replacement of machines and processes by new machines and processes causes insecurity and anxiety in all of us.

Loss of Control

Control is lost when machines interpose a distance between persons and their foci of interest. Even the rapidity of electronic aids may confuse and inhibit communication. The complexity of domestic and public devices bewilders us, and when they fail we are paralysed. Centralization and aggrandisement of authority is encouraged by the availability of large energy resources which require massive plant to exploit them. Each of us is at risk of being controlled by our domestic energy slaves.

Our household does not own a car, but when we borrow or hire one, we are surprised by the many tasks and opportunities which present themselves just because there is a car to use. Tom Gavranic, in a provocative abstract for this conference (2) suggested that beyond an average of eight energy slave equivalents per person, the social structure becomes distorted towards maintaining the slave system, rather than maintaining personal autonomy.

Disease

The circumstances outlined above arouse harmful, 'unhealthy' responses which are largely latent and sub-clinical, but which are increasingly recognised as inimical to well-being. Beyond these there are patent maladjustments which are defined as sickness. Some are called *organic:* bronchitis, emphysema, and the diseases of pollution; coronary artery disease and the diseases of gluttony and sloth; peptic ulcer; hypertension, and the diseases of stress. Others are called *non-organic:* anxiety states, psychosomatic and psychiatric disorders.

The way in which we approach the relationship between energy and health stems from two influences:

1. A medical background

In our practice, we become more and more impressed by the importance of social pathology in the causation of disease, and of the basic roles which anxiety, frustration, alienation and repressed or open violence play in determining 'health' for all of us. Anyone considering this subject will quickly associate toxic industrial smoke and chronic bronchitis, but even if all pollution is controlled, if "pure" energy becomes freely available, there are many undesirable and unhealthy effects of energy use to be faced.

2. A cross-cultural experience

For several years we lived in a Papuan village close to Port Moresby. It was an intimate, integrated community, in which all persons knew their shared past through the genealogies of their families, and knew in great detail their immediate environment — the inter-relationships of all the 1,000 villagers, the names and characteristics of each fruit, each fish, each little hill, each part of the reef. They knew the tides and the winds and the seasons, and geared their activity to fit with natural cycles.

Papuans live in close contact with their earth. They have experienced hunger, and they are mindful of their dependence upon the earth and aware of the limitations of its resources. They have never treated the land they garden or the sea they

Diesendorf, M. (ed.) (1979). *Energy and People*. Canberra, Society for Social Responsibility in Science (A.C.T.)

fish as commodities, but rather as things alive and sacred. So they never fished in excess of their daily needs. Their gardens were tended in a spirit of energetic cheerfulness with measured bursts of activity and ample time out to talk, cajole, romp, help, sleep or eat. This alternation of intense activity and rest is said to be a more efficient use of human energy than our relentless striving. Certainly New Guinea farming is very efficient — New Guinea farmers can produce up to 20kJ of food energy for every 1kJ expended. Our farming technologies demand about 5kJ of energy for each 1kJ equivalent of food produced (3).

Papuans share what they grow and what they make — *not* to do so goes against the natural order as they see it. They have no expectations of a "good bargain", of something for nothing — evil comes from such an inclination. Every obligation is repaid, which is an excellent way of keeping in touch with the effort expended.

The "advances" promoted in our culture claim to increase human efficiency by enabling people with greater slave energy and offering them cheap power — the equivalent of something for nothing. Buckminster Fuller sees this in a broader context of *Cosmic Costing,* and explores the very different concept of *"doing more with less"* aiming to obtain more useful units of life support (food, shelter, health care, education, recreation, transport, communication) from fewer resources. If you 'cost' the energy used down the millenia to make even a single gallon of petrol, then our system of transport is wildly inefficient. Until a century ago, 99% of the total energy used by mankind was consumed as food to power both humans and animals. In the U.S.A. in 1971, only 1% of the energy consumed was in the form of food to support muscle-accomplished work, while 99% was consumed by inanimate powered machines and processes. Highlighting this, is the delightful illustration that in the North American continent at any one time, two million cars are halted at stop lights with their engines running — the equivalent of 200 million horses jumping up and down and going nowhere (4).

How *can* you live with energy-conserving people and not become conscious of the gross folly of practices of your own culture? Our life style in Papua was dependent upon two motor cars, and foodstuffs almost totally imported from Australia, Europe or North America. Our practice of medicine, imported and imposed with powerful benevolence, took some rude shocks. Papuans saw disease as occurring when a natural law or accepted custom was transgressed, or when relationships between people were threatened by anger, guilt or fear. Sickness was met by family consultation to uncover the circumstances which had initiated it or the persons who were aggrieved or angry, so that the harmony which maintains health could be restored. Such consultation promotes the interdependence of the family group, and is a highly health-maintaining activity. Further, most people get well anyway, and in this context our pill-popping and plastering came to appear wasteful, even harmful — an assault on the human biological system. In that close-knit and secure setting, one was reminded that the human body is a truly magnificent piece of equipment, inherently capable, by itself, of withstanding many afflictions.

With Lewis Thomas (5) we began to look to a system for increasing awareness about human health, with more "curriculum" time for acknowledgement and celebration of the absolute marvel of good health which is the real lot of most of us most of the time. Our experience in Papua made it easy for us to endorse the statement by White (6) that "the type of social system developed during the human energy era was unquestionably the most satisfying kind of social environment that man has ever lived in" — meaning, for us, that the comprehensive knowledge and shared institutions available to the people of Pari gave them a confidence and security and meaning which are no longer accessible to us, who behave, as Schumacher notes (7), "as though we had no past and no future".

It is not surprising that in returning to Australia we have sought to capture some of the characteristics of Pari life for ourselves. We came to an urban community of two families and several adults who practised total economic sharing, part-time employment, active food production and voluntary simplicity which allowed no motor car or T.V. We sought to live lightly and harmoniously with our earth, and drew great strength from the group in breaking away as individuals from the usual expectations laid upon us. Self-actualization for the women was an important component of our change. Nancy Todd (8) has written that "as a woman it was devastatingly insulting to have an economy structured around the fact that (she was) malleable and stupid enough to be manipulated into buying whatever (she was) told in order to keep a small cog in the economic machine turning". So we aimed to become non-consumers, or at least critical and frugal consumers.

Through this experience of another culture and through our struggle to initiate some innovation and autonomy within our own culture we have come to see *Energy* and *Health* as words whose meaning is strongly influenced by culture. In our culture, these words have overtones of exploitation and chance. When we say 'energy', we mean energy available to *us,* a change from low entropy to high entropy which we can employ. When we say 'health' we too often mean the happy chance that the misfortune of disease is far away. Papuans had no word for energy — only words for heat, bigness, speed or personal strength which were used to express ideas of power. And they had no word for health. But they had a very strong sense of a natural order and the need to maintain it, and an understanding that there are causes and meanings to be discerned behind apparently chance events. For them, both energy and health were related to a holistic view of the universe and its harmony.

We are suggesting that this unifying view is appropriate to our own life situation. We are not suggesting a return to village life. Much that was there was brutal, cruel and hard. We are suggesting a non-exploitative and consistent approach to the universe which accepts the moral view which we were made aware of in Papuan life. The way ahead then seems to be not merely a matter of saving on fossil fuels or finding non-polluting power sources (important though those will be). Nor will it be a matter of manipulating funding mechanisms to reduce health costs. What we are looking for is a new paradigm and a new morality, changing from the view which applauds the good luck of success in doing things bigger or better or faster than the rest, to a view which acknowledges the limitations of our world and accepts responsibility for sustaining it, knowing that energy in nature and human well-being are of one kind:

The force that through the green fuse drives the flower
Drives my green age; that blasts the roots of trees
Is my destroyer.
And I am dumb to tell the crooked rose
My youth is bent by the same wintry fever (9).

References

(1) Newcombe, K. (1975) Energy Use and Human Well-Being in Urban Centres: The Case of Hong Kong. Paper presented to *13th Pacific Science Congress.* Cres, ANU.

(2) Gavranic, T. (1978) The Per Capita Energy Limits for a Sustainable Society. *Abstracts,* p. 10. *Energy and People Conference,* September 7-9, A.N.U., Canberra.

(3) Todd, J. (1976) The Dilemma Beyond Tomorrow. *Journal of the New Alchemists,* No. 1, 122.

(4) Fuller, R. Buckminster (1973) Geoview: Cosmic Costing, Part II. *World,* February 13th., p. 41.

(5) Thomas, Lewis (1974) *The Lives of a Cell. Notes of a Biology Watcher.* Future Publications Ltd., London.

(6) White, L. A. (1959) *The Evolution of Culture.* McGraw-Hill, New York. quoted by Newcombe (1).

(7) Schumacher, G. F. (1969) Industry and Morals. *Resurgence 2,* No. 5 reprinted in *Time Running Out,* Prism Press, 1976.

(8) Todd, Nancy J. (1976) Women and Ecology. *Journal of the New Alchemists,* No. 3, 107.

(9) Thomas, Dylan (1954) *Collected Poems,* 1934-1952. J. M. Dent and Sons, London.

The Fusion Alternative — How Clean?

Phillip Hart

Plasma Physics Dept
Sydney University 2006

A paper like this is perhaps a little early. About twenty years too early. It will be at least twenty years before fusion energy is commercially available.

Fusion seems to have something for everyone. It is "clean", which suits environmentalists; it is centralised, which suits industrialists; it is "inexhaustible", which suits us all.

Until now, a paper like this on fusion would have been merely conjecture. Recently, however, several trial designs have been completed for possible reactors. In a widely misinterpreted report, it was recently announced that Princeton University has demonstrated that one of the most feared theoretical blocks to fusion is, in practice, not a problem (1). There is tremendous excitement in the fusion research community.

So it *is* time to see whether fusion is all it's cracked up to be. If we start now, we have time to avoid environmental and design problems.

How are we going to go about all this? We can play a little game; first we can talk about what "they" do tell us about fusion, and then we can talk about what they don't tell us.

Exactly how "inexhaustible" and "clean" is fusion?

Let's first see how a fusion reactor will work. In this paper, we will not look as "laser fusion". It is about 10 years behind the "plasma fusion" program, and has entirely different problems.

A plasma fusion reactor has a very hot gas as the main component. This gas is so hot that the atoms get torn apart. The nuclei then combine to form new, heavier elements. A normal gas we can keep in a bottle. A very hot gas ("plasma") can be kept in a "magnetic bottle".

The type of magnetic bottle that currently looks most promising is the so-called "Tokamak". The diagram of a Tokamak (Fig. 1) shows its dough-nut shape, with the plasma at the centre of the "dough", surrounded by a "first wall", with a coolant flowing through it. This heated coolant produces steam to turn such common-place turbines as we use today in coal and fission systems, to produce electricity.

The fusion device simply supplies the heat.

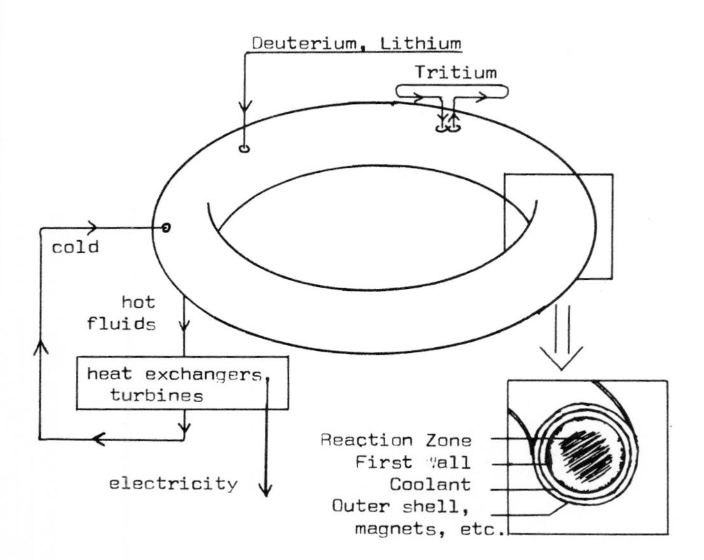

Figure 1
Schematic Diagram of a "Tokamak" Plasma Reactor

What is the fuel? Two isotopes of Hydrogen are used in the plasma, both of them very small fractions of the Hydrogen available to us. The isotopes are called "Deuterium" and "Tritium", of which Tritium is radioactive.

How much of this fuel do we have? Present-day U.S. electric power generation consumes, on average, about 180,000 metric tons of coal per hour. We could replace this by using just 10 kilograms per hour of Deuterium. This is the amount of Deuterium which we can separate per hour from water flowing through a 5 centimeter diameter pipe at normal pressures (2).

That's what they tell us.

What they don't tell us is that one of the fuels, Tritium, is so rare in nature that we have to make it ourselves, and to do this, we need Lithium. Unfortunately, Lithium isn't as plentiful as Hydrogen (which, incidentally, we can get from tap-water). If we were to supply all the U.S.'s electricity using fusion, we would have enough Lithium for three year's supply, at present prices (3).

This doesn't turn out to be as big a problem as you might first think. Several reassessments of Lithium reserves have proved favourable (4), and if we are willing to pay $100 per gram for Lithium, we can get it from sea-water. The Lithium reserves then are 100,000 million tons! (5). This is the first sign, though, that we have to be more careful in our claims about fusion.

Are there any other incidental resource problems that we don't hear about? Well, yes. Most present designs for reactors require Niobium for the magnets and Beryllium as a "neutron multiplier". Table 1 shows how much of a resource problem we have with these elements if we rely on them.

Table 1
WORLD AVAILABILITY OF SELECTED METALS
(Derived from Mitchell (3))
In Millions of Tons

Metal	Requirement for 1 year's operation at present US electricity consumption rate.	World reserves at: Present prices	3xPresent prices	10xPresent prices
Lithium	0.3	1.0	2.7 + ?	? + ?
Niobium	0.2	7.8	17	
Vanadium	0.1	26	100 +	
Aluminium	0.2	1500 +	2600 +	very large
Beryllium	0.04	0.02 + ?	0.04 + ?	0.25 + ?

So, using present designs, fusion power is not "virtually limitless".

Despite these observations, many physicists see no problem here. There are two reasons for this. Firstly, many don't even realise there is a problem. The work is still in very early stages, and it is very hard to get a clear perspective on the whole problem. Secondly, most of the physicists who do recognise the resource problems are confident that new solutions will be found later. The present emphasis is on just getting some reactors working; the second and third generations of reactors will benefit from what new things we learn about materials' properties and resource availability. Since fusion won't be supplying more than a tenth of our electricity until half-way through next century (4), we are not in danger of running out very soon, anyway!

One major advantage of fusion reactors is that they cannot

Diesendorf, M. (ed.) (1979). *Energy and People*. Canberra, Society for Social Responsibility in Science (A.C.T.)

have "core melt-down" as in fission reactors. Nor can they explode. Physicists have to work very hard to heat the plasmas, and whenever something goes wrong, the plasma touches the wall and instantly cools down. Very much like dropping a match into a bucket of water, the fragile heat source is extinguished within thousandths of a second.

This also means that terrorists cannot use the fusion process to make bombs. To make a Hydrogen bomb, you need to surround it with a Uranium/Plutonium fission-type bomb, so there is no diversion problem. Also, as the Tritium is produced on the reactor site, it does not have to be transported long distance (6).

But let us look at one of the other claims for fusion: "There are no radioactive wastes".

There are no high-level, "hot" radioactive wastes. There are, however, considerable volumes of low-level radioactive wastes. The walls, supports and magnets of the device are exposed to high-energy neutrons. These neutrons hit the walls each with fourteen times more energy even than those of Uranium-type fission plants.

Metz (7) suggests that a sizeable fusion program will generate some 250 tons of low-level radioactive wastes per year.

The story is this:— All nuclear energy reactors produce low-level wastes. The advantage of fusion (Hydrogen) over fission (Uranium/Plutonium) is that even the worst radioactivity from fusion is much less offensive than that from fission. Table 2 compares the two systems. The "Relative Biological Hazard" is a commonly-used measure of the "nastiness" of a particular waste. Let us compare a Tritium leak, which is the main radioactivity problem with fusion, with Plutonium waste, a major hazard with fission.

Table 2

RADIOACTIVE WASTES COMPARISON
(After Steiner (15), See also Coffman and Williams (14))

	FUSION			FISSION	
	Tritium	Niobium 93m	Niobium 95	Iodine 131	Plutonium 239
Half-life	*12 years*	*13 years*	*35 days*	*8 days*	*24,100 years*
Radioactivity (millicuries per watt (th))	12	87	155	32	0.03
Maximum permissible concentration (millicuries per centimetre cubed)	2000	40	30	0.001	0.0006
Relative biological hazard (millions)	*0.006*	2.2	5.2	23,000	*49*

Tritium, being a Hydrogen isotope, is a very small molecule, and hence easily slips through cracks, vacuum seals, and is even capable of migrating through hot metals! The leakage of Tritium under normal conditions is said to be "manageable" (4).

A thousand times more radioactivity of Tritium than of Plutonium is produced, but the chemical and physical properties of the two are such that Plutonium is three million times more dangerous to us. (The "Maximum Permissible Concentration" is that much smaller for Plutonium than for Tritium.) Hence, that measure of "nastiness", the Relative Biological Hazard, is ten thousand times higher for Plutonium than for Tritium. It is that much "worse".

All this also ignores the fact that Plutonium is with us for a much longer time after we've made it. Plutonium's half-life is 24,100 years, while Tritium's is just 12 years!

Sounds pretty good, eh?

One problem remains, though. By using Niobium in the magnet of the fusion machine, some rather vicious radioactive wastes are formed, amongst them, Niobium 93 and Niobium 95m. These are about a thousand times "nastier" than the Tritium, which is the next worst.

What is the point of all this? The point is that we should not be designing all these problems into our machines. Right now we should be remembering the advantages of fusion, and not be designing machines in which these advantages are minimised.

Low activity blankets have been designed, where none of the Niobium (or equally worrying, Stainless Steel) radioactive daughters are present. Vanadium and Aluminium structures have been designed which have thousands of times less radioactivity problem (8, 9, 10).

And so we come to the point. The reason why fusion energy is a nice energy is that the problems mostly come from how we design the reactor. There is a comparatively large number of ways we can design a fusion reactor.

Compare this with a fission reactor; there, the problems result directly from the type of fuel used. The radioactivity from fission is worst in the actual products of the reaction. In contrast, fusion energy's problems can, in principle, be largely designed away (11).

And this is why the whole topic is being raised now; if we realise *now* that we want to minimise certain advantages (viz. low radioactivity, abundant fuel) from fusion, we can devote effort in the right directions, and avoid putting a lot of time into answering the wrong questions. It may be too late in twenty years to turn around and say "If only we'd thought about this or that twenty years ago."

To better assess the environmental impact of fusion, we should start thinking *now*!

To close the paper, let me mention the third of the really big problems facing fusion.

Tokamaks may be too big and costly.

They are very complex, and need to be very large before we get a net energy gain. Typically, the reactors must be thousands of megawatts before they are economic. How many electricity authorities are going to be happy to buy such a technologically complex, mammoth plant to supply their base power? If the plant were to go down, the bulk of the production would go all at once (13). (However, it has been pointed out that one thousand megawatts will be only one percent of the U.K.'s electricity requirements in the year 2000, so perhaps it is actually relatively small (12)). Reactor designs are actually getting more complex as engineers learn more about what is required.

It is becoming clear that whether fusion power is economic will depend heavily on how much time is spent maintaining and repairing the materials exposed to the intensely radioactive environment at the centre of the reactor (4).

Summing up, we must admit that there are plenty of problems with fusion energy. There are, however, very great gains to be made in using fusion. Perhaps fusion will not be the best energy source. It will probably be the best centralised energy source.

References and Notes

(1) The media reported incorrectly that fusion conditions had been attained under experimental conditions. In fact, the Princeton experiment achieved the highest plasma temperature to date, but at the expense of plasma density, so fusion was not attained. The so-called "trapped-ion instability", which was thought to be the biggest problem still confronting fusion and was expected to appear at these very high temperatures, did not appear. See *Science* **201**, 792 (1 Sept 1978) and **202**, 370 (1978).

(2) Post, R. F. and Ribe, F. L. (1974). Fusion Reactors as Future Energy Sources. *Science* **186**, *397-407*.

(3) Mitchell, J. T. D. (1974). *Nuclear Fusion (Vienna)* Spec. Supp. p517; Workshop on Fusion Reactor Design Problems, Abingdon, Berks. U. K. 1974.

(4) Kulcinski, G. L. (1978). An Assessment of the Advantages and the Role of Fusion Power in Meeting Future Energy Needs. *Nuclear Fusion* **18**, 7, 988-992.

(5) Because fusion is such a capital-intensive energy source, using this more expensive Lithium as fuel will result in only a 0.05% increase in the cost of the electricity; if Lithium is used only as a fuel, it appears economic. However, many present designs also use Lithium as a coolant. This would be prohibitively expensive.

(6) Some considerable movement towards a so-called "hybrid" reactor has taken place in USSR, and to a lesser extent in USA. A hybrid reactor has both fission and fusion reactions taking place, and shows that it is possible to breed Plutonium using an adapted fusion reactor. The advantage of the

hybrid is that it reduces the constraints on the fusion part of the reactor, essentially making it easier to get the fusion reaction.

We have seen how carefully you need to tend a plasma to get fusion, unlike fission reactors, where it is a matter of trying to get a large enough chunk of fissile material together. Because the hybrid is very complicated, "back-yard" terrorists are incapable of getting their Plutonium this way.

Governments, however, are better equipped. But it may not be worth all that trouble, when a relatively simple breeder reactor does the trick.

(7) Metz, W. D. (1976).
Fusion Research (I); What is the Program Buying the Country? *Science* **192**, 1320-1323.
Fusion Research (II); Detailed Reactor Studies Identify More Problems. *Science* **193**, 38ff.
Fusion Research (III); New Interest in Fusion-assisted Breeders. *Science* **193**, 307-309.

(8) Powell, J. R. et al (1974). Minimum Activity Blankets for commerical and experimental power reactors. *Nuclear Fusion (Vienna)* Spec. Supp. 343-346. Workshop on Fusion Reactor Design Problems, Abingdon, Berks. U.K.

(9) Darvas, J. (1974). Blanket design with low Lithium and Tritium inventories. *Nuclear Fusion (Vienna)* Spec. Supp. 377-384. Workshop on Fusion Reactor Design Problems, Abingdon, Berks, U.K.

(10) Sako, K. et al (1974). Conceptual Design of a gas-cooled Tokamak reactor. *Nuclear Fusion (Vienna)* Spec. Supp. 27-50. Workshop on Fusion Reactor Design Problems, Abingdon, Berks. U.K.

(11) For example, it was originally thought that the temperature of a reactor would be 600-1000°C, resulting in big thermal pollution problems. Recent work (4) indicates that the operating temperature can be 300-500°C, so that there will only be about as much thermal pollution as from light-water fission reactors. (These temperatures are those of the walls, not of the plasma.)

(12) Gibson, A. et al (1971). Economic Feasability of Stellarator and Tokamak Fusion Reactors. *Nuclear Fusion (Vienna)* Spec. Supp. 375-392. Plasma Physics and Controlled Nuclear Fusion Research Madison, U.S.A., June.

(13) Forster, S. (1974). Three notes on engineering aspects of fusion reactors. *Nuclear Fusion (Vienna)* Spec. Supp. 287-295. Workshop on Fusion Reactor Design Problems, Abingdon, Berks, U. K.

(14) Coffman, F. E. and Williams, J. M. (1975). Environmental Aspects of Fusion Reactors. *Nuclear Technology* **27**, p 174-181.

(15) Steiner, D. (1971). Emergency Cooling and Radioactive-Waste-Disposal Requirements for Fusion Reactors. *Nuclear Fusion (Vienna)* Spec. Supp. 1971, p 447-456. Plasma Physics and Controlled Nuclear Fusion Research, Madison, U.S.A. June.

RENEWABLE ENERGY SOURCES AND THEIR RESOURCE NEEDS

Solar Cells and People

Some implications of the possible adoption of Solar Cells by a section of the world community as a means of generating electrical energy.

E. S. Trickett
47 Wood Road
Griffith NSW

1. Introduction

This is an attempt to present a specialised subject to a general audience. For this reason it should require no more than a knowledge of high school physics and english for comprehension. Those who wish to become embroiled in the details of the technology are urged to follow up some of the references. These lead to the work of specialists in the field who are at present world leaders in their chosen area. Our business here, as I understand it, is to take a broader view of the possible impact that the adoption of these devices may have upon our social structures. In Australia we have the potential for the deployment of solar photovoltaics*, but so far little widespread interest or enthusiasm. Indeed, it has until recently been the general assumption that the use of such devices would be very modest for some time ahead. However, the solar photovoltaic cell, be it based on silicon, gallium arsenide, or "Munjuncus Diplomica", now seems to be maturing as a real device. It has, above all things, that one essential property that distinguishes it from a mass of misty predictive assumptions: it works, in the here and now, producing electric power anywhere that the sun shines. Near home there is tangible proof of confidence in the announcement that the entire trunk telephone network of Papua New Guinea is being solar powered, and that the Commonwealth Railways have called tenders for a system to serve the Tarcoola-Alice Springs railway, while the Moomba-Sydney gas pipe line is already thus served (1). Initial impetus came from the American space programme which channeled funds into initial development. In the same way a political decision in the US has now provided funds for the enablement of commercial development of viable systems.

It would be possible for Australia to participate in this development — we have the skills and resources — but it would be a political decision. So far, the decision has been that we should let someone else do the development, and expect to pay our share of the development costs when we buy the finished devices. The pros and cons of this attitude constitute one area suitable for discussion by consortia of sociologists, economists and technologists. The "developing" countries, i.e. those without an advanced technology base, are generally assumed as having to take this sort of attitude, and import a finished product. Do they, should they? Do we, should we? How far is our attitude based on lack of enterprise and shortage of risk capital? How far on shrewd caution? How far is it the result of making short term financial returns and balanced budgets an overriding consideration? We have the sunshine and the raw materials and trained people, and if the level of skill and training is not adequate, it points to a questionable state of affairs in our training and research establishments. So far, one Australian University has struggled to take up this challenge, and has come up with a very fair showing on modest resources.

What the solar cell does *not* at present have is an assured immediate commercial future in competition with other power sources at present prices, except in certain special applications. If the success in economic terms was obvious, there would at once be rapid investment of capital and development. Even those of us who believe that there must be a change in the situation in the foreseeable future, we are not yet prepared to commit our capital to such projects. Indeed it is those with this faith who will be most concerned to wait and watch for the moment when the initial capital expenditure needed for membership of the club will reach its lowest point. The desired fall is at least in part dependent on the growth of the market, so that we have a feed back mechanism operating.

The other main restraint at present is the uncertainty of the possible life to be expected from the solar panels. As has often been remarked, one of the problems with solar energy is that no one can make a profit selling it . All the expenditure lies in the initial capital outlay, apart from very modest maintenance costs. This means that the difference financially between success and failure is critically dependent on the working life achieved, and the level of replacement and maintenance required in practice, not in theory, or as an initial promise. Indeed, any guarantee that does not provide for direct replacement of all failures for a measurable period ahead, say five years, will be of little value.

The call for contributions to this conference set out certain lines of attack on possible problems inherent in the choice of future energy sources. It stated very shrewdly that it was not the intention to promote any single energy future. Examples quoted for topics might have been taken to indicate a reluctance to get into technical detail, perhaps because people from other disciplines might have some difficulty in following such detail. Here, the attitude that has been adopted is that it is necessary to have some detail in order to obtain balanced understanding; but the task of the specialist is to translate as far as possible into plain language.

2. The Photovoltaic Cell

The specific property of this particular capture system is that when illuminated it yields direct electrical output, and in itself has no moving parts to require any mechanical maintenance. In fact, maintenance consists only of keeping the surface of the collectors clean, and replacing them if they fail. A solar collection system that uses an intermediate thermal path to run an engine and a generator will have much of the present character of a thermal power station, and is seen as a different animal, to be discussed separately. This approach is open to argument, especially in the long term, where hybrid systems generating both photovoltaically and thermally may well become the major development.

The solar cell consists of an illuminated surface with a depth very small compared to its area. To this surface, electrical contacts are made, currently at the front and back, although this may alter with certain manufacturing techniques. Cells stacked for edge illumination are being investigated by some workers in the USA. The critical requirement is for a semi-conducting layer in the which light is absorbed to release charge carriers, holes and electrons, which will eventually recombine.

*Direct conversion of solar energy to electricity.

Diesendorf, M. (ed.) (1979). *Energy and People*. Canberra, Society for Social Responsibility in Science (A.C.T.)

The provision of a potential gradient across the layer causes charge carriers to migrate apart and form an increased carrier density at each electrode. When the electrodes are connected via an external circuit, the charges can neutralise via this external circuit, manifesting as a current flow capable of doing work. As long as the illumination is maintained, the current flow is maintained at a level dependent on the illumination intensity. The ratio of energy supplied to the external circuit, to that supplied to the cell in the incident light beam, is taken as the efficiency of the cell. Experimental values of efficiency of the order of 20% have now been obtained and it is suggested that values of over 25% may be possible in the future. Commercial arrays are more like 10%, but rising. Anyone unfamiliar with the devices who really wishes to study the detailed theory and practice might start with the reviews by Kelly (2) and by Godfrey and Green (3) and the relevant chapter of the National Academy of Sciences report on energy for rural development (4). This can be reinforced with an overview in detail from the IEEE photovoltaic specialist conferences (5), and the detailed design procedures given by Angrist (6). Those not motivated to enquire in depth may take the solar cell as a "magic box" that converteth sunlight into unidirectional electric current. Beyond this it may help to know that maximum efficiency is only maintained at all levels of illumination if the correct electrical load is presented to the cell, and that this correct load is a function of the level of illumination. If the temperature of the cell rises too far, the material may be degraded or destroyed. There is a tendency for output to fall as temperature rises, but this depends on the materials used. For some materials the optimum operating temperature may be well above ambient. If the cell is exposed to weather unprotected, it will also be liable to damage. Shading parts of an array can cause problems of local heating in the shaded areas (7). In short, there are rules of use to be learnt. But very crudely, one could almost look upon the solar cell array as a solar powered battery. When the sun sets, the battery is no longer powered and one must rely on power stored during the day in some other device.

3. The Photovoltaic Cell as a System Component

While popular attention focuses on the solar cell as the sunlight to electrical energy converter, the realities of direct solar electrical power generation demand other parts to the complete supply system. The cells have to be mounted in some fashion and the individual cells have to be protected from the weather and interconnected electrically. The d.c. output from the system may be fed directly to a bank of storage batteries, or it may be inverted to given an alternating current supply. If no form of storage is used then any power available will inevitably fluctuate as the sunlight changes with time of day and with cloud cover. If the solar panels are stationary this daily change will be increased due to the changes in effective area of cells offered at right angles to the solar beam as the sun follows its natural path. (If the panels track the sun this effect is removed and the only reduction is that due to increase in atmospheric path length away from the noon condition.)

Changes in output of cells are changes in both current and potential (voltage). Ideally, the electrical load resistance presented to the cells should vary as the output level varies in order to maximise the power into the load at all times. If this is not done, one loses some of the efficiency of conversion possible from a given set of cells. One comes to see the need for a controller, if only a simple limiter, to prevent overcharge of the batteries during prolonged sunny periods. Remember that the capacity of the battery bank and the area of the cells have to be adequate to cope with the worse case situations *if* the requirement is for power at any time. The simpler system will be one where, as with the mediaeval windmill, the job only gets done when the power source is energised: when the wind blows; when the sun shines.

Initially the focus was on the solar cell itself, because of the very high cost of the individual elements. We have now reached a stage where the cost of the cells may turn out to be only a modest part of the overall system cost. The reason for this has been the rise in interest in the possibilities of concentration of sunlight, i.e. with mirrors or lenses. It is possible to run solar cells at more than one sun, i.e. at intensities above maximum clear sky terrestrial solar. In fact, there have been reports of experiments running cells at concentrations of over 1 kilosun. Obviously, at high concentration levels, it will be essential to track the sun as tightly as a telescope tracks a star. Where modest concentrations of the order of 50 to 100 suns are considered there are less stringent requirements, and it has even been suggested that there are ways of obtaining static concentration (8). When cells are run at high concentrations one starts to find the problem of cooling becoming apparent. There have been suggestions and designs for hybrid systems where the cells are used as dual collectors in a combined thermal/electrical system. If high temperature operation is envisaged, gallium arsenide or some other alloy may be preferred to silicon as the material from which to fabricate the active portion of the cells, because silicon cells fall rapidly in efficiency as their temperature rises above ambient. If high values of concentration are used, then the use of more exotic and expensive materials which are less common in the earth's crust becomes more admissable. It is fairly obvious that any concentration system based on reflection from a shaped surface is likely to do better with a point source; i.e. a sun not obscured by clouds. A technique of static concentration such as that suggested by Weber and Lambe already mentioned (8), may to some extent get over this problem, if it can be shown to work and be an economic solution. However, one has the feeling at present that solar cells are likely to be used more in areas of high solar input than in the higher latitudes and cloudy zones, and that there will be a higher probability of finding systems using tracking and high values of concentration in areas of low cloud incidence. In areas where overcast skies predominate, one may expect to find modest concentration factors and little tracking. Where concentration factors are high, then cooling will become a more important part of the system. Where concentration is absent or low in value, the use of ambient cooling with little or no assistance may be expected. In short, there is not one possible system: there is a design for each situation, just as architecture ought to be.

It is certain that at present one will only adopt solar cells in areas where provision of other power sources and their maintenance is prohibitively expensive. Costs currently quoted lie in the range of $10-30 per peak watt installed capacity. Just what this represents in terms of costs per kilowatt-hour will depend greatly on the life time of the panels. Some performance data has been presented by Carmichael et al. discussing the results of sealing tests (9). Encapsulant is a major key in most instance to survival except where the actual cell material degrades on exposure to radiation or by thermally accelerated migration of added material or impurity. It will not, on present evidence, be easy to ge a 50 year guarantee on a panel in the near future.

A basic limitation is the maximum power density in the solar beam. While purists may argue, for all practical purposes, we can accept that we shall have a maximum at sea level on the planet surface, when the sun shines directly overhead in a clear sky, of around 1kW per square metre. As the angle between the solar beam and the receiving surface decreases from a right angle, the absorbed energy must decrease as the cosine of angle between surface and beam. This a purely geometrical matter. Obviously when the collection surface is parallel to the beam only the diffuse sky radiation component will be collected. There is also a fall in absorption toward grazing incidence with some surfaces, due to increased reflectance at low angles. This loss is a function of the particular surface used.

To obtain a certain value of power input means providing a certain area of collection. However, use of lenses and reflectors may make it possible to use less area of actual solar cell than might otherwise be necessary. To obtain a specified number of kilowatt-hours, or units of electrical energy as charged by supply authorities, one must achieve a required aperture and intensity for a specified time. To do work the electrical current at a specified potential must flow for a specified finite time. It is essential that those from other disciplines keep clear view of the distinction between electric power, which is a rate of energy input, and energy, which is the power input integrated over a specified period — the difference between kilowatts and kilowatt-hours.

To obtain sufficient integrated input requires adequate area and time. If the capture were thermal, the potential level, the temperature at which the energy was available, would be set by the relationship between the income and losses from the collector. In the electrical case we are less limited on potential since the elements which are all in parallel with regard to the solar input may be connected in series-parallel to provide the required electrical potential and current capacity. Even then, of course, the voltage is not entirely maintained at low solar inputs, and definitely not at night. One does have a more open design situation than with a purely thermal capture system.

4. Applications: Possible, Probable, Debatable, Past, Present, Future

The first application that ensured the necessary development effort was for power supply to spacecraft. These systems are well established, if costly, and the success of the work done between about 1955 and 1965 speaks for itself. Although photovoltaic devices have been in commercial production from the 1930's, they were regarded as transducers, measurement sensors, rather than as power converters.

However, once the rejects from the space programme (still it should be noted quite good cells) become available, they were eagerly deployed — first on an experimental and later on a commercial basis. Obvious choices of application were in areas of difficult terrain and access where the expensive but lightweight and rugged panels could compete with alternatives such as diesel generator sets. It may well happen within the next decade that there will arise situations where it is actually economically attractive to use the solar panels in direct commercial competition with other sources, not in space craft or on the the tops of moutains, but in towns and villages.

Brown and Howe (10) suggest that within the next ten years this could be case for the villages of Tanzania. It is in the more remote areas and in those with higher insolation levels that we are likely to see initial use of these devices. Conversely, these are the areas least likely to make them. We have to be looking at an inherent export of the device from developed to developing areas. This means from an area where their price will be weighted by the advanced living standards, to an area where the funds for the procurement will be limited. This suggests that one of the initial markets could be the oil rich countries interested in using their present wealth to ensure their long term future. Initial applications are for telecommunications purposes in the broadest sense. This is because the requirement is inherent for reliable power at modest levels, and the cells meet this requirement ideally. Polgar (11) describes the development of the power supply systems for Nigerian schools to operate television sets used in a national education programme. The solar panels were seen as more economically attractive than the former battery based system or the use of generator sets.

Telecom Australia are planning a 580 kilometre microwave trunk system between Tennant Creek and Alice Springs, with each of the 13 repeater stations powered from solar panels. This system is scheduled to start working in 1979, and should provide valuable practical experience (12). There must be many examples over the planet. However, because they do not relate to large sums of capital or to disasters or other events beloved of newswriters, they happen as the grass grows, almost unnoticed.

Acceptance of the solar panel for communications has arisen from the requirement for high reliability in remote and unattended situations. The cost of running power lines or supplying fuel to diesel sets, to say nothing of getting them into position initially, is such as to make possible the choice on economic grounds. There is a better inherent reliability in any system not dependent on a national power grid, which is vulnerable in this day and age to strike damage, by both lightning and people. This element of reliability has led to the adoption of the solar panel in areas which might be grouped under the heading of security and safety. There are a number of systems that rely critically on the continuity of supply over extended periods and only require modest levels of power. Such systems are typified by buoys, beacons, lightships and lighthouses, traffic lights, railway crossing warnings and gate controls, even electric fences come under the same heading.

Indeed, electric fences are now commercially available in Australia and widely advertised. One may have confidence in a system powered from a collector placed within the perimeter of its own detection system. Any externally powered system must eventually be vulnerable if the external power is for any reason cut off.

The upsurge in privately controlled radio is to be noted as a social fact (e.g. "CB"). Such equipment with solar powered supply has interesting implications, especially in times of stress. Government policy all over the world has always been aimed in general at maintaining control of the means of individual communication over the airways. Is this system changing? Any person, group, community, that has control of a solar powered system is assured of a continuing modest electrical energy supply without the risks to which a supply authority is exposed: no strikes, no unpaid accounts. This makes such supply very attractive in situations where confidence must not be misplaced. Essential emergency lighting, be it in house, factory, or hospital can thus be assured with the same sort of confidence that one now places in a well made torch with a new and proven battery.

Under the heading of protection we might also include the cathodic protection of metal structures, e.g. pipelines, piling, even perhaps ships hulls. This is another place where reliable long term supply is needed at modest power levels. Low voltage d.c. output is well matched to direct connection to the protective network. An advantage of the use of a self-contained power supply is that of avoiding connection to any power system that may have unexpected earth leaks, which might later spoil the system of protection and lead to early failure of the structure.

There are other possible applications involving electrolysis that would be classifiable as modest power level applications. There is already a commercial unit run from the mains to produce chlorine at a preset rate for the purification of swimming pools and possibly drinking water. Such a unit could be equally well designed to run from solar panels either with or without battery storage. Electrolysis of water is of course a route to stored power and to an oxyhydrogen flame for special metal and glass working operations (13).

At higher power levels, electrolytic refining of metals such as zinc, copper, and manganese, are interesting possibilities. It should be feasible to site the collector at the site of the ore deposit and to avoid shipping out anything except high grade metals. Electrolytic refining of zinc would be one way to store solar energy in a form ready for insertion into air-depolarised zinc carbon batteries, with reuse of the spent zinc. While overall efficiency of the system would be unlikely to be economically attractive in the near future, it is a system that could be made to work without any advanced technology other than the panels themselves. The lead acid and other secondary cells may all be classified as uses of electrolysis. The twist of thinking that starts to ee the possible applications of electroforming as a way to avoid machining process costs in energy and dollars could lead someone into one of the major industries of the 21st century.

Application of solar photovoltaics to electrolytic refining is not a new thought, as is evidenced by a picture in a recent advertisement from Philips of an installation in South America. Power requirements for such operations are currently extracted where possible from hydropower, another form of stored solar power, as with the present aluminium industry. Any changes in metallurgical technology will have profound effects here. One wonders what exactly has happened to the alternative path to aluminium using manganese fluoride. Is it unsatisfactory, or does it pose a threat to invested capital of the existing industry?

One cannot envisage supplying the power requirements for an aluminium extraction plant of the present type from solar panels for some time yet, but any process that can be arranged to work at low temperatures via electrolysis for any metal could be feasible as soon as panels are in commercial production at low enough cost.

The next application at modest power levels is food preparation, especially at the domestic level, and in the more affluent communities. Provision of heat for cooking (sterilisation) of foodstuffs is probably more a matter for the thermal solar collector champions, but one might envisage a microwave or even a well insulated thermal cooking stove being run from solar charged batteries. The energy used each day by a household for thermal food treatment is of the order of a few kilowatt hours, and could well be achieved photovoltaically, *at a price*. The attraction of the microwave stove is the high thermal efficiency of the process which essentially heats only the food being processed. One might even start to take the "waste" heat from the r.f. generator and use this to preheat the water in the kettle! When one starts to consider operations such as running a "mixmaster", the possibility is quite definite. Energy consumption for such operations is not large, but the relief it affords the human arm is considerable.

In general the most likely applications will involve modest power levels of the order of 0.1 to 1 kW for short periods, probably related to turning a shaft or producing other motion. Where mains power is installed, solar cells will have little meaning at present prices, but could complement other power sources in areas where mains power is not available. In fact, one can run from batteries the sort of 32-volt equipment that is already well known to many Australians who have lived in rural areas.

Coming up the power scale, one may in the same way run a small workshop, where loads of up to a few kilowatts are required for short intervals. It is interesting to speculate on the possible implications of an increased interest in d.c. welding from special heavy duty batteries and the effect that it might have on the development of the technology. One can conceive performing the sort of work involved in the repair of farm machinery, even if the heavy duty cycles of large structural work were too demanding. Battery welding has been used in the past, but is currently considered as superseded by mains operated a.c. systems. The competition here could be between the electric weld, and the solar powered oxy-hydrogen torch.

Some of the larger photovoltaic proposals are almost literally pie in the sky. The proposals to put a solar power satellite in orbit and beam power to earth at micro-wave frequencies has received so much publicity in the popular media that almost every one is likely to be familiar with the general concept. The scheme is analysed by the CCIR working group 2A, and their report has recently been published (14). If it does get into orbit, this scheme will have the same effect as providing another large central power station, but it is difficult to see in what way it will be better than the same effort devoted to a somewhat larger array area on the planetary surface or to a number of smaller installations dotted over the country side. If serious thinking people will spend time and effort on the analysis of schemes of this novelty, surely some of the more modest earth based possibilities are worth a careful look?

At the other extreme from the concept of vast central satellite power stations is the "solar tile" concept. Here each of the roof tiles used on a house, or at least on the sun facing areas of roof and possibly the walls, would include in its outer surface a solar cell. As the roof tiling proceeds, each of these would be connected to busbars along the tiling battens, leaving the finished residence with its own power source independent of all others. This would remove the need for any distribution network but would provide a limited input. Not, however, an entirely insignificant one. At 10% efficiency, one might hope for a peak power output of some kilowatts from a roof area 5 metres by 10 metres. Cost and energy inputs are quite another matter. If there were no economic catch all the world's bright entrepreneurs would be now be well entrenched in a brand new industry with masses of glossy adverts. We would find getting technical data about as easy as getting real detailed data on batteries is now. This sort of situation is not likely to happen until after we start to pay considerably more for a litre of petrol than we are paying for a litre of orange juice.

5. Impact of Solar Cells

This can be a wide ranging matter and has here to be treated within the terms of reference of this conference. Attempt will be made to do just this, but the indulgence of readers and audience is requested in that the target selected is the essential meld of technical and sociological.

(a) Industrial impact in developed and developing countries.

At the outset there are some points to set straight. First, that the energy source under examination is unlikely to make in the near future any spectacular contribution to world energy requirements. However, somewhere between the present day and about 2100 AD, the production of solar photovoltaic collectors and associated equipment may have become a multimillion dollar industry. One expects a steady rise in fuel prices, with a slower rise in the production costs of the solar cells, since they are still benefiting from the learning process. Once they have reached a more stable technological base, their costs will level out, and must then rise as fossil fuel prices rise. This rise must continue until such time as the installed capacity is adequate to provide the power base for the manufacture of future equipment. At this point the industry takes off, and is thenceforth free from any ties with the problems of recovery of fossil hydrocarbons, except in so far as they may represent an essential chemical feedstock, e.g. for plastic insulation material etc.

Even once it has reached this point, it may well prove inadequate to cater for a major part of the essential power *needs* of the population of the planet. In fact, it probably cannot at any time achieve this status unless world populations fall. What it may well do, however, is provide a stable and reliable base to power certain essential services and protect us from the consequences of rapid breakdown that can occur with social instabilities.

The provision of a stable local base is discussed by Martin (15) in the context of solar energy, but he seems to consider the high price of solar cells a major impediment to their use. He does not seem to have great hopes of an early improvement and therefore looks to a fuel cell powered from reformed methane. The point is raised as to how much of the cost of a house in an affluent country should be devoted to the energy base. Martin is aiming at a peak electrical demand of 6 kW. The fuel cell is intended to meet this demand. It could be met by a storage battery bank and an inverter, but this and the cells to charge it would be costly. What proportion of the cost of the house should be considered reasonable? There is bound to be considerable argument and discussion between the electricity from the sun, and the gas from waste champions, in the next few years. In the LOPSOL Farmstead described by Martin, the anaerobic digester is a crucial item to the whole scheme. Such digesters can be made to work, but require sympathetic care. The solar cell is possibly less critical once it has been made, but does not yield fertiliser in operation.

The second important point that must be stressed is that the impact will depend very heavily on the type of society into which the technology is introduced. A recent article by Maggie Black (16) reports on the difficulties of transferring the technology of the developed world into third world countries, and of Kenyan attempts to meld traditional local technology with imported ideas. Agarwal (17) discusses similar ideas and the problems raised by the very low income levels which will effectively prevent any natural "market place" transfer of the advanced technologies where they may most be needed. At least initially the technological base for manufacture is unlikely in the Third World, although this may change over decades. It is in the developed and technologically highly developed areas of the planet that the equipment, especially the solar cells themselves, are likely to be made at first. This implies that the costing of the hardware will be loaded with the high costs of living of the developed country in which it is made, and that all the manufacturing techniques will be aimed at removing the labour content from the product as far as possible, since this represents a high cost factor: a factor which would be far less significant were the devices to be made with labour from the area that needs the device. In the case of the silicon cell, the large power requirement for the present production processes make it feasible to consider the possibility that the production of the solar grade silicon, if not the whole device, might appeal to the OPEC countries. Once the manufacturing skills and procedures are well established, will we see large factories springing up in, for example, Hong Kong, as happened with other semi conductor mass production lines, or even in the middle east?

What are the implications inherent in these possible developments, both for the people involved and for world trade? Application of the new technologies are likely to be delayed in the developed countries because they already have a power distribution network, and there is no immediate gain to be made by transferring the power base from this network, especially for the small loads that an individual might be able to afford to transfer to the high cost of solar panels. Conversely, the production techniques, especially for the present silicon cells, may involve large power loads that could be ill met in a developing country with embryonic power supply facilities. This will not hold for those countries with hydropower schemes. It is the technical expertise that is more likely to present barriers at present however.

If the developing countries wish to purchase into the new power system, how will they finance their development? By appeals for loans from the world bank? If they borrow funds, will their new power base assist them to repay these loans, or will they eventually become, de facto, a form of overseas aid from the nations making the equipment? The countries that make the equipment will have a new industry that will employ a number of people, but they will also expend megawatt hours that will be a very real cost to their economies. Perhaps the manufacturers will lease the equipment to the users, or arrange some form of hire-purchase.

One of the present pointers to the current state of the art is that technical information is still pretty freely available. Once there is any real chance of an industry being established with profits and long term future, details of production techniques will be less easily obtained.

It should be emphasised that the technical skills involved in the manufacture of solar cells, even if one has to use very special materials, are no more challenging than the multifarious skill levels involved in a modern motor car production unit, or in making microcircuits. The special skill and technology in the metal casting industry alone is often taken for granted, but represents a very finely honed situation developed over more than half a century. A similar input of development to the solar collector industry, once it can "take off" should ensure considerable development. The new factor in the semiconductor industry is purity and doping control. This may not, however, in the long term prove essential for the solar panel, depending on the success or otherwise of the present intense activity on *amorphous semiconductors* (19, 27, 28). Compared with single crystal semiconductors, amorphous materials can be prepared rapidly and cheaply, with low energy inputs.

What the industry seems likely *not* to need to the same extent as other engineering industries, is dimensional control, except perhaps in the production of the silicon ribbon crystal, if that be used, or in the thickness control of deposited films. The area of the elements and the general tolerancing on the support gear is of a low order of accuracy compared with much modern engineering practice. There may be a call for a fair degree of precision for the tracking gear if high concentration ratios are used, but nothing that industry would find hard to take in its stride.

Some experts have expressed a hope that the actual cells may be made by techniques of rolling and printing, rather akin to an offset litho plate. But it is as yet too early to have any hope of predicting the real details of development. This current uncertainty about the direction which events may take may be viewed as one of the impacts on our society. The reasons are essentially technological detail, which must not be slurred over, or brushed aside.

Ralph (18) gives some magnitudes that may be thought provoking. This was one of the earlier attempts to encourage thinking at a level previously considered as science fiction. To cover a square mile with silicon at 0.01″ thickness is given as requiring 1.5 x 10^6kg of solar grade silicon. Accepting a figure of 0.5 x 10^6kg for the US production in 1970, one is led to speculate that even today there would be some difficulty in meeting the demand for a few square miles of panel each year.

It would be the special purification and crystal growing facility that would need to be expanded. This would create a few jobs, demand capital for manufacture of plant, and, above all, would require a good deal of energy. Hunt (20) supplies some sobering figures in this respect for the energy required to make solar grade silicon. The conversion of the pure material to the finished cell takes only about 5% of the energy. The manufacture of the high grade polycrystalline material of sufficient purity takes about 95% of the quoted 3588 kWh/kg of silicon. Wolf (21) suggests an energy input of 3800 kWh/m^2 of array, but sees hope of a reduction of an order of magnitude in this figure.

It must be emphasised that this does not necessarily represent the only route to high efficiency solar cells. It is the route however, that has so far yielded the working cells for spacecraft and some terrestrial projects. Hunt emphasises these points. Further, he emphasises that improvements may well be possible and that it may not be a silicon cell that is eventually most successful. Possibilities inherent in a range of semiconductors, including the low-cost amorphous materials, have still to be evaluated. Some of these may well provide cells of acceptable efficiencies without the need for high purity of single crystal material.

What is already clear is that we may require a large amount of electrical energy in a relatively short period in order to transfer to our next energy base. It behoves mankind to refrain from squandering or misapplying too much of the current reserves of fossil fuels before the transfer has been effected. We may in effect need to use past solar inputs stored in oil and coal in order to buy the lead time for change over, be it to nuclear, solar, tidal, or windpower. If, for example, it takes 10 years to capture enough energy to make a solar cell system, then we need to have up our sleeves when we start that much energy in reserve, and to be able to get a life of at least 20 years from the cells if we expect them to yield us 50% of their total capture. If they only last 10 years, there will be little either of joy or energy to be had from the industry in the long term. It must be stressed that the figure of 10 years to repay the manufacturing energy is taken only as an example. One must hope that the figure will be more favourable than this. But a life of at least twice the payback period seems essential as a target (See 22, Table 28). One thing is certain: the days of our present throw-away outlook, so widespread in the developed world, are numbered. We shall look to increased life from all our "consumer" goods once the energy content starts to become a significant part of the initial production costs. This is the present situation with the silicon cell and we hope that we shall be able to reduce this component in the future. With most articles we have in no way reached this situation as yet. Any focussing to give higher intensities at the cell should also be a multiplier of the kilowatt hour yield per kg. of fabricated material.

If special materials which use elements not normally in demand become important, this could raise political issues of the control of supply of the raw materials. Materials like gallium are not necessarily found all over the planet as is silicon. We already have the pattern emerging with uranium becoming involved as an element in diplomacy in some minds.

We have not so far mentioned storage elements. While the use of hydrogen as a storage medium is well championed (23), it seems that batteries are likely to be an essential element in at least some solar cell based systems. If this were to prove the case we need to be aware of possible effects on patterns of endeavour, and consumption of resources. A considerable amount, possibly more than a third, of the present world lead production is used in the manufacture of lead acid accumulators. Most of these are for automobile use and will be discarded after only a few years use. From these discards some of the lead will be recovered and recycled.

To produce each kilogramme of lead requires an expenditure of 14 kWh of energy. It seems we shall have to look to recycling as much as possible, although it looks pretty cheap in energy units when compared with the silicon used in the cells.

More importantly, we shall be looking to ways to extend the working life of the batteries. So far, developments have been aimed at the short term economic gains resulting from a steady replacement market. There are currently a lot of development programmes in train on secondary cells. Most are taking place as they have done for years, within commercial boundaries, and are thus closed to external enquirers. The major developments were made many years ago and only recently have there been signs of any great changes. The most likely development in solar electric applications in the near future is an increase in the use of very pure lead for the manufacture of the plates, in place of the present antimonial lead, or even the lead calcium alloy favoured in the USA. Nickel cadmium cells seem to offer advantages mainly to those who make and sell them. The other cell that has given very faithful service in the past is the original Edison Nife cell. It is not clear why this cell is not now in extensive use. One is led to wonder if the very extended life was not in the long run an economic drawback to the firms who found they had saturated the market. An upsurge in the demand for nickel could have widespread ramifications.

One may ask if there is likely to be a resurgence of the production of the glass or ceramic battery case for fixed installations. Life for a glass container, if it is handled carefully, might be expected to be centuries. If one contrasts the figure of 6-8 kWh to make a kilogramme of glass with the 10 to 12 required for the average plastic, it looks as if in the future glass might win hands down. Where the energy expended is not a consideration, the light weight shock resistant, limited life plastic may have advantages and be preferred, basic raw material costs permitting. Once again however, we must consider both social and technical relationships in context, but we can be sure that whatever materials are used, the industry that makes the storage batteries is likely to be a growth area.

The larger the solar cell industry the more stable it can become, since it will then be able to use power from already manufactured units to process new ones, and will be less affected by the spiralling costs of fossil fuels.

Another matter for discussion is the impact on our choice of power supply conditions. At present we are accustomed to think of alternating current supply with a frequency of 50 or 60 Hz and a voltage of 240 or 110 volts as the world wide accepted standards. This means there are vast numbers of motors and appliances made to operate from this type of supply. This is one of the reasons for there being so much pressure on the research effort in solar photovoltaics to come up with solutions that will prop up the status quo, rather than attempting an entirely new approach. So far no one seems to have come up with a concept for the direct conversion of sunlight into an alternating current power source. Even Peter Glaser's panels in the sky are designed with klystrons in mind as the conversion units. This means that, if we start to meld our photovoltaics with existing power supply systems and devices, the inverter (which converts d.c. back to a.c. at an efficiency considerably less than 100%) will become an essential component of our system.

However, electric lighting and electric appliances with resistive heating will run off d.c. Appliances with "universal" electric motors (i.e. will run off d.c. or a.c.) used to be readily available, especially in the USA. Those appliances which depend critically on the specific a.c. frequency may require expensive conversions to d.c.

If we do not demand that the new technology has to fit in with existing devices, then there may be a whole new set of design criteria for electric motors for use with solar photovoltaic systems. If we wish to maintain the maximum overall efficiencies in order to minimise physical area and hence the cost of the arrays, we shall have to discard, at least temporarily, the idea of melding with existing power systems, and try to design optimised self contained systems, without imposed restraints. When we have found these optima, and proven them, we can re-examine the matter of coupling with existing neworks. One wonders if the concept of supply at a fixed voltage may not have to be abandoned in favour of the concept of maximum power into the load at various solar inputs.

Think for a moment of the mediaeval windmill, on which we once relied critically for our daily bread. They could be run only when the wind blew, and sails had to be trimmed and grainflow adjusted to meet the changes. They were in their time very effective tools of survival, but they do not meet the ideals of a society engrossed in a system where the dollar per hour and the fixed working week are almost sacred cows. Will it become necessary for our social outlooks to attain greater flexibility in order to meet the challenges of implementing a solar based economy one day? It could be that it is toward these goals that those concerned sincerely with "alternative cultures" are grouping.

(b) Environmental impacts.

Possibly the most important environmental impact may eventually be that they may compete for ground area with agricultural or other requirements. Another impact will be pollution resulting from the mining of the raw materials and the manufacturing industries for the cells.

In particular, the use of gallium arsenide might raise some questions. Although it takes from 10 to 100 milligramme of arsenic as the trioxide to kill an adult human, there is a query over the long term effects of low levels. With regard to gallium, no cases of industrial illness among workers with the metal seem to be on record up to 1960. In brief, there seems no serious problem here, and silicon obviously presents no problems. If, however, it were found desirable to use tellurides, there would be cause for some concern and a need for careful legislation and controls. An expansion of the storage battery industry, even if lead were used, does not present any problem we do not already have with us.

(c) Legal problems and implications

It has been pointed out that there is no escape from the need to provide a specific area of solar collection in order to capture energy at a required rate. This means that once an installation has been designed, fitted, and paid for, the owner will not take kindly to having it obscured either in part or in toto by trees or structures belonging to their neighbour. This problem is discussed in a recent note by Eisenstadt & Utton (24). It would seem that in the USA one might possibly expect to have to negotiate an easement to ensure that one had uninterrupted solar access. This unless a specific City Ordinance or State Law had been passed to ensure the rights of each individual to uninterrupted sunshine. This is not a simple matter, and the authors refer to a prior paper that deals with it in some depth.

A whole range of matters, including the one mentioned above, are discussed by Robbins (25). It may be valuable to quote here verbatim from this paper.

> *Though these eight problem areas exist I don't want to overstate them. Our work is the work of cautious lawyers. We want there to be no snags in this change in technology. We want to clear up every possible impediment that the legal system offers . . . We don't want users to be caught with a large investment in time, in money, in equipment or ideas just to find there's a law that bars solar use.*

This attitude appears very relevant to the present discussion and it may be summed up in two words: "legal enablement". This paper is recommended reading for anyone concerned with this area of social repercussion. Here it would suffice to point out that the problems arise from an interactino between a technical reality, the collection process for solar energy, and a human social structure.

Ideally the laws should be remoulded to meet the needs of people and the demands imposed by the nature of the process. What we must avoid is any attempt to remould the laws to meet the demands of some of the people at the expense of the needs of the rest. It is to be hoped that there will be a continuing dialogue through the publication of these issues, and that it will involve sociologists, lawyers and technologists with the relevant skills.

6. Conclusions

At present, except in the minds of a few people and in a few specialised laboratories and factories, the solar photovoltaic systems are a non-event. We know they work but there is no obvious large profit to be reaped easily at little risk. They have neither the great military significance nor the potential for aggression implied in the nuclear option. Where they have significance for their special properties they have already been deployed, as in communication satellites.

The next step will await the achievement of cost reduction targets and the rise in costs of other power sources. When "take off" point is reached there will be a choice: a choice between using the resources to try and prop up existing systems and using the special attribute of freedom from large distribution networks. The first road leads to modest power supplied in cities where it can be little more than supplementary, and the second offers modest power supplied in rural areas which may be very telling in the long run in encouraging decentralisation.

Barry Commoner in a recent interview has some interesting things to say about the two choices (26). It is to be hoped that his quoted values for the possible price reduction in solar cells happen soon. As he points out, this hinges largely in the USA on the political decision to back an infant industry against the known giants who also have known lifespans. This means that the world depends largely on this decision because no other country is currently so well poised to make the first steps.

The questions raised by the conference organisers can be turned round and spelt out as queries to those servants of the peoples of the world, the politicians: "Your masters would like to know what *you* intend to do about it, and when?".

References

(1) Anon. (1978). Telecommunications Solar Powered for Papua New Guinea. *Science and Technology* **15** (5), 21.
(2) Kelly, H. (1978). Photovoltaic Power Systems. *Science* **199**, pp 634-643.
(3) Godfrey, R. B. & Green, M. A. (1978). A Review of Current Solar Cell Technology. Monitor. *Proc. IREE Aust.* pp 87-91.
(4) Anon. *Energy for Rural Development, Renewable Resources and Alternative Technologies for Developing Countries.* Report of an Ad Hoc Panel of the Advisory Cttee. on Technology Innovation Board on Science and Technology for International Development Commission on International Relations. National Academy of Sciences, Washington, D. C. pp 27-29.
(5) Conference record of the Twelfth IEEE Photovoltaic specialists Conference 1976. Held November 15-18 at Baton Rouge, Louisiana, USA.
(6) Angrist, S. W. (1976). *Direct Energy Conversion,* 3rd Edition Allyn & Bacon Inc. Boston.
(7) Sayed, M. & Partain, L. (1975). Effect of Shading on CdS/CU_xS Solar Cells and Optimal Solar Array Design *Energy Conversion* **14**, 61-71.
(8) Weber, W. H. & Lambe, J. (1977). *Applied Optics* **15**, 2299. Reported in *New Scientist,* December 1, p 566.
(9) Carmichael, D. C. et al. Review of World Experience and Properties of Materials for Encapsulation of Terrestrial Photovoltaic arrays. ERDA/JPL 954328-76/4.
(10) Brown, N. L. & Howe, J. W. (1978). Solar Energy for Village Development. *Science* **199**, 651-7.
(11) Polgar, S. (1977). Use of Solar Generators in Africa for Broadcasting Equipment. *Solar Energy* **19**, 201-204.
(12) Holderness, A. L. (1978). Solar Power for Telecommunications. *Search* **9** (4) 143-147.
(13) Costogue, E. N. & Yasui, R. K. (1977). *Solar Energy* **19**, 205-210.
(14) Anon. *Microwave Transfer of Energy.* A new report from CCIR Working group 2-A on Characteristics and Effects of Radio Techniques for the Transmission of Energy (Question AV/2) Australian electronics engineering May 1978. pp 19-26. See also: Kistemaker, J. (1978). Solar Energy from Space. *Physics Bulletin,* February, p. 58 for a good brief analysis of the proposed system.
(15) Martin, J. H. (1976). Solar Energy in Providing for Alternate Lifestyles. *Solar Energy* **18** (6) 481-487.
(16) Black, E. L. (1978). Small is Difficult. *New Internationalist* **63**, 19-21. May.
(17) Agarwal, A. (1978). Solar Energy and the Third World. *New Scientist,* pp 357-359. February.
(18) Ralph, E. L. Large Scale Electric Power Generation. *Solar Energy* **14** (1) 11-20.
(19) Wilson, J. I. B., McGill, J., and Kinmond, S. (1978). Amorphous Silicon MIS Solar Cells. *Nature* **272**, 152-153.
(20) Hunt, L. P. 1976 in reference 4.
(21) Wolfe, M. (1975). Cost Goals for Silicon Solar Arrays for Large Scale Terrestrial Applications — Update 1974. *Energy Conversion* **14** (2) p 57.
(22) Weingart, J. M. (1977). The Role of Solar Energy in the Future of a Small Planet. *Science and Technology.* pp 4-18. November.
(23) Bockris, J. O'M. *Energy, the Solar Hydrogen Alternative.* Australia and New Zealand Book Co. Sydney 1975.
(24) Eisenstadt, M. M. & Utton, A. E. (1978). On the right to sunshine. *Solar Energy* **20** (1) 87-88.
(25) Robbins, R. L. (1975). Law and Solar Energy Systems: legal impediments and inducements to solar energy systems. *Solar Energy* **18** (5) 371-380.
(26) Commoner, B. (Interview) (1977). Solar Energy is for Now. *Mazingira,* no 3/4. pp 54-61. Pergamon Press.
(27) Wilson, J. (1978). Amorphous semiconductors in action. *New Scientist* **80.** 760-1.
(28) Ovshinsky, S. R. & Madan. A. (1978). *Nature* **276**, 482-3.

Note

This contribution represents the personal convictions of the author, and bears no relation to any official policy of the government or any government department.

Forests for Energy

J. Lejeune

Forestry Department
Australian National University
P.O. Box 4
Canberra, A.C.T. 2600

Introduction

Photosynthesis ensures that man has a source of food, fibre and fuel and, because of this, forests have always made important contributions toward meeting the needs of most human societies. These contributions have changed in their relative importance throughout history, starting with simple fuels followed by the use of solid wood in building and construction, culminating in more complex conversion processes such as those associated with some kinds of paper manufacture or the production of wood alcohol. Wood being what it is, further developments must occur in the future.

In recent times the development of industrial civilisations using concrete, steel and glass seems to have relegated wood to a secondary role. Nonetheless, the absolute consumption of wood in all countries in the last 20 years has increased rapidly, even though less and less wood is used in its original form and its most important contribution to the world's economy may be yet to come.

The oil crisis of recent years has focused attention on alternative sources of energy and has made us realise that oil, gas, uranium, coal and many other resources are finite and that we are consuming them at an alarming rate. Wood, however, has three characteristics that differ from all other materials, i.e. its versatility, which makes it suitable for many manufacturing processes; its renewability, so that, in contrast with minerals and petroleum, it need not be threatened by exhaustion when properly managed; and, finally, its potential for conversion to other products some of which we have barely begun to explore. For example, wood could become an even more important source of energy in most energy poor countries and certainly in developing countries which would like to become less dependent on oil and gas imports. Currently it contributes only 6% to world energy consumption and to reach this small percentage about half of all wood felled annually is used as fuelwood, i.e. 1184 x $10^6 m^3$ (1). Table I gives absolute quantities and percentages of roundwood produced in 1976 in three broad economic categories plus the percentages which fuelwood represents of these totals.

Table I.

Total world roundwood and fuel wood production in 1976 for developed, developing and centrally planned economies. (1)

	Roundwood[1] x $10^6 m^3$	Production % world total	Fuelwood[2] x $10^6 m^3$	Production as a % of roundwood production
Developed Market Economies	773	31	50	6.5
Developing Market Economies	1050	42	871	83
Centrally Planned Economies	691	27	263	38
World	2514	100	1184	47

1. Roundwood is wood in the rough, in its natural state, as felled or harvested.

2. Fuelwood is wood in the rough (from trunks and branches of trees) to be used as fuel for purposes such as cooking, heating and power production. Wood for charcoal and portable ovens is also included.

In the countries of the Third World, 83% of roundwood harvested in 1976 was burned for primary energy. The remainder, less than 200 million m^3, is considered as industrial wood. In the U.S., fuelwood is only 5% of the total cut compared to 10% around the 1950's and 90% during the mid 19th century.

The recent rises in oil prices in the developed countries may, ultimately, reverse the historical trend towards a decrease in the use of fuelwood as a major source of energy. Developing countries which are deficient in oil and coal and already have a high fuelwood consumption have little choice other than to use wood as their main energy source. A principal reason for the declining popularity of fuelwood has been its low calorific value (between 14 and 20 MJ/kg depending on moisture content) compared with oil and natural gas (between 30 and 40 MJ/kg). Moreover, coal, oil and natural gas are more convenient to use and easier to store and transport than wood. Now, however, with the mounting costs of oil, wood is being regarded in a different light.

Many scientists and industrialists believe that wood could be a major alternative source of energy. The concepts of complete tree utilisation and plantations for energy have become more attractive in many parts of the world. Existing sources of additional biomass include under-utilised standing forests, logging residues, forests killed by natural causes, mill residues and pre-commercial thinnings. Biomass from plantations, on the other hand, represents a substantial long-term investment which could contribute to a nation's energy supply.

Energy Farming

Principles

An energy plantation produces fuels by converting and storing solar radiation in plants grown specifically for their fuel value. Selection of high-yielding species, adapted to the site and to specialised cultural practices, is the key to producing fuels from plantations at an attractive cost. The spacing between trees is chosen to maximise the yield. Spacings of one meter or even less are commonly adopted. If growth is sufficiently rapid the stand can be clear-felled within two years of establishment. Thereafter, coppicing* will ensure that further harvesting can occur at 4 to 5 year intervals.

Advantages

The photosynthetic process is an inefficient means of converting and storing solar radiation compared to other proposed solar energy conversion systems, but the advantages of forests for energy are numerous:

— Fuels derived from trees provide an inexhaustible source of energy, assuming sufficient land is available on which to grow those trees.

— Many of the techniques for using wood as a fuel are already well defined; the limits of its use are determined only by economic considerations. Costs may be very substantially reduced by all the year round, large scale production as close to the main centres of population as possible.

— Wood may provide independence from unreliable or finite fuel sources.

— Higher yields per unit area ensure more complete use of land and resources. Areas which are not very useful at present but reasonably fertile can be brought under cultivation.

— Perennials can be harvested continuously throughout the year in response to fuel demand although, if demand is seasonally low, the harvesting rate can be temporarily reduced. This will, however, reduce the net financial return on the crop.

— Low planting costs are involved since the species referred to later in the text coppice readily after cutting.

— Operations on an energy farm include land preparation, species selection and/or production and selection of clones, planting, weed and disease control, harvesting and transport. All these can be done in the traditional way of agriculture and silviculture; hence most existing technology, equipment and skill can readily be incorporated.

— Efficient and cheap mechanised harvesting operations are possible because the age of the trees, their size, form and structure are kept uniform.

— The harvested material can be used directly as solid fuel to produce electricity, as raw material in the paper and other comminuted wood industries or can be converted to gas, ethanol, methanol, charcoal and methane. The technology to use wood for all these purposes is either well established or developing rapidly.

*Coppicing is the capacity of certain plant species to resprout spontaneously from the stump after cutting.

Useful species

A tree species suitable for energy farming must have the following features:

— Vigorous early growth with high biomass yields so that short rotations can be used.
— The ability to coppice rapidly after each harvest and to maintain its coppicing vigour through several rotations.
— The ability to reproduce vegetatively from clones and include the capacity for rapid genetic improvements.
— Disease resistance and freedom from major insect and fungal pests.
— Adaptability to varying site conditions, soil types and climates.
High basic density and calorific value (extractives high and moisture content low).

There are several high yielding agricultural crops such as sugarcane, sorghum, kenaf, sugarbeet, maize, cassava and elephant grass, but perennial woody species are to be preferred over annuals for energy plantations for the following reasons:

— Annuals must be planted and harvested in a well defined short period of time and the machinery and manpower for this are employed for only a short period of the year.
— Planting and management costs are lower because of the ability to coppice.
— There are no storage problems in times of low fuel demand.

Many conifers grow rapidly but they are not very suitable for energy farming because:

— Production rates on short rotations are usually less than from broadleaved species (2).
— They will not coppice. Furthermore, many of them cannot be reproduced vegetatively from clones.

The most promising species are found amongst the eucalypts (*Eucalyptus*), planes (*Platanus*), alders (*Alnus*) and poplars (*Populus*) because of rapid early growth, easy reproduction and a wide range of potential species from which to choose. The best results and the longest experience in tree farming for the whole range of energy products has, so far, been obtained with poplar hybrids (U.S.A., Belgium).

Yields

Short-rotation energy forests would usually be grown for their total above-ground biomass rather than just for stemwood. This means an increase in productivity in comparision to more usual rotations since bark and branches represent between 10 to 25 per cent of the total above ground tree biomass (3). Further increases in productivity may arise from the closer spacings of short rotation crops.

The biomass production rate depends upon the number of plants per hectare and the age of the stand at harvest time. Other factors include the species planted, the cultivation and fertilisation program followed and the local climate and soil conditions. Maximum insolation combined with a favourable soil/water balance is necessary for achieving the highest yields and all weeds must be cut back to ensure this.

Through recent advances in silviculture and plant genetics, annual increments can be increased significantly; however, even these increments could be doubled within 25 years (4) by concentrated research effort on species selection, genetic improvement and energy crop management.

The extensive literature dealing with wood increments summarised in Table II lists typical yields of between 15 and 55 dry tonnes per hectare per year (oven-dry basis) of above ground biomass. These yields were obtained with no irrigation and a minimum of weed control and fertilisation.

In a developed country a yield of about 16 dry tonnes per hectare per annum would be a minimum requirement but should preferably be higher. At 16 tonnes/ha Szego (2) suggests the minimum total area needed would be about 13,000 ha with an optimum planting density of 1 plant per 0.4 m². If higher yields per hectare can be achieved then a smaller area may be planted.

Table II

Examples of high, above-ground, biomass production

Species	Location	Biomass Prod dry tonnes/ ha/yr	Rotation period (years)	References
Hybrid poplar	Pennsylvania	20-30	—	(5)
Hybrid poplar	Belgium	36	4	(6)
Populus tristis	Wisconsin	16-19	4	(7)
Red alder	Washington	23	—	(5)
Red alder	West Coast U.S.A.	37	5	(8)
Eucalyptus spp.	California	54	—	(5)
Eucalyptus spp.	Ethiopia	49	—	(5)
Eucalyptus spp.	Portugal	40	—	(5)
Eucalyptus spp.	South Africa	28	—	(5)
E. globulus	India	43	15	(9)
E. globulus	Italy	27	6	(10)
E. grandis	Queensland	21	8	(11)
E. saligna	Brazil	17	10	(12)
E. regnans	Victoria	24	24	(13)

Environmental and Social Implications

Land requirements

Energy farming must be on easy terrain to allow mechanised crop management; on reasonably fertile soil with sufficient rainfall to allow high biomass production; and near to population centres or near installations requiring fuelwood.

The availability of land for energy forests depends on competition from other land uses such as grazing, food production, forests products, etc. These products are currently sold for the equivalent of 2 to 10 times their value as sources of energy according to R. S. Greeley (14). The same author concludes that an energy crop cannot compete economically with food and forest crops grown on prime agricultural and forest land at present but this situation can change in the future in specific locations.

Australia has 70 x 10^6 ha arable land but only 45 x 10^6 ha is currently under cultivation (15). Some of the uncommitted arable land may be suitable for new plantations. Ethanol produced from cellulose might provide half Australia's estimated fuel needs for transportation in the year 2,000 by planting half the uncommitted arable land (16). However, quite a few authors disagree with this figure, finding it unrealistic, largely because much of the uncommitted arable land is marginal and the productivity likely to be low.

Tree crops are not usually competitive with agricultural crops. Even in steep or broken terrain unsuitable for agriculture, forest plantations for fuel production would have to compete with traditional production forestry for the available land.

Any development of energy cropping has to occur within the existing pattern of agricultural settlement due to high capital costs of infrastructural development in remote areas. This limitation is especially relevant to a country such as Australia where some of the more suitable and available lands can be many kilometres from established centres.

In Europe, large areas of unused land (e.g. roadside strips, marginal soils) could be planted with several new poplar hybrids (yielding about 36 dry tonnes per hectare per year). However, except in special cases, the growing of biomass for energy is unlikely to be competitive with coal, oil or natural gas until the accessible resources are depleted or changing political factors force industrial countries into using alternative energy sources. Ultimately, irrespective of these factors, the ever increasing cost of conventional fuels may eventually necessitate the use of suitable agricultural lands for the sole purpose of photosynthetic energy production.

Most countries in the world are capable of producing significant amounts of alternative energy in a form that suits present needs. Energy farming could be developed to produce organic materials at low cost which could be converted to fuel.

As fuel costs rise, energy plantations will become increasingly competitive and could bring into use millions of hectares of land currently of low productivity.

Social implications

Growing trees for energy at close spacings over a short rotation is similar to ordinary farming. The only important differences are in planting and harvesting: energy forests are perennial whilst agricultural crops are usually annuals; and the methods and machinery currently being developed for energy forest harvesting may be somewhat different from those now in use for most crops. Nevertheless, farmers, with their knowledge of particular crops and farming techniques, will have to change their method of management in only a comparatively small way in order to produce energy crops. Also, as farmers become actively involved in forestry, there may well be some reconciling of the long standing dispute between agricultural and forestry interests in many parts of the world.

However, not all social problems are so easily solved. Farming different crop species and applying new management techniques can only become a reality if the alternative crop gives a significantly better financial return than the conventional crop.

Environmental considerations

Any adverse effects of energy farming on the environment could be an important impediment to its introduction. Environmentalists and all those who enjoy nature may fear that both large and small areas of native bush with their typical fauna and flora will be replaced. Conversely, it is simpler to plant pastures with trees than to convert forests into pastures. Regional and local planning is required and the public must be informed, made aware of, and given an opportunity to comment on, forestry proposals.

Monoculture with its known limitations and dangers is another aspect to be considered. However, energy farming could help to encourage planting of different species and, in the case of eucalypts and poplars, a variety of clones and hybrids could be used. Short rotation crops may take more nutrients and humus from the soil than a conventional timber harvest. As cropping depletes the soil, energy forests will require fertilizers. As intensive agriculture is typically an "energy loser" due to the inputs of fertilizers and other materials, as well as the direct fuel and power inputs for cultivation and processing, energy input/output analysis of tree farming is essential. As energy plantations aim to replace fossil fuels by photosynthesis, the cropping system must not depend on continual large inputs of a non-renewable mineral resource such as phosphate fertiliser.

Influencing public attitudes

Energy forests emphasize the use of materials grown specifically to produce energy. There are many organic 'wastes' from farms, forests and urban areas which could also produce energy. The public needs to become aware that energy is a product which, when not easily 'stolen' out of the ground, can be obtained only after hard work. The direct involvement of more people in "growing" energy may influence social attitudes towards energy so that energy saving, reducing energy demand in domestic and industrial life and the widespread use of solar heating systems may be important side effects of growing energy forests. Problems of energy supply and demand are unlikely to be solved by technological means alone; they have sociological and ecological as well as agricultural, engineering and forestry components.

Summary and Conclusion

This paper has dealt briefly with the social and industrial importance of wood, stressing its potential use as an energy source. World fuelwood consumption currently represents only 6% of total energy consumed annually but, in all the developing countries it is, nonetheless, still a principal source of energy.

Recently, in the developed world, wood has started to regain significance through the management of plantations specifically grown for their energy value. The advantages of such an energy source have been discussed, useful species and obtainable yields noted.

The limitations and problems of forests grown for energy with their environmental and social implications were outlined. In 'energy poor' countries, forests may already contribute significantly to the nation's supply of energy and, if oil prices continue to rise and if the development of the processing technology for using wood as a source of energy becomes more efficient and cheaper, wood may be used as an energy source universally.

Acknowledgements

I would like to thank Mr K. W. Groves for reading and commenting on the manuscript.

References

(1) F.A.O. (1978). *Year book of forest products 1966-1976,* Rome.

(2) Szego, G. (1976). Design, operation and economics of the energy plantation. Paper presented at a conference on Capturing the Sun through Bioconversion, Wash., D. C., March.

(3) Koch, P. (1972). Utilisation of the Southern Pines, U.S.D.A. Forest Service. Agricultural handbook No. 420. Southern Forest Experiment Station.

(4) Inman, R. (1977). Silvicultural Biomass farms. Mitre Technical Report No. 7347 — Mitrek Division and the Georgia-Pacific Corporation, Parkland, Oregon, U.S.A.

(5) Alich, J. and Inman, R. (1976). Energy from Agriculture. Paper presented at *Clean Fuels from Biomass, Sewage, Urban Refuge and Agricultural Wastes,* Orlands, Ela, January 26-30.

(6) Schalck, Professor J. (1978). Personal communication.

(7) Ek, A. R. and Dawson, D. H. (1976). Actual and projected growth and yields of *Populus tristis* # 1 "under intensive culture". *Can. Jour. of For. Res.*

(8) Evans, R. S. (1974). Information Report VP-X-129. Department of the Environment, Canadian Forestry Service, Western Forest Products Laboratory, British Columbia.

(9) Boden, R. W. (1964). Eucalyptus in Indian Forestry. *Aust. For. 28,* 234-241.

(10) Gemignani, G. (1967). Preliminary observations on the growth of some Eucalyptus species in the Pontine area. Presented at the F.A.O. World Symposium on Man-made Forests and their Industrial Importance, Canberra, 1968.

(11) Carter, W. G. (1974). Growing and harvesting Eucalyptus on short rotations for pulping. *Aus. For.* **36,** 214-225.

(12) Guimaraes, R. F. (1961). Experimentos regionalis de progenies. F.A.O. Second World Eucalyptus Conference Brazil 1: 369-374.

(13) Bevege, D. I. (1976). A green revolution in the high yield forest. Rational Silviculture of New Technology. *Aust. For.* **39** (1) 40-50.

(14) Greeley, R. S. (1976). Energy from land and fresh water farming. Proceedings of a conference on Capturing the sun through bio-conversion. Washington, D. C., March.

(15) Nix, H. A. (1974). Climate and crop productivity in Aust. Proc. Conf. at Int. Rice Res. Inst. Los Banos, Philippines.

(16) Morse, R. N. and Siemon, J. R. (1976). The role of forests as a source of solar energy. Paper represented at Institution of Engineers Aust. Conference, May.

(17) Nix, H. A. (1978). Availability of land for energy cropping in Australia, in Alcohol Fuels conference, Sydney, August.

A Community Use of Windpower

Mark Diesendorf

CSIRO Division of Mathematics and Statistics, Canberra

Introduction

If society wishes to continue to generate and distribute electricity on a large scale, then the following windpower projects, some of which may be already economically viable, are worthy of consideration:

(i) A large contribution to Britain's electricity supply could be provided by aerogenerators (i.e. windmills designed specifically to produce electricity) constructed around the coastline (1, 2) and offshore (3). In particular, if energy storage systems consisting of large hot water tanks were installed at places of end use, windpower may offer a higher availability (reliability) of energy for space heating than nuclear power (4) and may also be substantially cheaper (1, 2).

(ii) With the price of Australian petroleum being raised towards parity with world prices, windpower could replace most of that portion (20%) of Western Australia's grid electricity which is currently generated from oil (5).

(iii) The South Australian coastline, near the existing electricity grid, can provide far more wind electric power than could be utilized in South Australia in the forseeable future. Natural gas currently provides 60% of the primary energy for electricity generation in that State, and is running short. If the price of Australian natural gas were raised to its replacement value, then it may become economically realistic to export South Australian windpower via a new transmission line to Victoria and store it there for peakload use in new pumped hydro-electric storages. There is considerable untapped potential for pumped hydro in Victoria. By reducing Victoria's demand for peakload electricity from the Snowy Scheme, South Australian windpower could also, with the introduction of some institutional changes, indirectly benefit New South Wales (5) which shares the Snowy power with Victoria.

These three examples of the possible utilization of a system of large aerogenerators connected to centralized state or national electricity grids will not be treated further in this paper. They serve merely as a contrast to the alternative method described here of utilizing large aerogenerators as a local energy source for isolated communities. Since there are several Australian towns, e.g. situated on windy regions of the Western Australian coastline, which are isolated from the State electricity grid, the Danish example (below) could be relevant to Australia.

The Tvindmill

If size is significant, then it is worth noting that the aerogenerator at the Tvind Schools, which are situated near Ulfborg in Western Denmark, is at present the world's largest operating machine and, by a small margin, the largest ever built. The diameter of the circle swept out by the three blades is 54m, the height of the tower is 53m, and the alternator is rated at 2 megawatts in a windspeed of 15m/s.

The "Tvindmill" was built by the school teachers together with volunteer workers, both men and women. The purpose of the project was twofold:

(a) To provide energy in the form of heat and electricity for the three schools (6) which have a total population of 700 to 800 people;

(b) to produce a demonstration model of the community use of a renewable energy source.

Construction work on the Tvindmill commenced on 29 May 1975 when the teachers and students started digging the foundations. The project was financed from a common fund to which the teachers contribute their entire salaries. No funding was received from the government, industry or research organisations. The volunteer workers received food and shelter but were otherwise unpaid.

The group of workers on the Tvindmill decided from the beginning that they would work together co-operatively, taking mutual responsibility for all aspects of the project. During the construction period the whole "mill group" held a meeting each morning where spokespersons from various subgroups — e.g. dealing with the electrical system, the hydraulic system which changes the pitch angle of the wings, the minicomputer control system — could report, for general discussion, their progress and problems.

Very few members of the group had initially the necessary skills for constructing a large aerogenerator. Where possible, they sought advice from consulting engineers. They also developed many skills and ideas through enquiries with skilled workers on bridge works and similar projects. The members of the group doing the welding of the hub all obtained official certificates for welding; and all welds were tested with acoustic waves. The mill group devised a system of "licences" as a safety measure for all workers using potentially dangerous tools and equipment.

The wings of the Tvindmill are hollow and weigh only 5 tonnes each. They are constructed out of fibreglass and plastic. Since none of the members of the group had worked with fibreglass before, it was decided to build a small fibreglass fishing boat to gain experience, while the wing plans were being drawn up by a firm of aeronautical engineers. By the time the plans were finished the wing group had built three 10 metre fishing boats.

Apart from the construction materials of the wings, another special feature of the Tvindmill is that each wing-tip contains a small parachute which serves as an emergency speed control. Access to each wing tip is achieved from a small, demountable "flying fox" which runs on cables.

Contrary to the present trend in medium and large aerogenerators, the Tvindmill has a variable angular speed of rotation, up to a maximum value of 42 rpm at the rated windspeed v_r = 15m/s. Below v_r the rotational speed is controlled by adjusting the load on the alternator while above v_r the hydraulic pitch control of the wings keeps the rotational speed constant.

On 26 March 1978, the giant wings began to turn. For its initial test program, the Tvindmill is operating at low speeds on semi-automatic computer control. The data currently being collected from its operation will contribute substantially to the future development of large aerogenerators. At present surplus electricity is sold to the local electricity grid at 13ore (about 2c) per kWh, which is the value of the electrical energy as a fossil-fuel saver, as estimated by the local power company. During windless periods the Tvind schools buy off-peak power from the grid at 26ore (4c) per kWh.

Total cost of the Tvindmill was 5.5-6.0 million Dkr (about $900,000), which is $440-480 per rated kilowatt. This may be compared with the capital cost of a new 1000 MWe nuclear power plant, ordered in 1975, of around $1000 per kilowatt. Taking into account that power from a nuclear plant will be available for about 60% of the time while the rated (i.e. peak) power from the Tvindmill (without storage) will be available about 30% of the time (but there is also significant energy output when the aerogenerator is operating below its rated power), it is clear that windpower could be significantly cheaper than nuclear power in Western Denmark (7).

Diesendorf, M. (ed.) (1979). *Energy and People*. Canberra, Society for Social Responsibility in Science (A.C.T.)

In the past it used to cost the schools about ½ million DKr (about $80,000) per annum for oil and electricity. Thus the payback period for the economic investment will be at least 12.5 years. However, the payback period for the energy invested in the construction of such a large aerogenerator is expected to be much shorter — about one year (8, 9).

Since the main form of energy required by the schools is heating for the buildings, the mill group is considering whether to construct a large, insulated hot-water reservoir, heated resistively by electricity from the Tvindmill. It has been estimated that a reservoir of volume 3000 m³ would hold enough thermal energy to cover a calm period of 7 days. Since a calm of this length has never been recorded in the region, virtually complete thermal energy autonomy would be gained from the proposed storage.

Recent newspaper reports originating from the European electricity supply industry have claimed that the Tvindmill is "a failure", on account of the small quantity of electrical energy which it has produced to date. However, since the Tvindmill is still undergoing its initial test programme, under operating restrictions determined by the Danish Ministry of Housing, a large energy output could hardly be expected over this period. Therefore, it appears that the electricity industry, with its commitment to nuclear power despite the latter's growing economic problems, feels threatened by the successful construction of the Tvindmill.

References

(1) Ryle, M. (1977). Economics of alternative energy sources. *Nature* **267**, 111-7.

(2) Ryle's basic thesis has already survived considerable debate; see Clement, C. F. (1977). *Nature* **268**, 396; Ryle, M. (1977). *Nature* **268**, 482; Leicester, R. J., Newman, V. G. and Wright, J. K. (1978). *Nature* **272**, 518-21; Diesendorf, M. and Westcott, M. (1978). *Nature* **275**, 254; Anderson, M. B., Newton, K., Ryle, M. and Scott, P. F. (1978) Nature **275**, 432-4.

(3) Musgrove, P. (1978). Offshore wind energy systems. *Physics Educ.* **13**, 210-2.

(4) In Denmark, the addition of hypothetical short-term storage capable of delivering the average power for one day, makes an aerogenerator at least as reliable as a large nuclear power plant: see Sorensen, B. (1978). *Solar Energy* **20**, 321-31. Similar results have been obtained recently in Australia by J. van Leersum and by M. Diesendorf and K. Malafant.

(5) Diesendorf, M. (1979). Recent Scandinavian research and development in wind electric power: implications for Australia. *Search* **10** (5) 165-73.

(6) The three schools making up the community at Tvind are the Travelling Folk High School (for students over the age of 18), the Necessary Teacher Training Course and the Tvind After-School (for 14 to 18-year-olds).

(7) Indeed, this finding was already published in: Danish Academy of Technical Sciences (1975). *Vindkraft (Windpower)*, English summary report; see also Sorensen, B. (1976). Wind energy. *Bull. Atomic Sci.* (September) 34-45.

(8) U.K. Dept. of Energy (1977). *The prospects for the generation of electricity from wind energy in the U.K.*, Energy paper No. 21, London, HMSO.

(9) Sweden (1977). *Vindenergi i Sverige (Wind Energy in Sweden).* Report NE 1977:2. Nämnden för energiproduktionsforskning, Box 21048, 100 31 Stockholm.

Note

Any social interpretations expressed in this paper are those of the author and do not necessarily reflect the views of CSIRO.

Metals and Energy Options

Michael Wadsley

6 Primrose Crescent
East Brighton, Victoria

Introduction

During the 5000 years until 1550 A.D., people smelted metals from their ores using renewable resources. Human, animal, water and wind power provided the mechanical energy necessary for mining, transport, crushing and concentration of ore while charcoal derived from wood provided the necessary chemical energy. Soon after this date, an acute shortage of wood in Europe, particularly in Britain, caused this region to substitute coal for wood in the production of heat and metals (1).

A detailed, comprehensive record of metal production immediately prior to 1550 A.D. is provided in "De Re Metallica" by Georgius Agricola (1494-1555) (2). A comparison of metals produced during that era with those available to modern civilisation is instructive. Agricola describes the production of gold, silver, copper, tin, iron, antimony, lead, bismuth and mercury, and of the alloys bronze, steel and brass. Formation of zinc had been observed but it was not commercially produced in Europe. Modern society values metals such as aluminium, titanium, chromium, manganese, silicon, nickel, cobalt and cadmium which were not known to Agricola or to archeology. While non-recognition of nickel and cobalt, for example, may be attributed to the frequent co-occurence of iron with their ores and similarities of their physical and chemical properties with those of iron (e.g. see Table 1), the absence of metals such as aluminium must be attributed to other factors. Aluminium is a major component of soil, clay and rocks and the metal has physical characteristics which allow melting in a simple furnace. The failure of the skilled ancients to produce this metal may only be explained by studying smelting energy.

Minimum Smelting Energy

Many metals naturally occur in chemical combination with oxygen, examples being tin, iron, aluminium and titanium. Other metals occur in chemical combination with sulphur but are readily converted to oxides by roasting the sulphide ore or concentrate in air. Examples are lead, copper and zinc. The basis of smelting is the separation of a metal from its combination with oxygen. The energy required to make this separation varies from metal to metal and also varies with temperature for each metal (3). This information is illustrated

Table 1

Metals and their Physical Properties

Metals	Melting Point °C	Boiling Point °C
Known prior to 1550 A.D.		
Gold	1063	2966
Silver	961	2212
Copper	1083	2350
Tin	232	2300
Iron	1530	2730
Antimony	631	1380
Lead	327	1750
Bismuth	271	1560
Mercury	−39	357
(Zinc)	419	907
Also alloys: bronze, steel, brass		
Examples of modern metals		
Aluminium	660	2400
Titanium	1720	3200
Chromium	1890	2482
Manganese	1240	2097
Silicon	1410	2355
Nickel	1450	2840
Cobalt	1490	2900
Cadmium	321	767
Tungsten	3410	5927
Magnesium	651	1107

Diesendorf, M. (ed.) (1979). *Energy and People.* Canberra, Society for Social Responsibility in Science (A.C.T.)

in Figure 1. It is apparent that all the metals known to Agricola are, unlike aluminium, less strongly bound to oxygen than is carbon.

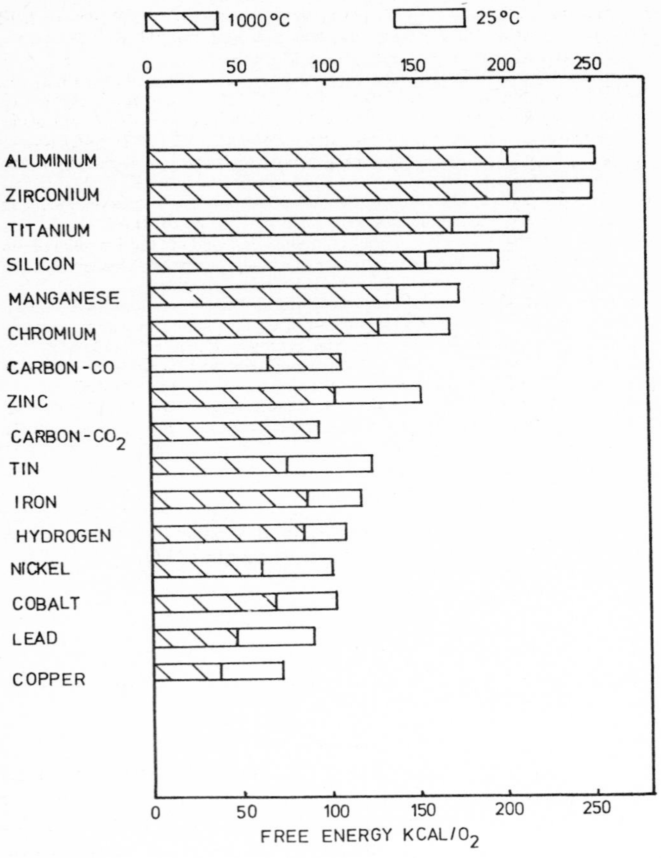

Figure 1
Energy to separate metals from oxides

This observation is important as it demonstrates the concept of energy quality. Only the ability of carbon to combine with oxygen and not the quantity of carbon determines whether metal is produced when smelting a particular ore. (If metal is formed then the quantity produced is determined by the quantities of carbon and ore, and by the smelting efficiency.) It was not until people learned to smelt by multistage processes or high grade energy, in the form of electricity, was harnessed, that other metals became available. The concept of energy quality, the Second Law of Themodynamics, is an expression of the intrinsic molecular, atomic or sub-atomic character of macro energy transfer or conversion processes and is fundamental to the understanding of the use of energy.

Current Commercial Smelting Energy

The production of metals in the modern world uses vastly different sources of energy to those described by Agricola. Energy input occurs at three readily identifiable stages of metal winning: mining, purification and smelting, and is either direct fossil fuel (natural gas, oil or coal) or electrical energy (derived from fossil fuel, hydro or nuclear sources). Conversion of fossil fuel or nuclear energy to electricity yields only about one third of the energy released by combustion or fission. The remainder is lost as waste heat — hot air or water.

Analysis of data in Table 2 indicates that the relative theoretical energy required to win metals from their ores bears little relation to the energy usage in the real world (4). Thus iron, which occurs as high grade ore, requires little energy for its conversion to metal while copper, whose minerals are highly diluted in its ores and are intimate associations of copper, iron and sulphur, requires disproportionately large energy input. Lead occurs as ores of intermediate grade which are readily concentrated and the metal is easily refined. Alluvial tin ores are very low grade but readily produce a pure concentrate

Table 2

Energy Use in Metal Winning
Units — btu/ton FF = Fossil Fuel EE = Electrical Energy

Metal	Mining + Transport		Purification: Physical		Purification: Chemical		Śmelting		Sub-Total		Total
	FF	EE	FF	EE	FF	EE	FF	EE	FF	EE	
Aluminium	4.47	0.32	—	—	36.0	6.64	27.8	168.9	68.3	175.8	224.1
Titanium	9.12	5.52	a	a	25.1	39.0	9.09	339.7	43.3	384.2	427.5
Iron	0.29		—	—	—	—	22.6	0.26	22.9	0.27	23.1
Tin Alluvial	3.61	157.4	a	a	b	b	16.8	1.98	20.4	159.2	179.6
Lode	3.54	129.3	a	a	b	b	109.9	12.2	113.4	141.5	254.9
Copper	13.3	8.28	11.3	31.0	6.84	3.30	24.9	4.68	56.3	47.3	103.6
Lead	1.45	2.90	0.83	3.97	6.27	1.74	8.67	0.97	17.3	9.56	26.8

class 17.4 Btu/ton Cement 7.6 Btu/ton
a. Included with mining and transport energy;
b. Included with smelting energy.

which is easy to smelt. Lode tin ores are low grade, difficult to concentrate and also difficult to smelt and refine. Aluminium and titanium ores are relatively easy to mine but require much energy for purification and smelting. Similar analysis may be made for all metals. Ores mined in the future will generally be less accessible, lower in grade and more complex than those mined at present so significant increases in energy efficiencies will be needed if requirements for mining and purification are not to increase. The energy requirements for smelting and refining have, in general, shown a decrease per unit production with time. Hopefully this trend will continue but it is the quantity of materials produced which determines the absolute energy input into metal production (Table 3) (4).

Table 3

Total Energy Requirements
For USA Consumed Materials

Metal	Energy/Ton Million Btu	Million Tons	Total Energy Btu $\times 10^{12}$
Aluminium	244	5.77	1408
Titanium	408	0.02	8
Iron	24	140	3350
Tin	190	0.07	12
Copper	112	1.97	221
Lead	27	0.89	24
Glass*	17	12.4	216
Cement*	8	90.5	688

*Metal substitute.

Future Alternatives

Future society will eventually have to be based on resources which are essentially limitless at the then rate of exploitation. I believe a morally sound policy is to educate society to routes to the future of smooth transition, not potential crisis. Metals have a critical role in producing the goods and chattels fo civilisation, particularly as heat transfer surfaces, electrical conductors and shafts and bearings in energy transfer and conversion processes (4a). Continued depletion of high grade resources and high quality energy on low efficiency and frequently frivolous pursuits with little regard to the future will not promote the founding of a stable society. The current trend might be slowed towards equilibrium by consideration of the following:

(a) Reducing the energy required for metal production.
(b) Substitution amongst metals and with other materials.
(c) Reduction of corrosion and wear of metals.
(d) Recycling of used metals.

Stating these aims is simple. Practical implementation will call for compromises which will have to be desired by society as a whole or the solutions will be unworkable.

(a) Production efficiency

If the price of fossil fuels continues to rise towards a value equivalent to the cost of deriving fuels from renewable resources, as it inevitably must, then metal producers will naturally counter by effecting greater energy efficiencies. A slowing of demand for the more costly products will also occur. Typical energy saving measures currently being effected include use of oxygen enrichment in combustion processes and direct use of fuels rather than electricity for process heat (5).

(b) Substitution

Substitution of one material for another as availability and costs change is considered to be a major potential solution to future problems (6). For example, aluminium, a high energy metal, might be replaced in some uses by iron. However, the extra energy required to accelerate the heavier weight of iron in transport applications might over-ride other considerations. It is generally found that such complex compromises will need to be thorougly evaluated before particular substitutions can be determined to be socially desirable.

(c) Corrosion

Corrosion is estimated to cost the U.S.A. some \$70 x 10^9 p.a. in terms of all the various inputs that are required to replace the lost metals. In addition, metals affected by corrosion and wear frequently become highly dispersed and/or revert to oxides thus preventing low energy recycling. Common answers to corrosion problems are to apply metal coatings such as zinc, zinc-aluminium alloy or tin, or to apply paint coatings containing titanium, zinc or lead oxides. An unfortunate consequence is that in protecting the bulk metal, the coating is necessarily thinly dispersed and recovery of the contained metals for recycling becomes costly in energy requirements. This is another area where the results of careful evaluation must be presented to society for decisions regarding priorities.

Table 4

Unit Energy for Production of Primary and Secondary Metals

10^6 BTU per Ton Metal

Metal	*Primary from ore*	*Secondary from scrap*	*Energy Saving from recycle*
Magnesium	358	10	348
Aluminium	244	12	232
Nickel	144	15	129
Copper	112	18	94
Zinc	65	18	47
Steel	32	13	19
Lead	27	10	17

Ref: Kellogg (8).

Table 5

Factors That Encourage Scrap Recycle

1. *High Metal Price*

	\$/lb
Silver	65.6
Nickel	2.0
Copper	0.64

2. *Ease of collection and sorting*
   Lead in storage batteries, cable covering, type metal
   Steel in heavy machinery, structural, rails, etc.
   Copper and brass by colour.
3. *Ease of refining to "primary" specifications*
   Silver
   Copper
   Steel
   Lead

Ref: Kellogg (8).

(d) Recycling

Recycling of metals has great intuitive appeal for it is apparent that if 100% recycling is approached and population size stabilizes, very little further mining and primary metal production will be needed. The energy savings from recycling are also potentially very large (Table 4) (8). However there are both social and technical considerations working both for (Table 5) and against (Table 6) effective recycling of used metals. I believe only effective re-education can solve the social problems while science can only provide a factual basis for decisions on the necessary technical compromises.

Table 6

Factors That Discourage Scrap Recycle

1. *Highly dispersed uses that inhibit collection and sorting*
   Copper wire in small magnets, cartridge brass and other brass items.
   Aluminium in packaging (cans, foil, etc.)
   Lead in solder, bearing metal, tetra-ethyl-lead, oxide and chemicals.
   Zinc in small die-casting, galvanised iron, oxide and chemicals.
   Nickel in low alloy steels, electroplate.
   Silver in photographic materials, electrical contacts, jewellery.
2. *Removal of impurities is difficult* (scrap must be cleaned and sorted)
   Aluminium: only magnesium can be easily removed by refining.
   Magnesium: can't remove baser metals, except by distillation of magnesium.
   Steel: copper, nickel and sulphur very difficult to remove.
3. *Abandonment of the Puritan "Thrift Ethic"*
   Design for convenience and appearance.
   The "disposable" product.
   The affluent society.

Ref: Kellogg (8).

Conclusion

The energy options available to society with respect to metals production and use are not clear-cut but will involve compromises which need to be acceptable to society as a whole if we are to achieve a smooth transition to a stable future. Desirable courses of action will need to be chosen and implemented in the near future before the current course to crisis becomes irreversible.

References

(1) Nef, J. U. (1977). An Early Energy Crisis and Its Consequences. *Scientific American,* **237,** 140-151.

(2) Agricola, G. (1556). *De Re Metallica.* Translated by H. C. and L. H. Hoover (1950). New York, Dover Publications, 638p.

(3) Kubaschewski, O., Evans, E. L. and Alcock, C. B. (1967). *Metallurgical Thermochemistry* 4th. Ed. Oxford, Pergamon Press, 495p.

(4) *Energy Use Patterns in Metallurgical and Nonmetallic Mineral Processing.* Phase 4 Report, NTIS PB 245 759/AS, Phase 5 Report, NTIS PB 246 357/759/AS, researched by Battelle Institute for USBM (1975).

(4a) Albers, J. P., Bawiec, W. J. and Rooney, L. F. (1976). *Demand for Non-fuel Minerals and Materials by the United States Energy Industry,* 1975-90. Geological Survey Professional Paper 1006-A. Washington, United States Government Printing Office.

(5) Kellogg, H. H. (1977). Conservation and Metallurgical Process Design *Trans. Instn. Min. Metall.* **86,** C47-57.

(6) Goeller, H. E. and Weinberg, A. M. (1976). The Age of Substitutability. *Science.* **191,** 683-9.

(7) Gray, A. G. (1978). What Corrosion Really costs. *Metal Progress.* **113,** 29.

(8) Kellogg, H. H. (1976). The Role of Recycling in Conservation of Metals and Energy. *J. Metals.* **28,** 29-32.

Excerpts from the Discussions

Most of the discussions at the *Energy and People* Conference were tape recorded with the intention of including them in this book. Unfortunately, the labour required for transcribing and editing them threatened to overwhelm all else. Therefore, with profound regret, we only include a tiny sample of discussions. I wish to apologise to the many participants who made excellent but unpublished contributions from the floor — *Editor.*

ENERGY, COLLECTING GUM, AND THE BASES OF ECONOMIC POWER

Allan Rodger

In one important way, our dependence on the present pattern of fuels and related resources, and the impediments to more appropriate, more responsible, and more equitable economic and social structures do not differ from the inequities and social inefficiencies associated with all processes of resource use. All are based on differential access to information.

Some time around about 1970 I had occasion to drive south westwards from Khartoum. The countryside is near to desert with only sparse vegetation. It is really rather thin scrub. But the Acacia trees are Acacia arabica and this is the tree from which gum Arabic is obtained. Gum Arabic was the first economic product to be developed in the Sudan following the establishment of British rule in the last years of the 19th Century and was for a long time the principal source of foreign exchange. While travelling through this light forest we came upon a man collecting the gum. It is obtained by a process rather similar to that used for the collection of latex from the rubber tree. The 'v' shaped scar on the tree oozes resin which forms into a hard lump about the size of a walnut. We saw the man passing from tree to tree collecting his harvest. He walked barefoot over ground laced with Acacia thorns wearing only a loin cloth. By any standards he would be considered poor; indeed, a poor man in Sudan is among the very poor of the World, at least in monetary terms.

The following day we were in a market town in which one of the principal goods being traded was gum Arabic. Here we again saw men from the Acacia arabica forest now come to town to sell their small bundles of gum to the local merchants. Observing this scene I began to wonder why it was that the price selected, and in some sense agreeable to both parties, was such that the gum collector returned in poverty to his endless life under the hot sun while the merchant, who perhaps not rich by many standards, enjoyed a very much higher material standard of living.

And then a few days later back in Khartoum, seeing the agents for these rural merchants trading with the national merchants in the capital city, and again seeing the difference between the return to the seller and the benefit to the buyer, and knowing something of the process which still lay ahead of these small resinous kernels, I was forced again to ask myself why it was that the poor man out in the forest could receive such scant recompense for his contribution to the high cost confectionery and edible gums for which his gum Arabic was an essential raw material.

Without having all the conventional answers already supplied to me, it seemed that the only explanation was that the hierarchy of income was directly related to a hierarchy of information. I wondered what thoughts might go through the mind of one of the buyers working for a major confectionery multi-national when deciding how much he was prepared to pay for refined gum Arabic, what range of alternatives would be open to him, to what miracles of the modern chemical industry could he turn in order to find a substitute if he thought the gum Arabic price too high, or with what alternative sources could he communicate through his telex or scrambled telephone. I thought of him, perhaps sitting in an airconditioned office in London or New York, effectively trading through a vast array of middle men with the original collector in that scrub to the South West of Khartoum, and I wondered what sort of fair and equitable deal could be possible in these circumstances. And turning to the gum Arabic collector, I wondered whether he had any idea of the complex, high technology industry of which he was a part, if he knew how little of the final value came to him, if he knew how many merchants lay between him and the consumer, and whether he could even imagine the possibility of operating a cooperative to sell the gum directly on the international market. What possibility was there for this man to consider alternative markets, perhaps to imagine new uses for his hard won gum, and then consider selling his produce in one market place only if the price was better than that available in another.

Knowing a little about the instability of the international commodities market, I wondered whether the sudden rises and falls ever communicated down through the long chain of middle men to the poor man who did the original collection, or whether through the tyranny of ignorance he was doomed to receive only that minimum recompense necessary to keep him alive and to ensure that he continued to serve the interests of all the other people engaged in the process.

It is a rather sad little story but one which gives a rather important insight into the facts of economic life. If we see it as our role in life to modify in any way the distribution of economic well-being in the world, then we must be concerned with the distribution of economic power, and an important aspect of economic power is economic information.

The system whereby control is exercised over the use of fuel is in some important respects not unlike that which controls the delivery and use of gum. It is equally brutal and no less effective in maintaining and even increasing current social stratification. To break out of the current pattern of pricing, distribution and use of resources — be it gum Arabic or oil we are dependent on the development of a vastly improved information base.

The idea is not one which will be readily accepted by the present power groups *because they know it to be true* and they know that, in important ways, their power depends on the continuation of ignorance. Alternatively, however, the idea presents us with an opportunity, because by helping to dispel ignorance we may make some contribution to the alleviation of economic oppression.

DISCUSSION FOLLOWING D. G. HILL'S PAPER

Roger Bartell: About 18 months ago, I represented the Society for Social Responsibility in Science in a series of workshop meetings with the National Capital Development Commission on transport planning in Canberra. During the course of these workshops we were given access to quite a wide range of information, some of which was labelled "confidential". An examination of the information we were

Diesendorf, M. (ed.) (1979). *Energy and People.* Canberra, Society for Social Responsibility in Science (A.C.T.)

given proved that some items were internally inconsistent. In the course of our lobbying, I wrote about these to the then Minister of the Capital Territory, Mr. Staley, who wrote back to me saying,

"Dear Dr. Bartell" (this is just the substance of what he wrote), "thank you for your letter and comments, and the information contained therein. I have read them with interest. However, you must understand that the National Capital Development Commission will exercise the same level of professional integrity as you would yourself in your own employment". (laughter)

Gerhard Weissman:: I used to work for the Electricity Trust in South Australia, where I experienced some of the same problems. An engineer works for two masters: himself and those who employ him. The things he must concern himself with in his employment are determined after he takes the job — and entirely by an organisation over which he has little or no control. The Institution of Engineers plays a part, as well. When I fell out with the Trust, I went to the paper and gave them some information which was public. Yet I was severely criticized by the Institution of Engineers for saying things critical of other engineers. Now, you have some professional prostitution going on there. For base profit you have to do what you don't want to do. If you don't do what you're expected to do — if you don't produce a report which favours the institution — then you're blacklisted and your professional ladder has no more rungs. After I questioned some data in the Trust, I was told that there was no more work for me in the department in which I worked, that if I wanted work elsewhere in the Trust there was a position in a power station where I could work on electric motors. Admittedly, it wasn't a C-3 job, but I would be paid as a C-3, anyway.

So you have to make the choice: do you want to be a prostitute, or don't you? If you don't, then you are in very hot water, indeed. So, unless you're independent and can thumb your nose at the institution, including the Institution of Engineers, you just can't live with the System. It's all right as far as information is concerned, but it's a question of which way to jump. I feel it's just not on for you people who aren't engineers to blame the individual engineer.

Doug Hill: I agree with that. I can understand the pressures on any individual. But it does seem to me the individual has to set limits on just how far he's prepared to prostitute himself. I can also understand an organization getting locked into something it's been several years developing. One thing that does surprise me in all this business is the role of the Institution of Engineers. I would have thought that they would have encouraged testing views and broadening the expanse of their professionals and the taking into account questions that hadn't previously been considered. I'm disappointed in their response.

In the cases of Lake Pedder and the Newport power station you have professional engineers near the top of their organisations building monuments to themselves — and they're the ones in control.

Louis Arnoux: Among those who apply for the job, you'll find it's the applicant who fits in psychologically with the project — everything else being equal — who will have the edge in the selection process.

Young woman: I'd like to comment on the type of comments that were made on the question of energy and conservation in the home. I find it quite incredible that you can talk about something like that for 30 minutes without ever mentioning women's role in the whole thing. You've been talking about hierarchies, yet you haven't even looked at the hierarchy that supports every other hierarchy, which is the hierarchy of men over women. The conservation of energy in the home is obviously something in which women would have to play a major part. The effect of this would be that women would be spending a lot of time saving energy, and this would keep them from spending much time on political issues, on efforts directed at getting industries to save energy. They would be dissipating their energies, directing them away from political areas, away from efforts directed at changing the system. In fact, they would be maintaining the basic structure underlying all the other hierarchical structures. There hasn't been any discussion of the role women should be playing. In fact, every mention in that last discussion was of "he", "man".

DISCUSSION OF PAPERS BY L. ARNOUX AND J. PRICE

Julie Dahlitz: I'm surprised that no-one has mentioned the military uses of energy. The United Nations has published figures showing that 25% of all scientific work is on warlike purposes; $400 billion per annum is spent on armaments; 40% of research and development goes to military purposes. The amount of actual energy has not been estimated, but extrapolating from the above, it must be a very substantial proportion. Where does this fit into the scale of things?

John Price: I'm not going to confront the question scientifically, head-on. But take the case of Japan — the Industrial Revolution. Increased energy use from centralized sources was a hidden cost in the process of disciplining the workers. Another hidden cost was the dependence on outside sources for expansion of the resource base necessary to industrialize. Therefore, the Second World War can be seen as a "hidden cost" as Japan tried to increase her resource base.

As we get more dependent on non-industrial production (and remember, lying on the beach is production, the satisfaction of a need), we have the chance of enabling the first and most important unit to be small and located close to us, and the others further away to become secondary. And so we have the prospect of all those bombs standing in silos, gathering cobwebs as monuments to man's past stupidity.

Michael Wadsley: John, I disagree with some of the generalisations in your paper. The Industrial Revolution came about 1550. Prior to that date, all the worlds energy needs were met by renewable resources. Then civilization found itself running short of wood — the main source of chemical energy; turned from wood to coal — which is not a very pleasant source as anyone knows who has worked with charcoal in a fire. However, because they were working with charcoal, they only had certain sources of metals available. Solar absorbers, flat plate collectors, for example, can only extend fossil fuel usage a minimal amount — produce water for example, only up to 60 - 100° centigrade. Heated water cannot supply all the things that society needs. There's no way that solar absorbers will do the trick — they can only be built by burning fossil fuels . . .

Mark Diesendorf: . . . Or by using electricity from wind generators!

Comment: Nepal is suffering from an energy crisis because it is running out of wood.

Price: The critical reason why Nepal is suffering is that wood as a resource is being exploited to gain capital for development.

Comment: Generation of energy is becoming economically less viable; energetically less viable as well. Deposits are becoming lower grade, more difficult to mine. Production is less efficient — less energy efficient.

Jack Mundey: I would like to comment on the importance of hours of work. Unions are mirrors of corporate power. I believe that we are going to have an industrialized society for some time to come. Growing urbanization of the globe bears testimony to that. Our task is to try to tame that industrialization; to concentrate on the improvement of working conditions. We should also be far more concerned about the consequences of the labour that we produce. I believe that this has been neglected by reformists and myriad brands of revolutionaries. I put it to you that maybe one of the reasons we don't attract industrial workers to many of the issues that you raise is, in fact, you can't go to them and say, well listen, you cut your wages, when they experience the daily grind and when real wages have actually fallen. I put it to you that even in the most idealized sophisticated form of capitalism, you'll never win real change amongst the lower echelon if you start with the idea of cutting wages, or if, as you put forward, working a number of hours without consideration from what we do during those hours, and which products we use, and which products we make. They seem the real gut issues — as well as the question of cutting the hours.

One thing I didn't like about your talk was this concept that the industrial society is about to come to an abrupt end.

Price: It seems to me that partly because of past Union activity — which I think is perfectly right and justifiable — the victories that workers have achieved for higher wages have not been victories for the workers themselves, but for the system, because money is the thing by which we are bound to that system. Money is what the system gives us in return for our labour. Money and goods are both parts of the system. If we are able, all of us, to satisfy more of our needs — apart from the system — this would give us greater independence to achieve the needs we want to achieve.

I had an experience when I attended a talk by Illich. A Unionist came up to me — a shop steward. He told me that many of his workers, not "hippie workers', normal, straight workers, had said:

"We would like to have more free time on a fixed wage basis."

The problem is that this has been a non-negotiable demand as far as management is concerned. This kind of proposition (i.e., more free time for less wages) should be discussed among workers and management.

Concerning the collapse of industrial society: look at the energy researchers who are going to save us — where is the money going?

Fission — the burner, the breeder
Fusion
Oil from coal
Centralized solar
Tidal power (in Britain)

All are more than 10-fold more capital intensive than the Middle Eastern oil. All of them require more lead times. *The capital expense is being imposed long before the time of the energy return.*

The drain in any one year of capital to maintain, and if possible, to make grow, the energy supply is increasing very rapidly. Now what this means is the economy is becoming unbalanced. More of your capital is going into producing the energy that is required to drive the goods that we buy in the home and the industrialized system. And this is likely to continue. The normal response to a capital shortfall is to blame past workers' increases in wages, and introduce cunning devices like partial indexation — reverses the competition.

To keep the system going, it is essential that the amount of capital and the amount of profits be increased. Subsidies are used to do this. Government and consumer subsidies of research and development, for example. The benefits go to the industrial-productive system, but the consumers pay for it; direct subsidies and investment allowances. Welfare and education subsidies are being reduced. The ordinary people are being made to pay for (in their private lives) the band aids of the system. How far can this go?

Real wages of consumers are dropping and yet it is out of the payment for the goods and services that the profit comes. It's self-defeating. This system could only have operated whilst capital prices of energy were reducing — it becomes totally unstable once they increase. How long could this go on? There are two things that could bring down the system: they probably won't happen, but they might.

1. There has been a run on the American dollar in countries overseas. What happens if it happens at home?
2. What happens if the Arabs have another war with Israel. Less likely now, but it still could happen. The system could come down within a month. We are incredibly vulnerable, as Ian Sykes pointed out.

Bruce Hamilton: Louis, do you have any sociological interpretation of our uses of leisure time?

Louis Arnoux: What is important is that we are earning time in order to spend it. We have to spend it in some way. We have to work to have our holidays — it's a vicious circle.

Price: There is a trend to go from complex, possession-oriented life to simpler holiday lifestyles — camping, visiting old Amsterdam or old Saigon rather than the Americanised strips.

Arnoux: Advertising is taking advantage of the back to nature trend. It offers us a nostalgic evocation of nature, a symbolic exchange, diffuse values which we have lost.

Young man: I was out of Australia for a while and had to ride a bicycle everywhere. When I returned and began driving again after only half an hour I had gained a feeling of power; I felt bigger. This has sexual connotations, I guess.

Arnoux: This feeling of power without effort is an important part of the consumer society.

Deborah White: D. H. Lawrence and Frieda were horseback riding in Mexico. Frieda, who wasn't noted for her intellectualism, said "Isn't it wonderful to feel this fine horse, vibrant with life, between my legs".

D. H. Lawrence looked around with alarm and said: "You've been reading my novels!" (laughter)

L. Endersbee: I'm disturbed about your sexual implications of energy use, because we in NEAC are trying to develop an energy conservation program. I'm horrified at the aberrations which may result from this. Can you suggest how we can avoid frustration?

Price: Why not withdraw? (laughter)

Arnoux: There is a lot of repression of the negative aspects of our society. We are obliged to be happy. The more we repress negative aspects, the more they are going to emerge. You say you are going to conserve energy . . . put people into boxes . . . put them into a childish position . . . tell them what to do. This will not achieve what we are trying to do.

Alan Roberts: Myths are propagated about the energy crisis by people who oppose the system, as well as by those who defend it. For instance, it has been said by some environmentalists that by the year 2010 we shall have run out of essential mineral resources. This does not take into account the use of substitute materials, e.g. for copper. If you look into it, a lot of these doomsday predictions are incorrect. They see the energy and resource crises as a set of isolated problems whereas in reality they are interacting.

The traditional scientific method is to tackle one problem at the time. I don't want the Institute of Physics to hear me, but physics isn't as hot as it looks. It looks terrific, but it's really a paper tiger. It has great difficulties in handling interacting systems.

Arthur Davies: This doesn't only happen in scientific areas. Much thinking about social, human problems is done in terms of separate little boxes.

ECONOMICS OF RESOURCE DEPLETION

Question: If I go to a local wood merchant and order a load of firewood, he goes out into the bush, cuts the trees and brings the wood in, and I pay him an amount which covers his costs of getting it out of the bush, plus a profit for himself. Where, in general, in the logging of forests, in the extraction of natural resources, is amortisation being paid on the time taken to create those resources? As far as I can see there is a gap somewhere. In the high energy scenario we seem to neglect sometimes to amortise the natural resource, while at other times, in making an investment in a timber plantation, proper account is taken of amortisation in estimating the likely profit thirty or so years hence.

Dr H. C. Coombs: I wonder if I might comment on that. Theoretically, in the collection of royalties, there is a way in which forestry resources can be replaced; good forestry practice is based upon obtaining from the users of the forests revenue to replace the forest over a forty or fifty year cycle or whatever the appropriate period may be. As far as I know there is nothing corresponding to that to provide for the replacement of fossil fuels. Theoretically, if the users of fossil fuels are to be put on the same basis as the users of timber they should be charged a royalty sufficient to replace the fossil fuels they have used by an alternative source of energy which they are using for ultimate. Whether that is practicable or not is an institutional question; where do the decisions lie?

ROLE OF THE MULTINATIONAL OIL COMPANIES

Lance Endersbee: One place where synthetic fuels from coal are likely to be developed fairly early is Australia, simply because of our high coal resources per capita. When we move towards synthetic fuels it will be by international investment, by the oil companies and others, who will be developing our coal resources for the world market.

We have to recognise that the international oil companies have huge resources, that they are in the game of supplying transportation fuels and that we can't just push them aside and say "You're not planning properly for the future. Governments are going to take over and do it for you". It just doesn't work that way. The international oil companies are huge operations and they have their own longterm future to think about. I see quite a lot of data which they produce and I have the impression that they are planning hard for the next generation.

They do tend to be rather secretive about it and there's a pretty good reason for that. They are all looking for what is going to be marginally the best synthetic fuel in the next generation — whether it's oil from oil shale, oil from coal, etc — and it's the relative costs of these which will determine the issue. When they move into these areas the oil companies will be investing tens of thousands of millions of dollars — investments which governments just could not make.

And the oil companies are ready to go. They are bidding for out top chemical engineers for these energy initiatives. The money is ready to go and they are doing their preparatory work.

These projects are going to take something like 8 years to build. There are problems with the present level of inflation. They don't want to invest too heavily when the final cost is uncertain. They don't want to be first up with synthetic oil at, say, $20 per barrel, when the Arabs are still charging $17.

NUCLEAR POWER

Price: I would like to ask Bruce Hamilton, why did Shell withdraw from the nuclear power industry?

Hamilton: Because it was too bloody uneconomic, that's why!

Recommended Background Reading

Schumacher, E. F. (1973). *Small is Beautiful.* Abacus.

Commoner, Barry (1976). *The Poverty of Power.* New York, Knopf; & Bantam.

Lovins, Amory B. (1977). *Soft Energy Paths.* Penguin.

Lönnroth, M., Steen, P. & Johansson, T. B. (1977). *Energy in Transition.* Secretariat for Future Studies (Fack, S-103 10 Stockholm)

Johannson, T. B. and Steen, P. (1977). *Solar Sweden.* Secretariat for Future Studies, Stockholm.

White, D., Sutton, P., Pears, A., Mardon, C., Dick, J. & Crow, M. (1978). *Seeds for Change — Creatively confronting the energy crisis.* Patchwork Press/Conservation Council of Victoria. (118 Errol Street, N. Melbourne 3051)

Chapman, P. (1975). *Fuel's Paradise.* Penguin.

Le Guin, Ursula (1974). *The Dispossessed* (a novel). London, Victor Gollancz.

Polanyi, Michael (1974). *Personal Knowledge.* rev. ed. University of Chicago Press or Routledge & Kegan Paul.

Hooker, C.A. and van Hulst, R. (1977). *Institutions, counter-institutions and the conceptual framework of energy policy making in Ontario.* Dept. of Philosophy, Univ. of Western Ontario.

List of Participants

ARNOUX, Louis	NZERDC, University of Auckland, Auckland 1, New Zealand
BARTELL, Roger	37 Woralul St, Waramanga, ACT 2611
BEARD, T. C.	Department of Health, PO Box 100, Woden, ACT 2606
BELYEA, Jane	c/- Animal Production Section, School of Agriculture & Forestry, University of Melbourne, Parkville, Vic 3052
BLIGH, Kevin	3 Windich Road, Bull Creek, WA 6153
BODDAM-WHETHAM, David	Energy Authority of NSW, Box 485 GPO, Sydney 2001
BOWMAN, Kaye	Human Ecology, Centre for Resource and Environmental Studies, ANU
deBURGH, Simon	Preventive Services, 18 King St, Rockdale, NSW 2216
CARSTAIRS, J. L.	Department of Zoology, ANU, Box 4 PO Canberra 2600
CASEY, Carlie	10 Cherry Street, Mexico, Maine 04257, USA
CHRISTIE, Marion	19 Henmant St, O'Connor, ACT 2601
COURTNEY, Mark	5 Sheehans Road, Blackburn, Vic
COWARD, Dan	Economic History RSSS, ANU
CREED, David	3/1 Scarborough Beach Rd, Scarborough, WA 6019
CROFTS, David	Toad Hall, ANU, ACT 2601
CROSSLEY, David J.	School of Australian Environmental Studies, Griffith University, Nathan, Brisbane 4111
CUMMINE, Alan	14 Brown St, Yarralumla, ACT 2600
CUMMING, Malcolm J.	CSIRO Div. Mineral Engineering, PO Box 312, Clayton, Vic
DAHLITZ, Julie	University House, PO Box 1535, 2601
DAVIES, Arthur	76 Sherbrooke St, Ainslie, ACT 2602
DAVIES, D.	30 Wandoo St, O'Connor, ACT 2601
DAY, Alice T.	Dept. of Sociology, ANU, Canberra, ACT 2600
DAY, Lincoln H.	Dept of Demography, ANU, Canberra, ACT 2600
DICK, John	Conservation of Urban Energy Group, Conservation Council of Victoria, 324 William St, Melbourne 3000
DICKINS, J. M.	14 Bent St, Turner, ACT 2601
DIESENDORF, M.	CSIRO Div. Maths & Stats, PO Box 1965, Canberra City, ACT 2601
DOBLE, Ann	Loc H Cres., Noojee 3833
DUNFORD, Richard	Sociology, SGS, ANU 2600
DUNN, R. J.	39 Macpherson St, O'Connor, ACT 2601
ENDERSBEE, Lance (Prof)	Dean, Faculty of Engineering, Monash University, Clayton, Vic 3168
EVANS, D. G.	Centre for Environmental Studies, Univ. of Melbourne, Parkville, Vic 3052
EVANS, Jeremy	Human Sciences Program, ANU, PO Box 4, 2600
EVANS, Margaret	2 Ridley St, Turner, ACT 2601
FISHER, Frank	Environmental Science, Monash University, Clayton, Vic 3168
FREEMAN, Terry	Centre for Environmental Studies, Macquarie Univ., North Ryde, NSW 2113
FURNASS, Bryan	ANU Health Service
GAVRANIC, Tom	58 Badenoch Cres., Evatt, ACT 2615
GERRITSEN, Rolf	c/-Political Science Dept., RSSS, ANU
GIFFORD, Roger M.	CSIRO Div. of Plant Industry, PO Box 1600, Canberra City, ACT 2601
GILDING, Jack	51 Nicholson St, Carlton 3053
GORRIE, Ian	Dept of Environment, Housing and Community Development, Box 1890, Canberra City, ACT 2601
GOTTLIEB, Kurt	46 Jennings Street, Curtin, ACT 2605
GREEN, Dorothy	18 Waller Crescent, Campbell, ACT 2601
HAMILTON, R. B.	Shell Australia, Melbourne
HART, Phillip J.	c/-Plasma Physics Dept., School of Physics, Sydney University 2006
HEACOCK, R. H.	Bureau of Transport Economics, PO Box 367, Canberra City, ACT 2601
HILL, Doug	Aust. Conservation Foundation, 672B Glenferrie Rd, Hawthorn, Vic 3122
HIND, Phillip	403 Station St, Bonbeach, Vic 3199
HUGHES, Ian M.	Human Sciences Program, ANU
HURLSTONE, C.J.	29 Cordeaux St, Duffy, ACT 2611
INALL, E. K.	Dept of Engineering Physics, RS Phys., ANU PO Box 4, Canberra, ACT 2600
JACKSON, Laurel	c/- Dept Environmental Design, TCAE, Box 1415P, GPO, Hobart 7001
JAMES, David	School of Economic and Financial Studies, Macquarie Univ., North Ryde 2113
JAMIESON, Marie	Adjungbilly via Tumut 2720
JENKIN, M.	70 Belconnen Way, Page 2614
JOHNSON, Judy	11 Love St, Flynn, ACT 2615
KALMA, Jetse D.	19 Cavill Close, Holt, ACT 2615
KING, R. J.	Centre for Environmental Studies, University of Melbourne, Parkville, Vic 3052
KROON, Dirk	37 Wilkinson St, Flynn 2615
LEJEUNE, J.	Forestry Dept. ANU, PO Box 4, Canberra, ACT 2600
LEYLAND, R.	Sec., National Energy Advisory Committee, PO Box 5, Canberra 2600
LUDOVICI, Charlotte	45 Nullagine St., Fisher, ACT 2611
LUSTIG, Eileen	15 Cottenham Ave, Kensington, NSW 2033
LUSTIG, Terry	15 Cottenham Ave, Kensington, NSW 2033
McALPINE, Geoff	PO Box 1890, Canberra City 2601
McCLYMONT, G. L.	University of New England, Armidale 2357
McCLYMONT, Vivien	140 Handle St, Armidale, NSW 2350
McKENZIE, W. M.	Div. Building Research, Highett, Vic 3190
McMAHON, Monica	35 Chermont St, Fisher, ACT 2611
MADDOCKS, Diana	215 Brougham Pl, North Adelaide, SA 5006
MADDOCKS, Ian	215 Brougham Pl, North Adelaide, SA 5006
MAHONY, M. J.	Agroindustrial Systems Program, Division of Chemical Technology, PO Box 1666, Canberra City, ACT 2601
MANNING, Alan	Box 1, PO Canberra
MARDON, C.J.	Conservation of Urban Energy Group, Conservation Council of Victoria, 324 William St, Melbourne 3000
MARSHALL, Peter	Ministry of Transport & Highways, 117 Macquarie St, Sydney 2000
MARTIN, J.	26 Beaurepaire, Holt
MILBOURNE, Raymond	22 Perkins Pl, Torrens, ACT 2607
MORGAN, I. G.	6 Hargraves Cres., Ainslie, ACT 2602
MOYAL, Ann	Director, Science Policy Research Centre, Griffith Univ., Nathan, Qld 4111
MUNDEY, Jack	Environmentalists for Full Employment, c/- ACF, 18 Argyle St, Sydney 2000
NAUGHTEN, Barry	c/-Dept National Development, PO Box 5, Canberra 2600
NEWMAN, I.	RMB 5323, Holgate, NSW 2251
NUZUM, Patrick	c/- Maum, 180 Brunswick St, Fitzroy 3065
O'NEILL, D. R.	c/- Dept of Foreign Affairs, Canberra 2600
PRESTON, Hugh	Dept of Science, Box 449, Woden 2606
PRICE, John	29B Mary St, Hawthorn, Vic 3122
PROUDLEY, Ray	36 Love Street, Black Rock, Vic 3193
RICE, John	School of Maths, Flinders Univ., SA 5042
RIDDETT, Ross	c/- Defence Facilities Div., Dept of Defence, 3-23 "K" Block, Russell 2600
RITCHIE, David	29 Bates St, Dickson, ACT 2602
ROBERTS, Alan	Physics Dept, Monash Univ., Clayton, Vic 3168
ROBERTS, Julie	29 Bates St, Dickson, ACT 2602
ROBERTS, Rhys A.	Australian Automobile Assoc., PO Box 1555, Canberra City 2601
RODGER, A.	Dept of Architecture and Building, Univ. of Melbourne, Parkville 3052
ROUTLEY, Richard	Philosophy RSSS, ANU
RUTHERFORD, Robert	Dept. Economics, Univ. Tasmania, Hobart, Tas

SADDLER, Hugh	CRES, ANU
SKENE, Phil	c/- SA Dept of Transport, GPO Box 1599, Adelaide, SA 5001
SMITH, G. D.	Dept of Biochemistry, SGS, ANU, ACT 2600
SOLIN, R. T.	27 McKean St, Fitzroy, Vic 3068
STEWART, G. A.	Agroindustrical Systems Program, Division of Chemical Technology, PO Box 1666, Canberra City 2601
SWAINE, D. J.	CSIRO Fuel Geoscience, PO 136, North Ryde, NSW 2113
SYKES, Ian	XL Petroleum, 317 Queensberry St, N. Melbourne
THOMSON, Michael	77 Miller St, O'Conner, Canberra 2601
TRICKETT, E. S.	47 Wood Rd, Griffith, NSW 2680
TYNDALE-BISCOE, Hugh	4 Steele St, Hackett 2602
WADSLEY, M.	6 Primrose Cres., East Brighton, Vic 3187
WALSH, D. T.	SMEC, PO Box 356, Cooma North, NSW 2630
WALSH, Patrick	CSIRO, Div. Building Research, PO Box 56, Highett, Vic 3190
WATSON, C. L.	79 Henry St, Latham, ACT 2615
WEISSMANN, Gerhard	c/- PO Hahndorf, SA 5245
WELLS, K. F.	RMB 1328, Sutton Rd via Queanbeyan, NSW 2620
WHEELER, Tone	School of Environmental Design, Canberra College of Advanced Education
WHITE, Deborah	Dept of Architecture & Building, Melbourne Univ, Parkville, Vic 3052
WILLIAMSON, Terry	Dept of Environment, Housing & Community Development, PO Box 1890, Canberra City 2601
WOOD, Ian	29 Greenough Circuit, Kaleen, ACT 2617
YONG, Mary	243 Carrington Rd, Coogee 2034